Macro- and Microemulsions

ACS SYMPOSIUM SERIES 272

Macro- and Microemulsions

Theory and Applications

Dinesh O. Shah, EDITOR
University of Florida

Based on a symposium sponsored by
the Division of Industrial and Engineering Chemistry
at the 186th Meeting
of the American Chemical Society,
Washington, D.C.,
August 28–September 2, 1983

American Chemical Society, Washington, D.C. 1985

Library of Congress Cataloging in Publication Data

Macro- and microemulsions.
 (ACS symposium series, ISSN 0097-6156; 272)

 "Based on a symposium sponsored by the Division of
Industrial and Engineering Chemistry at the 186th
Meeting of the American Chemical Society,
Washington, D.C., August 28–September 2, 1983."

 Includes bibliographies and indexes.

 1. Emulsions—Congresses.

 I. Shah, D. O. (Dinesh Ochhavlal), 1938– .
II. American Chemical Society. Division of Industrial
and Engineering Chemistry. III. American Chemical
Society. Meeting (186th: 1983: Washington, D.C.)
IV. Series.

TP156.E6M24   1985   660.2'04514   84-28358
ISBN 0-8412-0896-4

Copyright © 1985

American Chemical Society

TP156
E6M24
1985
CHEM

ACS Symposium Series

M. Joan Comstock, *Series Editor*

Advisory Board

FOREWORD

The ACS SYMPOSIUM SERIES was founded in 1974 to provide a medium for publishing symposia quickly in book form. The format of the Series parallels that of the continuing ADVANCES IN CHEMISTRY SERIES except that, in order to save time, the papers are not typeset but are reproduced as they are submitted by the authors in camera-ready form. Papers are reviewed under the supervision of the Editors with the assistance of the Series Advisory Board and are selected to maintain the integrity of the symposia; however, verbatim reproductions of previously published papers are not accepted. Both reviews and reports of research are acceptable, because symposia may embrace both types of presentation.

CONTENTS

MACROEMULSIONS

INDEXES

PREFACE

THE CURRENT STATE OF THE ART of various aspects of macro- and microemulsions is reflected in this volume. The symposium upon which this volume is based was organized in six sessions emphasizing major areas of research. Major topics discussed include a review of macro- and microemulsions, enhanced oil recovery, reactions in microemulsions, multiple emulsions, viscoelastic properties of surfactant solutions, liquid crystalline phases in emulsions and thin films, photochemical reactions, and kinetics of microemulsions.

This volume includes discussions of various processes occurring at molecular, microscopic, and macroscopic levels in macro- and microemulsions. I earnestly hope that this book will serve its intended objective of reflecting our current understanding of macro- and microemulsions, both in theory and practice, and that it will be useful to researchers, both novices as well as experts, as a valuable reference source.

I wish to convey my sincere thanks and appreciation to the chairmen of the sessions, namely, Professor K. S. Birdi, Professor S. E. Friberg, Dr. E. D. Goddard, Dr. K. L. Mittal, and Professor S. N. Srivastava. I also wish to convey my thanks to the Division of Industrial and Engineering Chemistry as well as various corporations: Alcon Laboratories Inc., Atlantic Richfield Company, GAF Corporation, Gulf Research and Development Company, Hercules Inc., Johnson & Johnson Inc., Phillips Petroleum Company, Procter & Gamble Company, SOHIO Corporation, Velsicol Chemical Corporation, Westvaco, and Syntex, U.S.A., Inc., for their generous support that allowed me to invite many researchers from overseas to participate in this symposium. I also wish to convey my sincere thanks to the Editorial Staff of the American Chemical Society's Books Department.

I am grateful to reviewers of the manuscript, as well as to my colleagues, postdoctoral associates, and students for their assistance throughout this project. Specifically, I wish to thank Dr. M. K. Sharma for his administrative and editorial assistance, as well as Ms. Melissa Maher, Mrs. Jeanne Ojeda, and Mrs. Derbra Owete for the typing of correspondence and manuscripts. I also convey my sincere thanks and appreciation to all the authors and coauthors of the papers without whose assistance this symposium proceedings would not have been completed.

I wish to convey my sincere thanks and appreciation to my colleagues and chairpersons, Dr. J. P. O'Connell in the Department of Chemical

ix

Engineering, and Dr. J. H. Modell in the Department of Anesthesiology at the University of Florida for their assistance. Finally, I wish to convey my sincere thanks and appreciation to my wife and children for allowing me to spend many evenings and weekends working on this volume.

DINESH O. SHAH
Center for Surface Science & Engineering
University of Florida

October 15, 1984

Introduction to Macro- and Microemulsions

M. K. SHARMA and D. O. SHAH

Departments of Chemical Engineering and Anesthesiology, University of Florida, Gainesville,
FL 32611

This paper reviews various aspects of macro- and microemulsions. The role of interfacial film of surfactants
in the formation of these systems has been high-lighted.
The formation of a surfactant film around droplets
facilitates the emulsification process and also tends
to minimize the coalescence of droplets. Macroemulsion
stability in terms of short and long range interactions
has been discussed. For surfactant stabilized macroemulsions, the energy barrier obtained experimentally
is very high, which prevents the occurrence of flocculation in primary minimum. Several mechanisms of
microemulsion formation have been described. Based on
thermodynamic approach to these systems, it has been
shown that interfacial tension between oil and water
of the order of 10^{-3} dynes/cm is needed for spontaneous formation of microemulsions. The distinction between the cosolubilized and microemulsion systems has
been emphasized.

Macroemulsions have been known for thousands of years. The survey of
ancient literature reveals that the emulsification of beeswax was
first recorded in the second century by the Greek physician, Galen
(1). Macroemulsions are mixtures of two immiscible liquids, one of
them being dispersed in the form of fine droplets with diameter
greater than 0.1 µm in the other liquid. Such systems are turbid,
milky in color and thermodynamically unstable (i.e. the macroemulsion
will ultimately separate into two original immiscible liquids with
time). Since the early 1890s, extensive and careful studies have
been carried out on macroemulsions and several excellent books have
been written on various aspects of formation and stability of these
systems (2,10). In addition, several theories and methods of macroemulsion formation have been discussed in the recent articles (11-
17). In spite of this progress, we still do not have good predictive
methods for the formation or breaking macroemulsions. For the formation of a stable macroemulsion from two immiscible liquids, there
is no reliable predictive method for selecting the emulsifier or

technique of emulsification for obtaining the optimum results. One
can use the concept of hydrophilic-lipophilic balance (HLB) for
initial screening (1,6) and most of the new macroemulsions are ultimately perfected by trial and error approach.

Macroemulsions are utilized in many applications and are very
important from the technical point of view. Many technologies and
processes involve production of stable emulsions, such as skin creams
(cosmetics), metal cutting fluids, fiber cleaning or removal of oil
deposits (detergency), mayonnaise (food industry), bitumen emulsions
(road construction), fuel (energy), herbicides and pesticides (agricultural sprays) and drug solubilization in emulsions (pharmacy). In
addition, it has been observed that some processes require emulsions
of long-term stability, whereas other require limited stability of
emulsions. There are processes such as formation of emulsions in oil
storage tanks and petroleum reservoirs where naturally occurring,
unwanted stable emulsions have to be broken down. In view of the
wide range of applications and technical importance of macroemulsions,
it is worth discussing various aspects of these systems.

Classification of Macroemulsions

In this section, we will briefly describe the classification of macroemulsions. Based on the dispersion of water or oil in continuous
phase and on the number of phases present in the system, macroemulsions can be subdivided into two categories.

Single Emulsions. These emulsions are formed by two immiscible
phases (e.g. oil and water), which are separated by a surfactant
film. The addition of a surfactant (or emulsifier) is necessary to
stabilize the drops. The emulsion containing oil as dispersed phase
in the form of fine droplets in aqueous phase is termed as oil-in-
water (O/W) emulsion, whereas the emulsion formed by the dispersion
of water droplets in the oil phase is termed as water-in-oil (W/O)
emulsion. Figure 1 schematically illustrates the O/W and W/O type
emulsions. Milk is an example of naturally occurring O/W emulsion in
which fat is dispersed in the form of fine droplets in water.

Double or Multiple Macroemulsions. These macroemulsions are formed
by two or more than two immiscible phases which are separated by at
least two emulsifier films. Multiple emulsions can also be subdivided as single emulsions in two categories (O/W/O) and (W/O/W)
emulsions (14). For a O/W/O system, the immiscible water phase
separates the two oil phases, whereas for a W/O/W system, the immiscible oil phase separates the two aqueous phases. These emulsions
are schematically shown in Figure 2.

The phase contained in the subdrops is often referred to as the
encapsulated phase. These systems are very relevant to transport
phenomena and separation processes (14), such as controlled release
of drugs in which the encapsulated phase can serve as a reservoir of
the active ingredient.

Mechanism of Macroemulsion Formation

Macroemulsions can be produced in different ways starting with two
immiscible liquids and by applying mechanical energy, which deforms

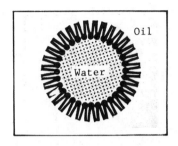

Oil-in-Water   Water-in-Oil

Figure 1.   Schematic illustration of oil-in-water (O/W) and water-in-oil (W/O) macroemulsions

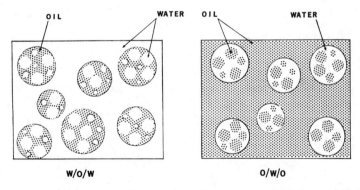

W/O/W   O/W/O

Figure 2.   Schematic illustration of multiple W/O/W and O/W/O macroemulsions

the interface to such an extent that it generates droplets. The formation of final emulsion droplets can be viewed as the stepwise process. Therefore, the disruption of droplets is a critical step in the process of emulsification. During emulsion formation, the deformation is opposed by the Laplace pressure. For spherical droplet of radius (r), the difference in pressure (Δp) at the concave side of a curved interface with interfacial tension (γ) is $2\frac{\gamma}{r}$. Further division of droplets leads to an increase in Δp as r decreases. In order to disrupt such a small droplet, the pressure gradient of the magnitude of $2\frac{\gamma}{r^2}$ must be applied externally. The viscous forces exerted by the continuous phase can also deform the emulsion droplets. The viscous stress ($G\eta$) should be of the same magnitude as the Laplace pressure to deform the droplets (9), where G is the velocity gradient and η is the viscosity of continuous phase. In any case, the pressure gradient or velocity gradient required for emulsion formation are mostly supplied to the system by agitation. The various methods of agitation to produce emulsions have been described recently (18). In addition, the emulsions of smaller droplets can be produced by applying more intense agitation to disrupt the larger droplets. Therefore, the liquid motion during the process of emulsification is generally turbulent (9) except for high viscosity liquids.

Energy Needed for Emulsion Formation. The total interfacial area (A) generated due to emulsification process is much larger because of the formation of the smaller droplets. Therefore, the increase in surface free energy of the system is given by $\gamma\Delta A$. It can easily be calculated that the energy needed to produce the emulsion of average droplet diameter (2μm) from the two immiscible liquids (1 ml of each) with interfacial tension γ = 10 dyne/cm would be about 3000 times higher than that of the surface free energy of the system. For the formation of some emulsions, the pressure gradient of 2×10^9 dynes/cm^3 and velocity gradient of 2×10^7/sec (assuming η = 1.0 cp) are needed (9).

The large excess of energy required to produce emulsions can only be supplied by very intense agitation, which needs much energy. In order to reduce the agitation energy needed to produce a certain droplet size, a suitable surfactant can be added to the system. The addition of surfactant reduces interfacial tension, which in turn decreases the surface free energy of the system. The formation of a surfactant film around the droplets facilitates the process of emulsification, and a reduction in agitation energy by a factor of 10 or more can be achieved (9). The nature and concentration of surfactant also affects the droplet size and energy requirement to form the emulsion. Besides lowering the interfacial tension, the surfactant film also tends to prevent the coalescence of droplets.

Interfacial Film of Surfactants. The droplets are surrounded by an interfacial film of surfactant in emulsion systems. The stability of such films can be increased by adding appropriate surfactants. The rate of change in interfacial tension with surface area from its equilibrium value is termed as the Gibbs elasticity E = 2dγ/d(ln A) (9). The factors which control E are the rate of transport toward or from the interface and the structure of surfactant as well as the

rate of compression and expansion of the interface. The film elasticity also plays an important role to stabilize emulsion droplets.
As the film is stretched, the local concentration of surfactant in
the film decreases. This causes a transient increase in interfacial tension. The highest elasticity in the presence of sufficient amount of surfactant, provides the greatest resistance against
stretching (9). Moreover, Prins (19) has shown that a stretched
thin film can also break if the interfacial tension exceeds a
critical value which depends on the system. For low surfactant concentration and rapid stretching, the critical value of interfacial
tension is attained rapidly. Based on the above mentioned factors,
it can be suggested that the surfactant is of primary importance
for the stability, or flocculation of the emulsions. In conclusion, the interfacial tension gradients are essential in emulsion
formation as suggested previously by Tadros and Vincent (20).

Stability of Macroemulsions

The emulsions are complex systems which present major challenges to
the scientists working in this field. Previous investigators applied
various theoretical approaches at the droplet level and also at
the molecular level to explain the behavior of these systems.
The forces such as electrical double layer, forces between emulsion
droplets, hydrodynamic inertial forces, entropic (Diffusional)
forces and the dispersion forces which act on the droplets or between the droplets separated at tens or hundreds of nanometers.
Sedimentation and flocculation processes involve the forces such
as the centrifugal force, applied electrostatic force and gravitational force. Before discussing the emulsion stability in terms
of these forces, we would like to explain the thermodynamics of
emulsion stabilization.

Thermodynamic Approach to Emulsion Stability. In this section, we
would like to discuss thermodynamic approach to emulsion stability.
Let us assume that the total free energy of the emulsion can be
separated into several independent contributions. Considering
hypothetically the formation or coalescence of emulsion of two immiscible liquids (e.g. oil and water), such that external field forces are
absent. The total free energy (G_B) of the system just before emulsification process can be expressed in the form (10)

$$G_B = G_I + G_E + G_{IE} + G_S \tag{1}$$

where G_I, the free energy of the internal phase; G_E, the free energy
of the external phase; G_{IE}, the free energy of the interface between
two liquids and G_S, the free energy of the interface between the
liquids and the surface of the container. In general, the solid/
liquid interfacial area will be small and therefore, G_S can be neglected. The free energies, G_I and G_E, will remain almost the same
before and after emulsification, whereas, G_{IE} will be at a minimum
before emulsification. The interfacial free energy, G_{IE}, can be
expressed in the form

$$G_{IE} = \gamma_{IE} \, A \tag{2}$$

where γ_{IE}, the interfacial tension and A, the interfacial area. After emulsification, the interfacial area greatly increases, therefore, G_{IE} is larger than that before emulsification. The free energy of emulsion formation can be written in the form (10)

$$\Delta G_{Emul.} = \gamma_{IE}\Delta A - T\Delta S \tag{3}$$

where ΔS is the change in entropy due to the process of emulsification. In general, the free energy of interface ($\gamma_{IE}\Delta A$) term is much larger than the $T\Delta S$ term. Therefore, the change in free energy of emulsion formation ($\Delta G_{Emul.}$) will be positive. This indicates that most of the macroemulsions are thermodynamically unstable or metastable. Moreover, the free energy of emulsification is positive; this means that the free energy of demulsification ($\Delta G_{Demul.}$) is negative. This implies that an extenral supply of free energy is needed for the formation of macroemulsions, and once formed, they are unstable. Tadros and Vincent (20) have shown that the variation in free energy change as a function of the demulsification processes (e.g. flocculation, coalescence) is continuous, and there is no free energy barriers to the processes until the drops are close enough for short range repulsive and attractive forces.

Short Range Interactions and Emulsion Stability. The stability of macroemulsions in terms of short range (e.g. inter-droplet) interactions will be discussed in this section. The dispersion (London) forces arise from charge fluctuations within a molecule associated with the electronic motion (21). Therefore, these forces can operate even between nonpolar molecules. London (21) reported an equation for mutual attractive energy between two molecules in vacuum in the form

$$V_A = \frac{3}{4} h \nu_o \frac{\alpha}{d^6} = \frac{\beta_{11}}{d^6} \tag{4}$$

where h is the Planck's constant; ν_o, the characteristic frequency of the molecule; α, the polarizability of the molecule and d, the distance between the molecules. The London forces between two molecules are short range as the V_A is inversely proportional to the sixth power of their separation. Assuming these forces between molecules could be summed for all the molecules in a particle of radius (r), the Equation (4) can be expressed (22) in the form

$$V_A \approx -\frac{Ar}{\pi}\left(\frac{2.45\lambda}{120H^2} - \frac{\lambda^2}{1045H^3} + \frac{\lambda^3}{5.62\times10^4 H^4}\right) \tag{5}$$

valid for $H > 150$ A^o, and

$$V_A \approx -\frac{Ar}{12H}\left(\frac{\lambda}{\lambda+3.5\pi H}\right) \tag{6}$$

valid for $H < 150$ A^o

where A is the Van der Waals constant; λ, the wavelength of the intrinsic electronic oscillations of the atoms ($\sim 10^{-5}$ cm); H, the interparticle distance between droplets and r is the average radius of droplets.

The so called Hamaker/Van der Waals constant, A, required to evaluate energy of attraction between two droplets (Equations 5 and 6) in vacuum is defined by

$$A = \pi^2 B_i^2 L_i \tag{7}$$

where L_i is the London constant, B_i, the number of atoms (molecules with ith kind contained in cm^3 of the substance). If the droplets are suspended in a continuous medium, then the net interaction is reduced and the Equation 7 can be rewritten in the form

$$A \approx (A_{o-o}^{1/2} - A_{w-w}^{1/2})^2 \tag{8}$$

and generally lies in the range of 10^{-13} to 10^{-14} ergs (15, 17).

The repulsive (e.g. electrical double layer) forces have been discussed in detail in literature (23,25). The repulsive energy derived by Derjaguin and Kassakov (26) is given by

$$V_R = \frac{\varepsilon r \psi_o^2}{2} \ln(1 + e^{-\kappa H}) \tag{9}$$

where ε is the dielectric constant of the medium; ψ_o, the surface potential and κ is the reciprocal "thickness" of the electrical double layer, given by

$$\kappa = \left(\frac{8\pi n z^2 e^2}{\varepsilon kT}\right)^{1/2} \tag{10}$$

where z is the valency of counter ions; e, the electronic charge; n, the number of ions per cm^3 in the solution; k, the Boltzmann's constant and T, the absolute temperature.

The total interaction energy is obtained by summing the V_A and V_R contributions. A schematic representation of these energies (3) is given in Figure 3. This curve shows a primary minimum at very short distances between the droplets (e.g. close contact), a maximum at intermediate interdroplet distances and a secondary minimum at large interdroplet distnaces. For irreversible flocculation into the primary minimum to occur, the energy barrier has to be surmounted. The height of the barrier is primarily controlled by the electrolyte concentration (15, 17). The reversible flocculation may occur in the secondary minimum as reported previously by several investigators (15,17,27). For surfactant stabilized emulsions, it has been reported that the energy barriers obtained experimentally are very high, which prevents the occurrence of flocculation in primary minimum (15,27).

Long Range Interactions and Emulsion Stability. The processes such as sedimentation, creaming and flocculation can be controlled by external forces, e.g. centrifugal, gravitational or applied electrostatic forces. These forces are considered to be essentially long range. In this section we would like to discuss processes controlled by long range forces.

Sedimentation and Creaming. The creaming and sedimentation processes occur in emulsion systems mainly due to the density difference between the dispersed and continuous phases. Assuming a steady state,

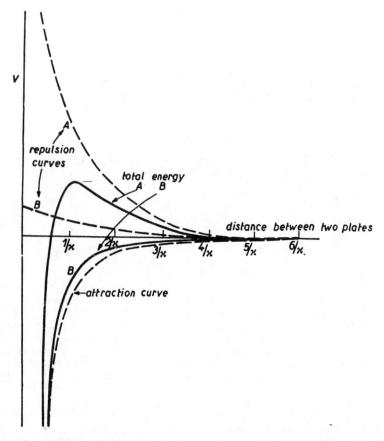

Figure 3.   Schematic illustration of interaction energies as a function of interparticle distance between two droplets.

the sedimentation or creaming rate (V) of non-interacting spherical droplets of radius (r) can be determined by equating two oppositely acting gravitational and hydrodynamic forces as given by Stockes' Law (28) in the form

$$\frac{4}{3} \pi r^3 \Delta\rho g = 6\pi\eta_o rV$$

and

$$V = \frac{2\Delta\rho g r^2}{9\eta_o} \tag{11}$$

In various emulsion systems, the main objective is to decrease the rate of sedimentation or creaming rather than to increase it. Equation 11 indicates that this can be achieved by increasing η_o or decreasing $\Delta\rho$. The former may be achieved by the addition of a structuring or gelling agent such as polymers, silica, etc., whereas, the latter by adding a suitable solvent. Other factors, such as deformation, polydispersity, droplet size, flocculation and coalescence also affect the processes of sedimentation and creaming. Both these processes can be studied employing ultracentrifuge (29,30).

Flocculation and Coalescence. Flocculation being the primary process, the droplets of the dispersed phase come together to form aggregates. In this process, the droplets have not entirely lost their identity and the process can be reversible. Since the droplets are surrounded by the double layer, they experience the repulsive effect of the double layer. Kinetically, flocculation is a second order reaction since it depends in the first instance on the collision of two droplets and is expressed in the form (31)

$$-\frac{dn_1}{dt} = K_1 n_o^2$$

or

$$1/n_1 - 1/n_o = K_1 t \tag{12}$$

where n_o and n_1 are the number of droplets present initially and after time t. The rate constant (K_1) depends upon the frequency of the collision between droplets and is governed by the repulsive forces between them. This constant is also known as Smoluchowski's constant. The values of K_1 reported in the literature are of the order of 10^{-13} (15).

Coalescence being the secondary process, the number of distinct droplets decreases leading to a stage of irreversibility and finally complete demulsification takes place. Coalescence rate very likely depends primarily on the film-film repulsion, film drainage and on the degree of kinetics of desorption. Kinetically, coalescence is a unimolecular process and the probability of merging of two droplets in an aggregate is assumed not to affect the stability at other point of contact (32).

$$-\frac{dn_2}{dt} = K_2 n_o$$

or

$$\ln n_o - \ln n_2 = K_2 t \qquad (13)$$

where n_2 is the number of individual droplets after time t. The rate
of coalescence depends on the lateral adhesion properties of the
surfactant film at the interface. The coagulation process comprises
the flocculation and coalescence of the system. These processes can
be determined experimentally by light scattering, droplet counting
and centrifugal methods. The theoretical and experimental discussion
of this topic is given by the previous investigators (33-35).

Microemulsions

Microemulsions were first introduced by Schulman et. al (36) in
1943. The various properties of these systems were studied during
the following years (37-40), and in 1955 (39), the systems were
called both swollen micellar solutions and transparent emulsions.
This ambiguity in the microemulsion terminology remains today (41).
The microemulsions are defined as the clear thermodynamically stable
dispersions of two immiscible liquids containing appropriate amounts
of surfactants or surfactants and cosurfactants. The dispersed
phase consists of small droplets with diameter in the range of
100-1000A°. Because of these properties, such systems have several
advantages over macroemulsions for industrial applications.
        The small droplet size in microemulsions also leads to a large
surface-to-volume ratio in an oil-water system. This is important
for chemical reactions in which the rate of reaction depends on the
interfacial area. The microemulsion can also be classified as W/O
or O/W similar to macroemulsion systems.

Mechanism of Microemulsion Formation

During the last four decades, several investigators have proposed
various mechanisms of microemulsion formation. The following is a
brief description of these mechanisms.

Interfacial Tension in Microemulsions. Schulman and his collaborators
(42) have postulated that the transient interfacial tension has to
be negative for the spontaneous uptake of water or oil in microemulsions. During the process of microemulsion formation, one phase
breaks up into the maximum number of droplets. The diameter of
these droplets depends upon the interfacial area produced by the
surfactant molecules. The transient interfacial tension (e.g.
the spontaneous tendency of the interface to expand) produced by
the mixing of the components became zero or a very small positive
value at equilibrium. Schulman and his co-workers have specifically
mentioned that the negative interfacial tension is a transient
phenomenon and that at equilibrium, the oil/water interface in a microemulsion has either zero or a very small positive interfacial
tension. However, Schulman's explanation of transient interfacial
tension has been misquoted and misunderstood by various investigators. Schulman et al. in 1959 suggested various possible ways of
producing transient negative interfacial tension and, therefore, the
formation of microemulsions.

The concept of transient interfacial tension has been further extended by Davis and Haydon (43). They described an experiment by Ilkovic (44) in which a negative potential was applied to a mercury drop in an aqueous solution of a quaternary ammonium compound. At -8 v/cm applied potential, the spontaneous emulsification of mercury occurred. The spontaneous emulsification was observed for surfactant concentrations which exhibited negative values for interfacial tensions upon extrapolation. These results indicate that for spontaneous emulsification, the dynamic interfacial tension may approach transient negative values. Moreover, this does not mean that at equilibrium, the dispersed droplets will have a negative interfacial tension.

Gerbacia and Rosano (46) have determined the interfacial tension at oil-water interface after alcohol injection into one of the phases. They observed that the interfacial tension could be temporarily lowered to zero due to the diffusion of alcohol through the interface. They concluded that the diffusion of surfactant molecules across the interface is an important requirement for reducing interfacial tension temporarily to zero as well as for the formation of microemulsions. They further claimed that the formation of microemulsions depend on the order in which components are added.

It has also been shown from thermodynamic consideration (Equation 3), that if the interfacial tension is very low, the thermodynamically stable emulsions can be formed. Previous investigators (20,45,47,48) have calculated that for a situation likely to occur in microemulsion formation, the interfacial tensions would need to be in the order of 10^{-4} to 10^{-5} dynes/cm for thermodynamic stabilization and for spontaneous formation of microemulsions.

Double Layer Interactions and Interfacial Charge. Schulman et al (42) have proposed that the phase continuity can be controlled readily by interfacial charge. If the concentration of the counterions for the ionic surfactant is higher and the diffuse electrical double layer at the interface is compressed, water-in-oil microemulsions are formed. If the concentration of the counterions is sufficiently decreased to produce a charge at the oil-water interface, the system presumably inverts to an oil-in-water type microemulsion. It was also proposed that for the droplets of spherical shape, the resulting microemulsions are isotropic and exhibit Newtonian flow behavior with one diffused band in X-ray diffraction pattern. Moreover, for droplets of cylindrical shape, the resulting microemulsions are optically anisotropic and non-Newtonian flow behavior with two difused bands in X-ray diffraction (9). The concept of molecular interactions at the oil-water interface for the formation of microemulsions was further extended by Prince (49). Prince (50) also discussed the differences in solubilization in micellar and microemulsion systems.

Scriven (78) proposed the role of the electrical double layer and molecular interactions in the formation and stability of microemulsions. According to them, the total interfacial tension (γ_T) can be expressed in the form

$$\gamma_T = \gamma_p - \gamma_d \qquad (14)$$

where γ_p is the phase interfacial tension which is that part of the
excess tangential stress which does not arise in the region of the
diffuse double layer and $-\gamma_d$ is the tension of the diffuse double
layer. This equation suggests that when γ_d exceeds γ_p, the total
interfacial tension (γ_T) becomes negative. For a plane interface,
the destabilizing effect of a diffuse layer is primarily due to a
negative contribution to the interfacial tension.

Adamson (51) proposed a model for W/O microemulsion formation
in terms of a balance between Laplace pressure associated with the
interfacial tension at the oil/water interface and the Donnan Osmotic
pressure due to the total higher ionic concentration in the interior
of aqueous droplets in oil phase. The microemulsion phase can exist
in equilibrium with an essentially non-colloidal aqueous second phase
provided there is an added electrolyte distributed between droplet's
aqueous interior and the external aqueous medium. Both aqueous media
contain some alcohol and the total ionic concentration inside the
aqueous droplet exceeds that in the external aqueous phase. This
model was further modified (52) for W/O microemulsions to allow
for the diffuse double layer in the interior of aqueous droplets.
Levine and Robinson (52) proposed a relation governing the equilibrium
of the droplet for 1-1 electrolyte, which was based on a balance
between the surface tension of the film at the boundary in its
charged state and the Maxwell electrostatic stress associated with
the electric field in the internal diffuse double layer.

In addition, Shinoda and Friberg (53) have summarized their
extensive studies on the formation of microemulsions using nonionic
surfactants. They proposed the following conditions to form microemulsions with minimum amount of surfactants:

1. Microemulsions should be formed near or at the phase inversion
    temperature (PIT) or HLB temperature for a given nonionic sur-
    factant, since the solubilization of oil (or water) in an aqueous
    (or nonaqueous) solution of nonionic surfactant shows a maximum
    at this temperature.
2. The larger the size of the nonionic surfactant, the greater is
    the solubilization of oil in water.
3. The mixing ratio of surfactants should be such that it produces
    an optimum HLB value for the mixture.
4. The closer the phase inversion temperature (PIT) of two surfac-
    tants, the greater is the solubilization and therefore, the
    minimum amount of the nonionic surfactants is needed.

Stability and Structural Aspects of Microemulsions

Several attempts have been made to explain the stability and structural aspects of various microemulsions (54-60). In this section,
we would like to describe some of the important aspects of microemulsion stability.

Stability of Microemulsions. The first attempt to describe the microemulsion stability in terms of different free energy components
was made by Ruckenstein and Chi (55) who evaluated the enthalpic
(Van der Waals potential, interfacial free energy and the potential
due to the compression of the diffuse double layer) and entropic

components. The qualitative results of this model (55) for the
given free energy charge (Δg) as a function of droplet size (R) are
shown in Figure 4. This figure shows three curves depending on
the values of uncharged surface free energy (f_s). If it is very
small, a stable dispersion of small droplets (microemulsion) can
exist. If it is too large, the dispersion cannot exist. Moreover,
for intermediate values, the metastable emulsions of large droplets
can exist. These three cases exist in nature which indicates the
validity of the model. This model also predicts the actual size
range of stable droplets. This treatment can also predict the
occurrence of phase inversion. Phase inversion occurs at that
volume fraction for which the values of change in free energy for
both kinds of microemulsions are the same. In several recent
papers, Ruckenstein and his co-workers (58, 61-63) have discussed
the thermodynamic stability of microemulsion systems. Eicke (64)
has also explained the effect of cosurfactants on the thermodynamic
stability of microemulsion systems in the presence of an additive.
The presence of such an additive will decrease or increase the
miscibility of the two-component system. An attempt has also been
made by Miller and Neogi (65) to explain the thermodynamic stability
in terms of chemical potential of the two phases, the interparticle
potentials, the entropic contribution and the interfacial free energy.

The kinetic stability of microemulsions has also been described
by Eicke (66,67) using fluorescence technique to detect a rapid
exchange of electrolyte solutions and water between inverse micelles
in iso-octane arising from collisions. Gerbacia and Rosano (46)
experimentally observed that some W/O microemulsions consist of
surfactant-plus-cosurfactant concentration of 16% were not stable on
a long term basis even in the presence of a small amount of water
(2.6%). A theory was presented (46) to explain the kinetic stability
using microemulsion droplets as a model. The theory is essentially
limited to the energy changes occurring in the layers of surfactant
and cosurfactant during coalescence of water droplets of microemulsions. The evaluation of the free energy changes in the interfacial
film was based on the regular solution theory, both for entropic
and enthalpic components.

Structural Aspects of Microemulsions. Several investigators have
studied the structure of microemulsions using various techniques
such as ultracentrifugation, high resolution NMR, spin-spin relaxation
time, ultrasonic absorption, p-jump, T-jump, stopped-flow, electrical
resistance and viscosity measurements (56-58). The useful compila-
tion of different studies on this subject is found in the books by
Robb (68) and Shah and Schechter (69). Several structural models
of microemulsions have been proposed and we will discuss only a
few important studies here.

Based on various physical techniques, Shah et al. (70) have
proposed structures for the microemulsion and cosolubilized systems
(Figure 5). Two isotropic clear systems with identical compositions,
except that one contains n-pentanol and the other n-hexanol, are
structurally quite dissimilar systems. The proposed structure for
the pentanol containing system is a cosolubilized system in which
one can visualize the surfactant and the cosurfactant forming a
liquid which can dissolve both water or oil as a molecular solution,

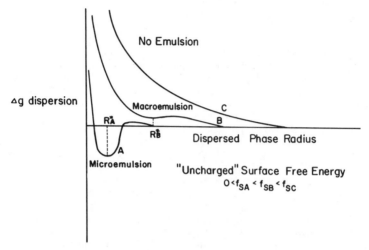

Figure 4.   Schematic illustration of free energy change (Δg)
            as a function of droplet size (R)

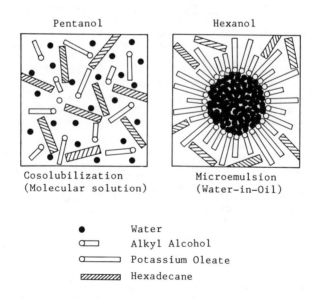

Figure 5.   Schematic illustration of the structure of cosolubilized
            and microemulsion systems

whereas hexanol containing system is a true water-in-oil microemulsion in which water is present as spherical droplets. The
structures shown in Figure 5 are schematic and should not be considered rigidly. These authors have also mentioned that there may
be small aggregates of water molecules or surfactant and cosurfactant
molecules in the cosolubilized system. However, the structure of
cosolubilized system shown in Figure 5 is consistent with the change
in electrical resistance upon addition of water to such systems.

For microemulsion system (hexanol system), since it contains
water spheres in a continuous oil medium, the addition of water
forms more spherical droplets. The continuous medium is still an
oil phase and therefore, the electrical resistance is maintained
at a high value in the range of 10^{-5} ohms (70). However, for cosolubilized system (pentanol system), as the amount of water is increased,
the average distance between alcohol molecules as well as between
water molecules would change and this consequently would influence
the hydrogen bonding ability of water and alcohol molecules, which
in turn would influence the chemical shift of the resonance peak.
Moreover, as one adds more and more water in cosolubilized system,
it becomes more electrically conducting and, hence exhibits a
continuous decrease in the electrical resistance (70).

In summary, these authors (70) have proposed that the transparent, isotropic, clear, stable systems prepared from oil/water/emul-
sifier can be classified into one of three main categories: Normal
or reversed micelles, water-in-oil or oil-in-water microemulsions
or cosolubilized systems. One can distinguish these classes of
structures by using a combination of physical techniques to study
the properties of such systems.

From the results of self-diffusion, Lindman et al. (71) have
proposed the structure of microemulsions as either the systems have
a bicontinuous (e.g. both oil and water continuous) structure or
the aggregates present have interfaces which are easily deformable
and flexible and open up on a very short time scale. This group has
become more inclined to believe that the latter proposed structure
of microemulsion is more realistic and close to the correct description. However, no doubt much more experimental and theoretical
investigations are needed to understand the dynamic structure of
these systems.

The ultrasonic absorption in relation to the transitions and
critical phenomena in microemulsions has been studied by Lang et al.
(72). The ultrasonic absorption is very sensitive to the concentration fluctuations which occur near the critical temperature or
composition in binary liquids. Similar absorption maxima were also
expected as the composition of the systems was varied in the vicinity
of composition where water-in-oil microemulsions convert into the
oil-in-water microemulsions. However, the most puzzling feature of
these data is probably the very continuous change of the relaxation
parameters with composition even in the range where W/O microemulsions turn into O/W microemulsions.

Friberg et al. (73) have proposed a random structure of microemulsions with varying curvatures. Taupin and co-workers (74) have
considered the presence of hard oil and water droplets with a relatively sharp transition between these, while Shinoda (75,76) has
proposed a lamellar structure with alternating water, amphiphilic

and hydrocarbon layers. Talmon and Prager (77) have suggested a hard randomly arranged hydrophobic and hydrophilic polyhedra, whereas Scriven (78) has viewed the middle phase microemulsions as a complex periodic three dimensional networks with both hydrocarbon and water continuity. It is certain that future investigations on the structure of microemulsions will reveal many interesting structural characteristics and diversity of phenomena in these systems.

Literature Cited

1. Becher, P. "Emulsions: Theory and Practice", Krieger Pub.: New York, 1977; p. 95.
2. Manegold, E. "Emulsionen, Chemie und Technic", Strassenbau: Heidelberg, 1952.
3. Kruyt, H.R. "Colloid Science", Vol. 1, Elsevier Pub.: New York, 1952.
4. Clayton, W. "Theory of Emulsion and Emulsification", Churchill: London, 1923.
5. Bancroft, W.D. "Applied Colloid Chemistry", McGraw-Hill: New York, 1932.
6. Sherman, P. "Emulsion Science", Academic Press: New York, 1963.
7. Beeker, P. "Emulsions: Theory and Practice", 2nd Ed., Van Nostrand Reinhold: New York, 1965.
8. Lissant, K.L., ed. "Emulsions and Emulsion Technology", Part I, Dekker: New York, 1975.
9. Becher, P., ed. "Encyclopedia of Emulsion Technology," Dekker: New York, 1983; Vol. 1
10. Lissant, K.L., ed. "Emulsions and Emulsion Technology", Parts II and III, Dekker: New York, 1976 & 1984.
11. Gouda, J.H.: Joos, P. Chem. Eng. Sci. 1978, 30, 521.
12. Sharma, M.K.; Sharma, M.; Jain, S.P.; Srivastava, S.N. J. Colloid & Int. Sci. 1978, 64, 179.
13. Torza, S.; Cox, R.G.; Mason, S.G. J. Colloid Int. Sci. 1972, 38, 395.
14. Straeve, P.; Varanasi, P.P. Separation & Purification Methods 1982, 11(1), 29.
15. Sharma, M.K.; Srivastava, S.N. Colloid & Poly. Sci. 1977, 255, 887, Agra Univ. J. Res. (Sci.) 1974, 23, 35.
16. Overbeek, J.G. J. Colloid Interface Sci. 1977, 58, 408.
17. Sharma, M.K.; Bahadur, P. and Srivastava, S.H., Indian J. Technol., 1075, 13, 419.
18. Walstra, P. in "Encyclopedia of Emulsion Technology", P. Becher, Ed.; Dekker: New York, 1983.
19. Prins, A. in "Foams"; R.J. Ackers, Ed.; Academic Press: London, 1972.
20. Tadros, T.F. and Vincent, B. in "Encyclopedia of Emulsion Technology"; P. Becher, Ed.; Dekker: New York, 1983.
21. London, F. Z. Phys. 1930, 63245.
22. Schenker, J.H.; Kitchener, J.A. Trans. Faraday Soc. 1960, 56, 161.
23. Verwey, E.W. and Overbeek, T.G. "Theory of Stability of Lyphobic Colloids", Amsterdam, 1948.
24. Bell, G.D.; Peterson, G. J. Colloid Interface Sci. 1972, 42, 542.

25. Devereux, O.F. and de Bruyn, P.L. Ph.D. Thesis, "Interaction of Plane Parallel Double Layers", MIT, Cambridge, Mass., 1963.
26. Derjaguin, B.; Kussakov, M. Acta Phys. Chem. USSR, 1939, 10, 25 and 153.
27. Sharma, M.K.; Srivastava, S.N. Indian J. Technol. 1977, 15, 82.
28. Stokes, G.G. Phil. Mag. 1851, 1, 337.
29. Levich, V.G. "Physicochemical Hydrodynamics", Prentice-Hall: New York, 1962.
30. O'Brien, R.N.; Echer, A.I.; Leja, J. Zh. Fiz. Khim. 1947, 21, 1183.
31. Smoluchowski, M.V. Z. Phys. Chem., 1917, 92, 129.
32. Van den Tempel, M. Ind. Intern. Cong. Surf. Act. Vol. 1, 1957, p. 439.
33. Muller, H. Kolloid-Z. 1926, 38, 1
34. Spielman, L.A. J. Colloid Interface Sci. 1970, 36, 562.
35. Van den Tempel, M. Rec. Trav. Chem. 1953, 72, 433 and 442.
36. Hoar, T.P.; Schulman, J.H. Nature 1963, 152, 102.
37. Schulman, J.H.; McRoberts, T.J. Trans. Faraday Soc. 1946, 42B, 165.
38. Schulman, J.H.; Riley, D.P. J. Colloid Sci. 1948, 3, 383.
39. Bowcott, J.E.; Schulman, J.H. Z. Electrochem. 1955, 59, 283.
40. Schulman, J.H.; Friend, J.A. J. Colloid Sci. 1949, 4, 497.
41. Prince, L.M., Ed. "Microemulsions", Academic Press: New York, 1977.
42. Schulman, J.H.; Staeckenius, W.; Prince, L.M. J. Phys. Chem. 1959, 63, 7716.
43. Davis, J.T.; Haydon, D.A. Ind. Intern. Cong. Surf. Activity, Vol. 1, 1957, p. 417.
44. Ilkovic, D. Coll. Trav. Chim. Tchecosl. 1932, 4, 480.
45. Hsieh, W.C.; Manohar, C.; Shah, D.O. J. Colloid Interface Sci. (IN PRESS).
46. Gerbacia, W.; Rosano, H.L. J. Colloid Interface Sci. 1973, 44, 242.
47. Reiss, H. J. Colloid Interface Sci. 1975, 53, 61.
48. Ruckenstein, E.; Chi, J.C. J. Chem. Soc., Faraday Trans. II. 1975, 71, 1690.
49. Prince, L.M. J. Colloid Interface Sci. 1969, 29, 216.
50. Prince, L.M. J. Colloid Interface Sci. 1975, 52, 182.
51. Adamson, A.W. J. Colloid Interface Sci. 1969, 25, 261.
52. Levine, S.; Robinson, K. J. Phys. Chem. 1972, 76, 876.
53. Shinoda, K.; Friberg, S. Advances in Colloid and Interface Sci. 1960, 4, 281.
54. Lang, J.; Djavanbakht; Zana, R. J. Phys. Chem. 1980, 84, 1541.
55. Ruckenstein, E.; Chi, J.C. J. Chem. Soc. Faraday Trans., II. 1975, 71, 1690.
56. Shah, D.O.; Hamlin, R.M. Science 1971, 171, 483.
57. Bansal, V.K.; Chinnaswamy, K.; Ramachandran, C.; Shah, D.O. J. Colloid Interface Sci. 1979, 72, 524.
58. Lindman, B.; Kamenka, N.; Kathopoulis, T.M.; Bilsson, P.G. J. Phys. Chem. 1980, 84, 2485.
59. Bansal, V.K.; Shah, D.O.; O'Connell, J.P. J. Colloid Interface Sci. 1980, 75, 462.
60. Sjoblom, E.; Friberg, S.E. J. Colloid Interface Sci. 1978, 67, 16.

61. Ruckenstein, E.; Narayanan, R. J. Phys. Chem. 1980, 84, 1349.
62. Ruckenstein, E. Chem. Phys. Letters 1980, 98, 573.
63. Ruckenstein, E. in "Surfactant in Solution-Theoretical and
    Applied Aspects"; K.L. Mittal, Ed.; Plenum Press: New York,
    1983; Vol. 3, p. 1551.
64. Eicke, H.F. J. Colloid Interface Sci. 1979, 68, 440.
65. Miller, C.A.; Neogi, P. J. Amer. Inst. Chem. Eng. 1980, 26, 212.
66. Eicke, H.F. "Topics in Current Chemistry", Springer-Verlag:
    Berlin, 1980: Vol. 87.
67. Eicke, H.F. J. Colloid Interface Sci. 1975, 52, 65.
68. Robb, I.D., Ed. "Microemulsions", Plenum Press: New York, 1982.
69. Shah, D.O. and Schechter, R.S., Eds. "Improved Oil Recovery by
    Surfactant and Polymer Flooding", Academic Press: New York, 1977.
70. Shah, D.O.; Walker, R.D., Jr.; Hsieh, W.C.; Shah, N.J.;
    Dwivedi, S.; Nelander, J.; Pepinsky, R.; Deamer, D.W. SPE
    5815 presented at Improved Oil Recovery Symposium, 1976.
71. Lindman, B., Kamenka, N., Brun, B., Nilsson, P.G. in "Micro-
    emulsions"; I.D. Robb, Ed.; Plenum Press: New York, 1982; p. 115.
72. Lang. J., Djavanbakht, A., Zana, R. in "Microemulsions";
    I.D. Robb, Ed.; Plenum Press: New York, 1982; p. 238.
73. Friberg, S.; Lapczynska, I.; Gillberg, G. J. Colloid Interface
    Sci. 1976, 56, 19.
74. Lagues, M.; Ober, R.; Taupin, C. J. de Physique Letters 1978,
    39, 487.
75. Shinoda, K.; Saito, H. J. Colloid Interface Sci. 1968, 26, 70.
76. Saito, H.; Shinoda, K.; J. Colloid Interface Sci., 1970, 32, 647.
77. Talmon, Y.; Prager, S. J. Chem. Phys. 1978, 69, 517.
78. Scriven, L.E. in "Micellization, Solubilization and Microemul-
    sions"; K.L. Mittal, Ed.; Plenum Press: New York, 1977;
    Vol. 2, p. 877.

RECEIVED December 20, 1984

MICROEMULSIONS

Onset of Chaos in the Structure of Microemulsions

E. RUCKENSTEIN

Department of Chemical Engineering, State University of New York at Buffalo, Buffalo, NY 14260

A thermodynamic treatment of microemulsions is developed to explain: (1) the transition from a single
phase microemulsion to one which coexists with an excess dispersed phase, (2) the transition from a two
to a three-phase system in which a middle phase microemulsion is in equilibrium with both the excess phases,
and (3) the change in structure which occurs near the
latter transition point. In addition, the same treatment is employed to explain observations concerning the
ill-defined, fluctuating, interfaces which can arise
inside some single phase microemulsions. Because two
length scales characterize a microemulsion, two kinds
of pressures are defined: the micropressures, defined
on the scale of the globules, and the macropressure,
defined on a scale which is large compared to the size
of the globules. In contrast to the former pressures,
the macropressure is based on a free energy which also
accounts, via the entropy of dispersion of the globules
in the continuous phase, for the collective behavior of
the globules. The macropressure constitutes the thermodynamic pressure of the microemulsion and is equal to the
external pressure p. An excess dispersed phase coexisting with a microemulsion forms when the micropressure
p_2 inside the globules becomes equal to p. A third

phase, the excess continuous phase, appears when, in
addition, the micropressure p_1 in the continuous phase

becomes equal to p. The change in structure inside
some single phase microemulsions as well as near the
transition from a two to a three phase system occurs
near the point where $p_2 = p_1$. This happens because,

under such conditions, the spherical interface between
the dispersed and continuous media of the microemulsion
becomes unstable to thermal perturbations. These
fluctuations in the shape of the dispersed phase

0097-6156/85/0272-0021$06.00/0

lead to chaotic variations of ill-defined structures
of the microemulsion. The possible existence of a
cascade of continuous changes from a chaotic microemulsion to molecular dispersion is also noted.

Since Schulman and his coworkers discovered microemulsions (1-3), the
preferred pathway to prepare them was to start from an emulsion stabilized by the adsorption of surfactant molecules on the surface of
the globules. The addition of a cosurfactant – a medium length alkyl
alcohol – generates an emulsion, containing globules of almost uniform size, lying between 10^2 and 10^3Å. In contrast to the conventional emulsions, this new kind of emulsion is, in general, optically
transparent and thermodynamically stable. Because of the small size
of its globules, it was called a microemulsion. The spherical shape
attributed to the globules of the dispersed phase is a result of not
only our image about the conventional emulsions, but also of numerous
experimental investigations (2-7).
 An alternate pathway, which does not generally use a cosurfactant (8-10), starts from micellar solutions and involves a nonionic
surfactant with polyoxyethylene head group or a double chain ionic
surfactant, such as Aerosol OT. It is well known that surfactants
dissolved in water form, above the critical micelle concentration, a
substantial number of large aggregates (micelles) (11,12). Hydrocarbon molecules, though insoluble in water, can be solubilized either among the hydrocarbon chains of the micelles (13-15) or for some
surfactants, such as those with polyoxyethylene head group, the solubilized molecules can form a core surrounded by a layer of surfactant
(a microemulsion). Micellar aggregates are spherical for small sizes
and cylindrical for large sizes (11). When the solubilization occurs
among the hydrocarbon chains, spherical or cylindrical shapes are
also expected to occur (14,15), depending upon the size of the
aggregate. When the aggregate consists of a core of hydrocarbon molecules protected by a layer of surfactant, intuition suggests that a
spherical structure will result for aggregates of any size (14,15).
However, since the interfacial tension at the surface of the globules
is very low, the entropic freedom which a non-rigid, non-spherical
shape can provide may overcome the effect of the increase in area. As
a result, non-spherical globules could be preferred thermodynamically.
 Surfactants which are dissolved in oil form small, non-spherical
aggregates (16,17). However, the presence of water favors the formation of "swollen micelles", containing a core of solubilized water
molecules. Again, intuition suggests a spherical shape for these
globules.
 For completeness, let us note that if the volume fraction of the
dispersed phase becomes sufficiently large, it is expected that the
interactions between the globules will affect their shape. As these
interactions become greater, a percolation threshold occurs. The
electrical conductivity of the system increases steeply at the
threshold because of the transient interconnections between an infinite number of globules of water.
 More recent experiments have indicated that the structure of the
dispersion could be much more complex. The first of these observations resulted from a study of the phase behavior of microemulsions.

This is best illustrated by what happens when the amount of salt is increased, for fixed amounts of oil, water, surfactant and cosurfactant (18-20). For relatively low amounts of salt, an oil in water microemulsion coexists with excess oil phase; at sufficiently high salt content, a water in oil microemulsion is in equilibrium with excess water phase, while at intermediate salinities, a (middle phase) microemulsion coexists with both water and oil excess phases. Of course, the excess phases contain some surfactant, cosurfactant and even water or oil at concentrations determined by the equilibrium between the phases. In the present context, it is important to note that, in general, the microemulsion contains spherical globules whose size distribution is fairly uniform when it coexists with a single excess phase, but that its structure becomes very complex as soon as it coexists in equilibrium with both excess phases.

Very recently, a novel Fourier transform NMR method was employed by Lindman, et al. (21) to obtain multicomponent self-diffusion data for some single phase microemulsion systems. Because of the large values obtained for the self-diffusion coefficients of water, hydrocarbon, and alcohol, over a wide range of concentrations, the authors concluded that there are no extended, well-defined structures in these systems. In other words, the interfaces which separate the hydrophobic from the hydrophilic regions appear to open up and reform at a short time scale.

The scope of the present paper is to show that the thermodynamic theory developed by the author (22,23) can explain the change in structure which occurs near the point of transition from two to three phases as well as in some single phase microemulsions. In essence, the thermodynamic considerations that follow demonstrate the existence of a transition point in the vicinity of which the spherical shape of the globules is no longer stable and is replaced by a chaotically fluctuating interface. For this reason, it is appropriate to call this transition the spherical to chaotic transition. The next section of the paper contains a qualitative explanation of the above phenomena. The corresponding thermodynamic equations are then derived and an explanation concerning the origin of the middle phase microemulsion as well as the interpretation of the NMR experiments for some single phase microemulsions follow. A discussion regarding the "transition" from macroscopic to molecular dispersion concludes the paper.

Qualitative Considerations

The dispersion of one phase into a second phase in the form of globules leads to an increase in the entropy of the system and results in the adsorption of surfactant and cosurfactant on the large interfacial area thus created. This adsorption decreases the interfacial tension from about 50 dyne/cm, characteristic of a water-oil interface devoid of surfactants, to some very low positive value. In addition, the concentrations of surfactant and cosurfactant in the continuous and dispersed phases are decreased as a result of the adsorption, thereby reducing their chemical potentials. This dilution of the bulk phases leads to a negative free energy change, which we call the dilution effect. Dispersions that are thermodynamically

stable are generated when the negative free energy change due to the
dilution effect and to the entropy of dispersion overcomes the positive product of the low interfacial tension and the large interfacial
area produced. This explains the thermodynamic stability of microemulsions (24).

Let us now consider that the microemulsion contains spherical
globules. These globules are, however, macroscopic bodies. Consequently, two macroscopic length scales characterize a microemulsion.
One of them is the scale of the globules, while the other, which is
much larger than that of the globules, is the scale of the microemulsion. The thermodynamic pressure is defined at the scale of the
microemulsion, on the basis of a free energy which includes the
"macroscopic" behaviour reflected in the entropy of dispersion of the
globules in the continuous phase. In addition, one can, however, define pressures at the scale of the globules. Let us call the pressure
p_2 inside the globule and the pressure p_1 near the globule in the continuous phase micropressures, to stress the fact that they are defined on the scale of the globules and to contrast them to the macro-
(thermodynamic) pressure which is defined on the scale of the entire
microemulsion. A particular globule which has the pressure p_2 inside
feels outside, in its vicinity, the micropressure p_1. Only a "macroscopic" part of the microemulsion, which of course should be large
compared to the size of the globules, feels the thermodynamic pressure. In other words, the micropressures are defined on the basis
of a free energy which does not include the entropy of dispersion of
the globules in the continuous phase (an effect which manifests itself at a scale of the order of the microemulsion, which is large
compared to the size of the globules). Because of the condition of
mechanical equilibrium, the thermodynamic pressure equals the pressure p of the environment (external pressure). It is important to
emphasize the p_2 and p_1 are real pressures at the scale of the globules, and that the thermodynamic pressure is a real pressure in a
volume which contains a large number of globules.

Let us consider one globule at the interface between microemulsion and environment. As long as the micropressure $p_2 < p$, the
globule will stay in the microemulsion and no excess dispersed phase
will appear, because the chemical potential at the pressure p_2 is
smaller than that at the pressure p. The mechanical equilibrium condition between microemulsion and environment is still satisfied because the macro (thermodynamic) pressure is equal to the external
pressure. For $p_2 > p$ the globules will disappear from the continuous phase, i.e., the microemulsion will give way to separate oil and
water phases. An excess dispersed phase in equilibrium with the
microemulsion will appear when $p_2 = p$. Similar considerations show
that a third phase, the excess continuous phase, appears when $p_1 = p$.

Consequently, a three-phase system composed of both excess phases and a microemulsion will form when $p_2 = p_1 = p$. However, the equality

$p_2 = p_1$ is not compatible with the existence of spherical globules,

because, near such a point, the interface becomes unstable to thermal perturbations and, therefore, it fluctuates. Thus, the change in structure observed experimentally occurs near the transition point from a two to a three-phase system, as a result of the equality of the micropressures at that point. Of course, the micropressures p_2

and p_1 can become equal without any phase separation (i.e., without

their common value being equal to p). For this reason, fluctuations of the spherical interface are also expected to arise in some single phase microemulsions. This explains the NMR experiments of Lindman, et al. (21).

The above simple considerations explain the origin of the middle phase, the change in structure associated with its occurrence, as well as the fluctuations of the interface between the continuous and dispersed medium which arise in some single phase microemulsions. While it is difficult to obtain detailed quantitative information on the above behaviour, the thermodynamic equations derived in the following section provide a framework for further theoretical development as well as some additional insight concerning the micropressures and various physical quantities involved.

Basic Thermodynamic Equations

Let us assume that the microemulsion contains spherical globules of a single size. For given numbers of molecules of each species, temperature and external pressure, we consider an ensemble of systems in which the radii and volume fractions of the globules can take arbitrary values. Because the equilibrium state of the system is completely determined by the number of molecules of each species, the temperature and external pressure, the actual values of the radius and volume fraction will result from the condition that the free energy of the system be a minimum.

We start with the observation that the dispersion of the globules in the continuous phase is accompanied by an increase in the entropy of the system and denote by Δf the corresponding free energy change per unit volume of microemulsion. However, let us consider, for the time being, a "frozen" system of globules inside the continuous phase, ignoring Δf. At constant temperature, the variation of the Helmholtz free energy per unit volume, f_0, of the frozen microemulsion can be written, if the actual physical surface of the globules is used as the (Gibbs) dividing surface, as follows (25,22,23):

$$df_o = \gamma dA + Cd(1/r) + \Sigma_i \mu_i dn_i - p_2 d\phi - p_1 d(1 - \phi) \qquad (1)$$

where γ is the interfacial tension, C is the bending stress due to the curvature $1/r$, A is the interfacial area per unit volume of microemulsion, n_i is the number of molecules of species i per unit volume,

μ_i is the electrochemical potential of species i, p_2 and p_1 are the micropressures inside the globules and in the continuous phase, respectively, and ϕ is the volume fraction of the dispersed phase. Of course, the radius r, the area A, the volume fraction ϕ, the interfacial tension γ and the bending stress C correspond to the selected (Gibbs) dividing surface. The area A and the radius r of the globules are related via the expression

$$A = 3\phi/r \qquad\qquad\qquad (2)$$

We note that Equation 1 contains the micropressures p_1 and p_2, since they represent the pressures in the continuous and dispersed media of the frozen state, the collective behaviour expressed via Δf being ignored for the frozen state.

One can remark that the interfacial tension γ, which is defined by a variation in the frozen free energy with area at constant ϕ, n_i and T, includes also those interactions between globules such as the van der Waals, double layer and hydration force interactions, which change with surface area A. However, a change in A at constant ϕ changes also the shortest distance, 2h, between the surfaces of two neighboring spherical globules. Thus, the virtual change used to define γ changes the spatial distribution of the globules and not just their surface area. Therefore, γ includes, in addition to the effect of the above interactions at a given distance 2h, the effect of the reversible work done against these forces, as h changes. Considering the force τ per unit area to be positive for repulsion, the variation 2dh produces, per unit volume of microemulsion, the work $A\tau dh$. A more familiar interfacial tension γ_h, from which the effect of the variation of h is eliminated, can be therefore defined by

$$\gamma \, dA = \gamma_h dA - A\tau \left(\frac{\partial h}{\partial A}\right)_{\phi, n_i, T} dA \qquad\qquad (3)$$

Since the equations written in terms of γ are more compact than those in terms of γ_h, the treatment which follows uses Equation 1 as the starting point.

The free energy per unit volume of microemulsion is given by the sum

$$f = f_0 + \Delta f \qquad\qquad\qquad (4)$$

which, when combined with Equation 1, leads to:

$$df = \gamma \, dA + Cd(1/r) + \Sigma_i \mu_i dn_i - p_2 d\phi - p_1 d(1 - \phi) + d\Delta f \qquad (5)$$

While all the variables r, n_i, ϕ, T and p are necessary to specify an arbitrary state, the equilibrium state of a microemulsion is completely determined by n_i, T and p. The values of r and ϕ will therefore emerge from the condition that the microemulsion be in internal equilibrium, i.e., f be a minimum with respect to r and ϕ (or A and ϕ). This leads to the equations:

$$\gamma = - \left(\frac{\partial \Delta f}{\partial A}\right)_{\phi} - C \left(\frac{d(1/r)}{dA}\right)_{\phi} \quad \text{(constant } n_i \text{ and T)} \tag{6}$$

and

$$P_2 - P_1 = \left(\frac{\partial \Delta f}{\partial \phi}\right)_A + C \left(\frac{d(1/r)}{d\phi}\right)_A \quad \text{(constant } n_i \text{ and T)} \tag{7}$$

For spherical globules, A and r are related via Equation 2. Then Equations 6 and 7 become:

$$\gamma = \frac{r^2}{3\phi} \left(\frac{\partial \Delta f}{\partial r}\right)_{\phi} - \frac{C}{3\phi} \tag{8}$$

and, since $\left(\frac{\partial \Delta f}{\partial \phi}\right)_A = \left(\frac{\partial \Delta f}{\partial r}\right)_{\phi} \left(\frac{dr}{d\phi}\right)_A + \left(\frac{\partial \Delta f}{\partial \phi}\right)_r$

$$P_2 - P_1 = \left(\frac{\partial \Delta f}{\partial \phi}\right)_r + \frac{r}{\phi} \left(\frac{\partial \Delta f}{\partial r}\right)_{\phi} - \frac{C}{r\phi} \tag{9}$$

It is instructive to write Equation 9 in another equivalent form. At constant m, where m $(= \phi \frac{4}{3} \pi r^3)$ is the number of globules per unit volume of microemulsion,

$$\left(\frac{\partial \Delta f}{\partial r}\right)_m = \frac{3\phi}{r} \left(\frac{\partial \Delta f}{\partial \phi}\right)_r + \left(\frac{\partial \Delta f}{\partial r}\right)_{\phi}$$

Combining this equation with Equations 8 and 9, one obtains:

$$P_2 - P_1 = \frac{2\gamma}{r} - \frac{C}{3\phi r} + \frac{r}{3\phi} \left(\frac{\partial \Delta f}{\partial r}\right)_m \tag{10}$$

Equation 10 reveals more obviously that Equation 9 constitutes a generalized Laplace equation.

The mechanical equilibrium condition between microemulsion and the environment (which is at the constant pressure p) provides a second relation between p_2 and p_1. Indeed, the variation of the Helmholtz free energy F of the entire microemulsion can be written as

$$dF = \gamma d(AV) + VCd(1/r) + \Sigma \mu_i dN_i - p_2 d(V\phi) - p_1 d(V(1 - \phi))$$

$$+ d(V\Delta f)$$

where V is the volume of the microemulsion and N_i is the number of molecules of species i in the entire microemulsion. Considering a variation dV of the volume at constant T, p and N_i, the mechanical equilibrium condition with the environment yields the equation:

$$\gamma d(AV) + VCd(1/r) - p_2 d(V\phi) - p_1 d(V(1 - \phi)) + d(V\Delta f)$$

$$- pdV_e = 0 \qquad\qquad (11)$$

where $dV_e = - dV$ is the variation of the volume of the environment. Combining Equation 11 with Equations 8 and 9, one obtains:

$$\frac{3\phi}{r}\gamma - (p_2 - p_1)\phi + (p - p_1) + \Delta f = 0 \qquad\qquad (12)$$

The system of Equations 8, 9 and 12, lead to the following expressions for p_2 and p_1:

$$p_2 - p = \Delta f + (1 - \phi) \left(\frac{\partial \Delta f}{\partial \phi}\right)_r + \frac{r}{\phi} \left(\frac{\partial \Delta f}{\partial r}\right)_\phi - \frac{C}{\phi r} \qquad\qquad (13)$$

and

$$p_1 - p = \Delta f - \phi \left(\frac{\partial \Delta f}{\partial \phi}\right)_r \qquad\qquad (14)$$

Equations 13 and 14 relate the micropressures p_2 and p_1 to the external pressure p,C and to Δf and its derivatives.

It is obvious that the mechanical equilibrium condition requires the macro (thermodynamic) pressure of the microemulsion to be equal to p. Equations 13 and 14 reveal that the macropressure is equal to p_2 or p_1 plus terms which arise as a result of the collective behavior of the globules reflected in Δf and also due to the curvature effect.

The Origin of the Middle Phase Microemulsion and of its Structure

The chemical potential in the microemulsion phase is defined as $\frac{\partial f}{\partial n_i}$ at constant T, A, ϕ and n_j (with $j \neq i$). Because the free energy Δf due to the entropy of dispersion of the globules in the continuous phase can be considered a function of only r and ϕ and independent of n_i (26,27), it follows that $\frac{\partial f}{\partial n_i}$ is equal to the chemical potential μ_i in the frozen state.

Assuming the concentrations of various components to be the same in the globules and the excess dispersed phase, the equality of the chemical potentials is equivalent to the equality of the pressures. The chemical potentials μ_i in the dispersed phase are expressed in Eq. (5) at the micropressure p_2. Since in the excess dispersed phase the pressure is equal to the external pressure, an excess dispersed phase forms when

$$p_2 = p \qquad\qquad (15)$$

which, when introduced in Equation 13, provides the expression

$$\Delta f + (1 - \phi) \left(\frac{\partial \Delta f}{\partial \phi}\right)_r - \frac{C}{\phi r} + \frac{r}{\phi} \left(\frac{\partial \Delta f}{\partial r}\right)_\phi = 0 \qquad (16)$$

Equations 8 and 16 provide the basic thermodynamic equations which can predict the dependence on salinity, of the equilibrium radius r and the volume fraction ϕ at the transition between the region in which a microemulsion phase forms alone and that in which it coexists with an excess dispersed phase. Any addition to the system of excess dispersed phase having the same composition as the globules will change neither ϕ nor r in the microemulsion, as soon as the transition point is reached. To carry out such calculations, explicit expressions are needed for the interfacial tension γ as a function of the concentrations of surfactant and cosurfactant in the continuous phase, of salinity and radius r, as well as expressions for C and for the free energy Δf. The interfacial tension depends on the radius for the following two reasons: If the radius were increased at constant ϕ, the surface area becomes smaller; this decreases the total amount of surfactant and cosurfactant adsorbed, but, because the system is closed the concentrations in the bulk become larger and the amount adsorbed per unit area increases. Thus, the interfacial tension decreases. In addition to the above mass conservation effect, there is also a curvature effect, due to the following relation which exists between γ and C (the generalized Gibbs adsorption equation (25,23)):

$$\frac{\partial \gamma}{\partial r} = - \frac{C}{3\phi r} \qquad \text{(constant } \mu_i \text{ and T)}$$

It is difficult to derive an equation for γ, particularly at high electrolyte concentrations. However, for a planar interface, with only a single surfactant, such an equation was recently established (28). It is also difficult to establish a relation for C. As for Δf, expressions have already been derived, either on the basis of a lattice model (26), or on the basis of the Carnahan-Starling approximation for hard spheres (27). These expressions could be introduced in Equations (8) and (16) to relate r and ϕ to γ and C, or perhaps even more meaningfully, to relate γ and C to r and ϕ.

Similarly, from the equality of the chemical potentials in the continuous and excess continuous phases one concludes that a third phase, the excess continuous phase, appears, when, in addition to $p_2 = p$, we also have

$$p_1 = p \qquad (17)$$

which, when combined with Equation 14, leads to

$$\Delta f - \phi \left(\frac{\partial \Delta f}{\partial \phi}\right)_r = 0 \qquad (18)$$

A change in structure is expected to occur near the transition to the three phase system, since the equality $p_1 = p_2 = p$ is not

compatible with spherical globules. Large fluctuations of the interface between the two media of the microemulsion are expected to
occur in the vicinity of such a point. This behaviour, which is similar to that occurring near a critical point, explains the results
of the light scattering experiments of Cazabat et al. (7), obtained
near the transition to the three-phase system and in the middle phase
region. These authors note the increase in turbidity, the decrease
of the diffusion coefficient, the large thicknesses of the two interfaces, as well as the very small values of the interfacial tension
between the middle phase and each of the excess phase, near the transition point. It is also of interest to note that the interfacial
tensions between the microemulsion and each of the excess phases could
be represented by expressions which are valid near a critical point
(29). While the above observations are typically valid in the vicinity of a critical point, they are, in the case of microemulsions, a
result of the fluctuations of the spherical interface, which occur
in the vicinity of the point where $p_2 = p_1 = p$. The above thermody-

namic equations are no longer valid in their initial form for the
middle phase microemulsion, because, in this case, we no longer have
spherical globules of only one size. Additionally, chaotic breakup
and coalescence are also probably taking place.

   To transform the above equations into a predictive tool is not an
easy task, for reasons already outlined. They constitute, however, a
thermodynamic framework on the basis of which further development
could follow.

Discussion of The NMR Experiments in Single Phase Microemulsions

As already noted, a particular globule, which has the pressure p_2 in-

side, feels outside, in its vicinity, the micropressure p_1. The con-

dition of mechanical equilibrium of the spherical interface of the
globules requires $p_2 > p_1$. However, it is important to realize that,

in contrast to the case of a liquid droplet surrounded by its vapors
(which is treated in Reference 25), this inequality does not require
$\frac{2\gamma}{r} - \frac{c}{3\phi r}$ to be a positive quantity, since the additional term $\frac{r}{3\phi}\left(\frac{\partial \Delta f}{\partial r}\right)_m$

which appears in Equation 10 is always a positive quantity. Indeed,
the increased volume exclusion (which arises when the radius r increases at constant number m of globules) decreases the disorder and
hence the entropy of dispersion of the globules in the continuous
phase. As a result, the corresponding free energy Δf increases with
increasing radius, at constant m. As the qualitative discussion of a
previous section demonstrates, one may also note that $p_2 < p$ in single

phase microemulsions and $p_2 = p$ when a microemulsions coexists with an

excess dispersed phase. Therefore, in order to examine the mechanical stability of the spherical interface of the globules, it is not appropriate to compare p_2 with the external pressure as one might be tempted to do.

The spherical shape of the globules is stable to thermal perturbations as long as $p_2 > p_1$ and becomes unstable in the vicinity of the point where $p_2 = p_1$. At this point, Equations 9 and 10 lead to the expressions:

$$\left(\frac{\partial \Delta f}{\partial \phi}\right)_r + \frac{r}{\phi}\left(\frac{\partial \Delta f}{\partial r}\right)_\phi - \frac{C}{r\phi} \equiv \frac{2\gamma}{r} - \frac{C}{3\phi r} + \frac{r}{3\phi}\left(\frac{\partial \Delta f}{\partial r}\right)_m = 0 \qquad (19)$$

The instability of the spherical shape occurs near the point where $\frac{C}{3\phi r} - \frac{2\gamma}{r} = \frac{r}{3\phi}\left(\frac{\partial \Delta f}{\partial r}\right)_m$. Because $\left(\frac{\partial \Delta f}{\partial r}\right)_m$ is always a positive quantity, a necessary condition for the instability to arise is: $\frac{C}{3\phi} > 2\gamma$.

Equations 8 and 19 allow one to calculate, in principle, the values of ϕ and r (as a function of the concentrations of the components involved) in the vicinity of which the above fluctuations arise. However, such calculations could be performed only if explicit expressions become available for γ and C. (Of course, γ and C could be related to r and ϕ by using Equations 8 and 19 as well as one of the expressions already derived for Δf.) Therefore, one can conclude that, near the "critical values" of ϕ and r, (provided by Equations 8 and 19), the interface of the spherical globules starts to fluctuate. As a result, the interface does not keep its identity, the dispersed units breakup and coalesce and the system acquires the chaotic characteristics of turbulence. For this reason, we are tempted to label this structure as chaotic.

Concluding Remarks

In the previous sections, the emphasis was on what we called the spherical to chaotic shape transition. While the dispersed phase maintains its macroscopic dimensions, it possesses an extremely labile, fluctuating interface, which leads to its breakup and coalescence. It is natural to suspect that, at least in some cases, the above transition is the beginning of a cascade of changes during which the macroscopic length scale of the dispersed phase decreases progressively until the molecular scale is reached. To better illustrate this point, let us consider a microemulsion in which the amount of alcohol is continuously increased and assume that the alcohol is soluble in the continuous as well as in the dispersed phases. Because this alcohol is compatible with both oil and water, it can, at least in principle, continuously increase the compatibility of the two. In addition, at a relatively low amount of alcohol, the transition from spherical globules to a chaotic structure occurs. An increase in the amount of alcohol could thus progressively increase both the compatibilitiy of the oil with water as well as the lability of the interface of the macroscopic dispersion until a molecular dispersion of oil, water, surfactant and cosurfactant is achieved.

This work was supported by the National Science Foundation.

Literature Cited

1. Hoar, T. P.; Schulman, J. H. Nature 1943, 152, 102.
2. Schulman, J. H.: Riley, D. R. J. Colloid Sci., 1948, 3, 383.
3. Stoeckenius, W.; Schulman, J. H.; Prince, L. M. Kolloid Z. 1960, 169, 170.
4. Bowcott, J. E. L.; Schulman, J. H. Z. Electrochem. 1955, 59, 283.
5. Falco, J. W.,; Walker, R. D. Jr.; Shah, D. O. AIChE J. 1974, 20, 510.
6. Hwan, R.; Miller, C. A.; Fort, T. Jr. J. Colloid Interface Sci. 1979, 68, 221.
7. Cazabat, A. M.; Langevin, D.; Meunier, J.; Pouchelon, A. Adv. Colloid Interface Sci. 1982, 16, 175.
8. Shinoda, K. J. Colloid Interface Sci. 1967, 24, 4.
9. Shinoda, K.; Ogawa, T. J. Colloid Interface Sci. 1967, 24, 56.
10. Kon-No, K.; Kitahara, A. J.; J. Colloid Interface Sci. 1970, 34, 22; 1971, 37, 469.
11. Tanford, C. "The Hydrophobic Effect", Wiley, New York, 1973.
12. Ruckenstein, E.; Nagarajan, R. J. Phys. Chem. 1975, 79, 2622.
13. Elworthy, P. H.; Florence, A. T.; McFarlane, C. B. "Solubilization by Surface Active Agents", Chapman and Hall: London, 1968.
14. Nagarajan, R.; Ruckenstein, E. Sep. Sci. Technol. 1981, 16, 1429.
15. Nagarajan, R.; Ruckenstein, E. In "Surfactant in Solution - Theoretical and applied Aspects"; Mittal, K. L., Ed.; Plenum Press, New York, 1983; Vol. 2, p. 923.
16. Kertes, A. S.; Gutman, H. M. In "Surface and Colloid Sciente"' Matijevic, E., Ed; Interscience, New York 1976.
17. Ruckenstein, E.; Nagarajan, R. J. Phys. Chem. 1980, 84, 1349.
18. Winsor, P. A. "Solvent Properties of Amphiphilic Compounds", Butterwoth, London, 1954.
19. Healy, R. N.; Reed, R. L.; Stenmark, D. G. Soc. Pet. Eng. J. Trans., AIME 1976, 261, 147.
20. Bourrel, M.; Koukounis, C.; Schechter, R.; Wade, W. J. Dispersion Sci. Tech. 1980, 1, 13.
21. Lindman, B.; Stilbs, P.; Moseley, E. J. Colloid Interface Sci. 1981, 83, 569.
22. Ruckenstein, E. Chem. Phys. Letters 1983, 98, 573.
23. Ruckenstein, E. In "Surfactant in Solution - Theoretical and applied Aspects"; Mittal, K. L., Ed.; Plenum Press, New York, 1983; Vol. 3, p. 1551.
24. Ruckenstein, E., Chem. Phys. Letters 1978, 57, 518; Faraday Discussions of the Chem. Soc. 1978, No. 65, 141.
25. Buff, F. R., J. Chem. Phys. 1951, 19, 1591.
26. Ruckenstein, E.; Chi, J. C. J. Chem. Soc. Faraday Trans. II, 1975, 71, 1690.
27. Overbeek, J. Th. G. Faraday Discussions of the Chem. Soc. 1978, No. 65, 7.
28. Beunen, J. A.; Ruckenstein, E. Advan. Colloid Interface Sci., 1982, 16, 201.
29. Fleming, P. D.; Vinatieri, J. E.; Glinsman, G. R. J. Phys. Chem. 1980, 84, 1526.

RECEIVED June 8, 1984

Stability of Premicellar Aggregates in Water-in-Oil Microemulsion Systems

STIG E. FRIBERG[1], TONY D. FLAIM[1], and PATRICIA L. M. PLUMMER[2]

[1]Chemistry Department, University of Missouri at Rolla, Rolla, MO 65401
[2]Physics Department, University of Missouri at Rolla, Rolla, MO 65401

The stability of premicellar association structures in W/O
microemulsion systems was calculated using the CNDO/2 method.
The results revealed the importance of directed interaction
of water molecules with the polar groups of the amphiphilic
compounds. Even extremely strong hydrogen bonds such as
those between ionized carboxylate and nonionized carboxylic
groups could be intercalated by water molecules with energy
conservation.

The knowledge about microemulsions has reached an advanced state
(1,2); especially so about the fundamentals for their stability (3).

However, some problems remain unsolved. One of them, which we
have found intriguing, is the fact that the systems at the lowest
water concentrations do no show the presence of inverse micelles;
in fact, the W/O microemulsions will tolerate rather large amounts
of water before any colloidal association takes place. This fact
was early pointed out by Shah (4) and the variation of particle size
with water content has been investigated using dielectric (5) methods,
light scattering and electron microscopy (6). These results strongly
indicate the size of the primary aggregates at low water concentrations not to be significantly different from the size of the solvent molecules.

Our interest in this phenomenon is mainly the role of the water
molecules for the stability of such aggregates; an interesting problem against the suggestion by Eicke (7) that small amounts of water
are essential for the stability of inverse micelles of aerosol OT.

In this article we evaluate interactions in a system stabilized
with an ionic surfactant and with a carboxylic acid as the cosurfactant. Such a system is distinguished from the common soap/alcohol
stabilizer combinations by the fact that the soap/acid system does
not require a minimum water concentration to dissolve the soap.

0097–6156/85/0272–0033$06.00/0

Calculation Method

In brief, the CNDO (the acronym stands for complete neglect of
differential overlap) approach is an all valence electron, self-
consistent field calculation in which multicenter integrals have
been neglected and some of the two electron integrals parameterized
using atomic data. Slater type atomic orbitals are used as the basis
2s, $2p_x$, $2p_y$, $2p_z$ for carbon and oxygen. In these calculations two-
electron integrals are approximated as

$$(\mu\nu/\lambda\sigma) = \delta_{\mu\nu} \, \delta_{\lambda\sigma} \, (\mu\mu/\lambda\lambda) = \gamma_{\mu\lambda}$$

where μ etc. stands for Slater Orbitals ϕ_μ, ... centered on the
nuclei. The electron interaction integrals are assumed to depend
only on the atoms to which ϕ_μ, ϕ_ν belong, and not on the specific
orbitals, e. g.

$$\gamma_{\mu\nu} = \gamma_{AB} = \iint s_A^2 \, (1) \, (r_{12})^{-1} \, s_B^2 \, (2) \, d\tau_1 \, d\tau_2.$$

Further

$$(\mu/V_B/\nu) = \delta_{\mu\nu} \, V_{AB}$$

where $-V_B$ is the potential due ot the nucleus of charge Z_B and the
inner shell of atom B and

$$V_{AB} = Z_B \int s_A^2 \, (1) \, (r_{1B})^{-1} \, d\tau_1$$

In addition, the off-diagonal core matrix elements, $H_{\mu\nu}$, are set
proportional to the overlap integral, $S_{\mu\nu}$.

$$H_{\mu\nu} = \beta_{AB}^\circ \, S_{\mu\nu}$$

where β_{AB}° is a parameter determined from atomic spectral data for
Atoms A and B. The specific parameterization used is called CNDO/2.
    The computer codes used for these calculations are modifications
of Dobash's program supplied by QCPE (8). The modifications principally consisted of increased dimensions to handle the large systems,
and a matrix extrapolation routine incorporated into the SCF portion
of the program to enhance convergence.

Results and Comments

The basic unit to be studied was the sodium formate/formic acid complex with two soap molecules and four acid molecules. This number
of molecules in the soap/acid complex has been experimentally
determined (9) for octanoic acid/sodium octanoate. In the present
calculations, the shorter chains are used in order to save the labor
of mapping the geometrics. Earlier calculations (10) have shown the

contributions of the hydrocarbon chain electrons to the head group
interactions to be negligible and in addition, the role of the chains
in mundus realiter is only to provide a hydrophobic shell for the
head group structure. Since we postulate the geometry a priori from
experimental data and since the hydrocarbon chains do not contribute
to the head group interaction to a significant degree, the formic
acid/formate combination model is justified.

The 2/4 soap/acid molecular complex may be structured with two
binding patterns (10). The acid carboxylic groups may be aligned
horizontally, Fig. 1a, or vertically, Fig. 1b. For our present
evaluation, only the vertical alignment will be examined; it offers
distinct advantages for the inclusion of water (11).

When the water molecules are attached to the soap/acid association structure, structural changes may be expected. In the present
investigation, three of these are examined.

1. The water molecules are hydrogen bonded to the polar
    groups with no accompanying change of geometry of the
    association complex.
2. The two soap molecules are separated from each other
    along the horizontal axis bisecting the carboxylic
    groups. The acid molecules retain their position
    relative to the soap molecule, Fig. 2.
3. The acid/soap hydrogen bond is broken and the $-0^-...H..0$.
    distance is increased.

The experimental evidence at hand (12,14) shows a maximum of 14
water molecules to be attached to the 2/4 soap acid before a phase
transition to a liquid crystalline structure occurs. The addition
of water is accompanied by a linear reduction of the number of carboxylic acid/carboxylate hydrogen bonds (13,14).

With this information at hand, an examination of the energy
changes for all the alternatives 1-3 were considered useful in order
to understand the energy foundation for the solubilization of water
into a soap/acid complex.

The energy needed to enhance the horizongtal distance between
the ionized carboxylate group oxygens to a sufficient degree to enable
water molecules to be inserted into the center of the structure was
25.6 Kcal/mole; the oxygen-oxygen distance now being 4.205 Å against
3.570 Å for the original structure. This low value was obtained
through repositioning of the carboxylic acid groups and by adjustment
of the vertical carboxylate group/sodium ion distance to its optimum
value at 3.265 Å.

This expanded structure allowed two water molecules to bind by
two hydrogen bonds to the two ionized carboxylate groups and by a
oxygen/metal ligand bond to the sodium ion. The position of this
water molecule is displayed as H_2O^1 in Fig. 3. Water molecule #2
was located symmetrically and has been omitted in the figure for
reasons of clarity. These water molecules were maximally occupied
in strong bonds which is reflected in their binding energy 37.9 Kcal/
mole water.

It should be noted that the energy released by these two bonds
more than compensates for the energy input to obtain the necessary
expansion to accommodate the two water molecules.

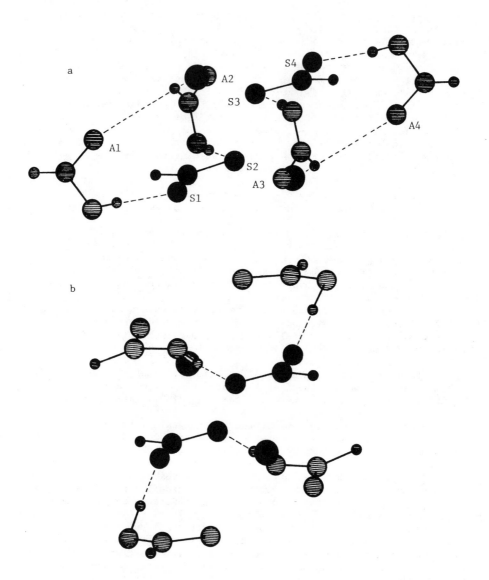

Figure 1.   The 4:2 acid—soap dimer with the acids bridging
horizontally (a) and vertically (b) after Bendiksen et al. (10).
Key: ⊜, acid; ●, soap.

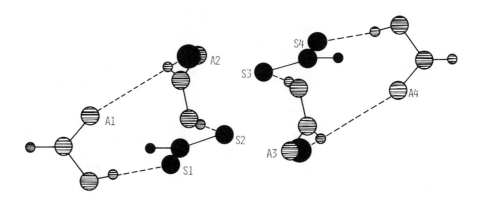

Figure 2. The 4:2 acid–soap dimer with the acids bridging vertically expanded to accommodate two water molecules in between the two main groups. A1 and S1 denote different positions in the <u>a</u>cid and <u>s</u>oap molecules. Key: ⊖, acid; ●, soap.

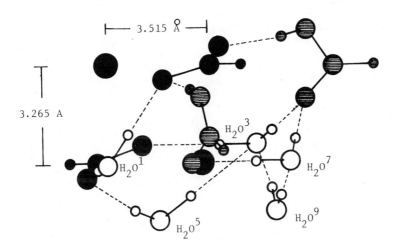

Figure 3. The position of 14 water molecules added to the expanded soap/acid association structures. Water molecule 2 is the symmetric identical to #1, #4 the corresponding pair to #3 and so forth. Key: o, water; ⊖, acid; ●, soap.

In addition, water molecules marked 3 on Fig. 3 and its symmetric location 4 (not included) gave binding energies of 27.6 Kcal/mole water. These molecules may be bound to the unexpanded soap/acid association structure with similar energies involved. The bonding of these two water molecules obviously would provide sufficient initial energy for the expansion of the soap/acid association complex to accommodate water molecules #1 and 2.

Additional water molecules were added according to Figure 3, to a maximum of 14. The energies released are given in Table I with binding sites according to Fig. 1b.

Table I. Association Energies and Binding Sites for Water 1-7

Water Molecule	Binding Site (Fig. 1b)	Association Energy (Kcal/mole-water)
1	S1, S3	-37.9
2	S2, S4	-37.9
3	S2, A4	-27.6
4	S3, A1	-27.6
5	S1, H_2O^3	-18.1
6	S4, H_2O^4	-18.1
7	A3, A4	-22.3

The numbers show an overwhelming stability for inclusion of the water molecules and encouraged the evaluation of energies involved in breaking the strong carboxylic acid/carboxylate hydrogen bonds. In the calculation the water molecules were added without breaking the carboxylate/carboxylic group hydrogen bonds, but experimental evidence (13,14) shows the hydrogen bonds to be reduced to one half of their original number at the point of transition to a liquid crystalline phase and an evaluation of the energies involved was considered useful.

The carboxylate/carboxylic acid group hydrogen bond energies in the expanded structure totalled 95.6 Kcal. Adding this number to the expansion energy means an input of 121.2 Kcal/mole to the soap/acid complex in order to accommodate the water molecules. A comparison of this value with the energy released by 14 added water molecules is illustrative. Addition of the 14 molecules will release an energy of 112.2 Kcal/mole [248 Kcal/mole (Table I) - 14 x 9.7 (evaporation heat of water)] = 112.2 Kcal/mole.

This number is slightly lower than the value for the energy of all the hydrogen bonds to be broken, which is in good agreement with the experimental results showing one half of the hydrogen bonds to be disrupted. The energy needed for that to be accomplished plus the energy for the structural expansion amounts to 73.4 Kcal/mole; a value well below the 112.2 Kcal/mole released by the water bonding according to our present calculations.

So far, the results of the attempts to calculate the magnitude of these interactions are encouraging; the efforts will be continued using more exact methods.

Experimental results (12) showed a transition to a lamellar
liquid crystal for 14 added water molecules. Our calculations (to be
reported at a later occasion) showed no discontinuity or any other
indication of instability of the soap/acid water complex for the subsequent water molecules added in excess of 14. It appears reasonable
to assume that the isotropic liquid/liquid crystal transition does
not depend on the energy levels of the polar group interactions. The
phase transition probably depends on the hydrophobic/hydrophilic
volume ratio and estimations according to Israelachvili/Ninham (15)
approach may offer a better potential for an understanding.

Literature Cited

1. Robb, I. D. (Ed.), "Microemulsions"; Plenum: New York, 1982,
   p. 259.
2. Friberg, S. E. and Venable, R. L., "Microemulsions", Encyclopedia
   of Emulsion Technology,; Becher, P., Ed.; Vol. 1, Chap. 4, pp.
   287-336 , 1983.
3. Ruckenstein, E., J. Dispersion Science & Technol., 2, 1 (1981).
4. Shah, D. O. and Hamlin, R. M., Science , 1971, 171, 483
5. Clausse, M and Rayer, R., "Colloid and Interface Science II",
   Academic: New York, 1976, p. 217.
6. Sjöblom, E. and Friberg, S., J. Colloid Interface Sci., 1978,
   67, 16.
7. Eicke, H. F. and Christen, H., Helvetica Chemical Acta, 1978,
   61, 2258.
8. Dobash, P. A., QCPE, 1974, 10, 141.
9. Söderlund, G. and Friberg, S., Physik. Chem. Neue Folge, 1970,
   70, 39.
10. Bendiksen, B., Friberg, S. E. and Plummer, P. L. M., J. Colloid
    Interface Sci., 1979, 72, 495.
11. Flaim, T., Thesis, University of Missouri at Rolla, Missouri,
    1983.
12. Ekwall, P., "Advances in Liquid Crystals", Academic: New York,
    1975, Vol. 1, p.1.
13. Friberg, S., Mandell, L. and Ekwall, P., Kolloid Z.u.Z Polymere
    1969, 233, 955.
14. Bendiksen, B., Thesis, University of Missouri at Rolla, Missouri,
    1981.
15. Israelachvili, J., Mitchell, D. J. and Ninham, B. W., J. Chem.
    Soc. Faraday Trans II, 1976, 72, 1525.

RECEIVED June 8, 1985

Viscoelastic Detergent Solutions from Rodlike Micelles

H. HOFFMANN, H. REHAGE, K. REIZLEIN, and H. THURN

Institute for Physical Chemistry, University Bayreuth Universitätsstraße 30, D-8580 Bayreuth, Federal Republic of Germany

Results of small-angle-neutron-scattering (SANS), static and dynamic light scattering, NMR and rheological
measurements on solutions of Alkylpyridiniumsalicylates
(RPySal) and Alkyltrimethylammoniumsalicylates (RTASal)
are presented and discussed. The data indicate the existence of rodlike micelles whose lengths L increase linearly with increasing concentration, until a critical
concentration c* is reached at which L approaches to
the mean distance D between the rods. Above c* the L-
values decrease with increasing concentration. Addition
of an electrolyte (NaCl) shifts c* towards lower values.
Below c* the unsheared solutions are not elastic and
have low viscosities; the viscosity and the elasticity
rise drastically when c* is surpassed. In order to explain this elasticity, the formation of a threedimensional network above c* must be assumed. The dynamic nature of this network can be seen from the storage modulus G' which increases with increasing angular frequency ω and reaches a rubber plateau at a characteristic ω-value.

Many surfactant solutions are normal Newtonian liquids even up to rather high concentrations. Their viscosities are very small as compared with the viscosity of the solvent water. This is particularly the
case for micellar solutions with concentrations up to 20% W/W in
which spherical micelles are present. Even in the presence of rodlike micelles the viscosities can be rather low. Systems with rodlike micelles have recently been studied extensively. Missel et al.
(1-2) studied alkylsulfate solutions and showed how the lengths of
the rods can be varied by the addition of salt or by the detergent
concentration. Under all these conditions the solutions are of rather low viscosity. On the other hand, we have studied a number of
cationic detergent solutions in which rodlike micelles were formed
and which all became quite viscous at rather low concentrations. In
addition some of these solutions had elastic properties. The phenomenon of viscoelasticity in detergent solutions is not new. Exten-

0097–6156/85/0272–0041$07.50/0

sive studies on such solutions have been carried out by a number of
groups. Bungenberg de Jong (3) already reported measurements on the
system Cetyltrimethylammoniumsalicylate. This system a few years ago
was extensively studied with different techniques by Lindman (4),
Wennerström (5) and Gravsholt (6). Other viscoelastic detergent systems which were formed from mixed anionic and betain type surfactants were studied by Tiddy (7). Kalus et al. (8) have reported on
viscoelastic perfluordetergents. The most detailed studies by Hoffmann et al. (9-11) were done on Cetylpyridiniumsalicylate. Many different techniques were used in these investigations to characterize
the system and in particular to look at the size of the rodlike micelles as a function of concentration and other parameters and the
influence of these parameters on the rheological behaviour of the solutions. Most of these results have been reported in the symposium in
Lund by Hoffmann (12) last year. In the meantime we have carried out
more measurements on viscoelastic systems and in particular we have
studied the influence of the chain length, ionic strength and temperature on the aggregates. We shall now report some of these new data.

At the beginning, however, we would like to summarize some of
the older results in order to establish a base for the discussion of
the new results. The cationic Cetylpyridinium- or Cetyltrimethylammo-
nium-ion combined with a weakly solvated counterion form rodlike micelles at very low concentrations even in the absence of an inert excess salt. Often there is only a very small concentration range in
which spherical micelles exist and when the concentration is increased
above a threshold value rodlike micelles begin to grow. The threshold
value is sometimes referred to as the cmc_{II}. We prefer the name transition concentration c_t. It is even possible that a concentration
range for spherical micelles does not exist at all and the system begins to form rods already at the cmc. The intermicellar interaction
between the micelles is still repulsive under these conditions. This
is noteworthy to mention because it was recently postulated that micellar rods can be formed only under attractive intermicellar conditions (13). The growth of the spherical micelles to rods is therefore
in these systems determined by the intermicellar interaction and the
monomer-micelle interaction. The reason for the growth of the rods
under these conditions is the strong binding power of the weakly solvated counterions. As a consequence these micelles have a much smaller surface charge density than micelles of the same surfactant ion
but in combination with more hydrophilic counterions like Cl^- od Br^-.
This effect is already evident in the rather low cmc-values of the
systems. Upon a further increase of the concentration above c_t the
rods grow approximately linearly in length with concentration. As
long as the lengths of the rods are smaller than their mean distance
of separation the viscous resistance of the solution is only moderately increased above the one of water. However, in this concentration
range the solutions are rheopectic. Their viscosity increases drastically upon shearing for an extended period of time which depends on
the shear rate.

As soon as the rotational volumes of the rods begin to overlap,
the viscosity increases abruptly several orders of magnitude within
a very small concentration range. At high concentrations even the un-
stirred solutions are viscoelastic. The onset of the elasticity occurs at a rather well defined concentration c*. For the detergent sy-

stem CPySal the threshold value is about 7,5 mM. Both characteristic
concentrations c_t and c* depend strongly on temperature and increase
rapidly with temperature. In the concentration range between c_t and
c* the length of the rods is mainly determined by these two concentrations. The elastic properties of the solution above c* are probably due to a temporary threedimensional network which is formed from
the rods. The dynamic properties of the network determine the rheological behaviour of the solution. The rods seem to be rather stiff
with persistence lengths above 1000 Å. At present we know very little
about the network points or contacts between the rods. Regarding this
problem it should be mentioned that the rods in the network can no
longer freely rotate while the translational diffusion as measured
from dynamic light scattering techniques can proceed almost unhinder-
ed and is influenced by the intermicellar interaction. Of particular
theoretical interest is the result that the lengths of the rods decrease again with increasing concentration above c*. The system ad-
justs the lengths of the rods in such a way that the rotational overlap of the rods is small. For concentrations above c* the dimensions
of the rods seem to be determined by the intermicellar interaction
energy. This will become evident again from the data which will be
presented.

Materials and Methods

The RPySal and RTASal solutions were prepared as previously described by ion exchange procedure from the corresponding chlorides or by
dissolving the salicylates which have been synthesized before. Both
methods gave identical results. The solutions were left standing for
two days in order to reach equilibrium (9-11).

Two types of viscometers were employed for measuring the rheological properties of the detergent solutions. The viscosities at low
dertergent concentrations were determined with a modified Zimm-Crothers viscometer. From these values the size parameters of the aggregates could be calculated. According to Doi and Edwards (14) the viscosity of a dilute solution of rodlike molecules is given by

$$\eta_o \simeq \eta_s \cdot (1 + \hat{c} \cdot L^3) \qquad (1)$$

Here η_o means the viscosity of the solution at zero shear rate, η_s
the viscosity of the pure solvent, $\hat{c}$ the number of rods per unit volume and L the lengths of the rods. As long as L is still shorter
than the mean spacing between the aggregates, the viscosity of the
detergent solution is still very low and always small in comparison
with the solvent viscosity. As a consequence, very accurate measurements are necessary to get some informations on the dimensions of the
rodlike micelles. The Zimm-Crothers viscometer is a very sensitive
instrument and it was possible to measure the viscosity of the dust-
and airfree solutions with an accuracy of 0,2% at very low shear rates (15).

The lengths of the rodlike aggregates can be calculated with
Equation 1 when the concentration $\hat{c}$ is known. The value for $\hat{c}$ can be
obtained either from the electric birefringence measurements or it
can be calculated by the simple expression

$$\hat{c} = c_M \cdot M / (\pi \cdot r^2 \cdot \rho \cdot L) \qquad (2)$$

In this expression r means the short radius of the rodlike micelles, c_M the concentration of the aggregated detergent, M the molecular weight of the surfactant monomers and ρ the density of the rodlike aggregates.

Inserting Equation 2 into Equation 1 leads to an Equation with the only unknown variable L:

$$\eta_o \simeq \eta_s \cdot (1 + K \cdot L^2) \quad \text{with} \quad K = c_M \cdot M/(\pi \cdot r^2 \cdot \rho) \tag{3}$$

Equation 3 is only correct for the dilute solution region.

Interference arises when the rotational volumes of the rods actually touch one another or even more when the rods begin to overlap. The condition for overlapping rods is $\hat{c} \gg 1/L^3$.

Because the normal viscosities change with time, dynamic experiments have been carried out with a rotary viscometer in the oscillating mode (Contraves Low Shear 30 sinus). In this way it was possible to obtain informations without disturbing the internal supermolecular structures. In the general case a sinusoidal deformation or strain is applied. The response of the liquid to the periodic change consists of a sinusoidal shear stress p_{21} which is out of phase with the strain by the phase angle δ. The shear stress is made up of two components. The first one is in phase with the deformation and the second one is out of phase with the strain. From these quantities the storage modulus G' and the loss modulus G'' can be calculated according to the Equations

$$G' = (p_{21}/\hat{\gamma}) \cdot \cos \delta \tag{4}$$

$$G'' = (p_{21}/\hat{\gamma}) \cdot \sin \delta \tag{5}$$

Here p_{21} means the amplitude of the shear stress, $\hat{\gamma}$ the amplitude of the deformation and δ the phase angle between stress and strain.

It is convenient to express the periodically varying functions as a complex quantity which is termed the complex viscosity $|\eta^*|$. This quantity may be calculated by the expression

$$|\eta^*| = \sqrt{(G'^2 + G''^2)}/\omega \tag{6}$$

It can be shown that for most dilute solutions there exists a simple correlation between dynamic and steady state flow characteristics (16). For most detergent solutions the magnitude of the complex viscosity $|\eta^*|$ at a certain angular frequency ω coincides with the steady state viscosity η_∞ at the corresponding shear rate $\hat{\gamma}$ (12, 17).

The simplest mechanical model which can describe a viscoelastic solution is called Maxwell element. It consists of a spring and a viscous element (dashpot) connected in series. The spring corresponds to a shear modulus G_o and the dashpot to a viscosity η. The behavior of the Maxwell element under harmonic oscillations can be obtained from the following equations:

$$G'(\omega) = G_o \cdot \omega^2 \cdot \tau^2/(1 + \omega^2 \cdot \tau^2) \tag{7}$$

$$G''(\omega) = G_o \cdot \omega \cdot \tau/(1 + \omega^2 \cdot \tau^2) \tag{8}$$

From these equations we see that for $\omega \cdot \tau \gg 1$ the storage modulus G' approaches to a limiting value which is identical with the shear modulus G_o. Under such experimental conditions the solution behaves as an elastic body. At low frequencies is $\omega\cdot\tau \ll 1$ and G' becomes proportional to ω^2. This region is called the terminal zone and the solution behaves as a liquid. Viscoelastic solutions can be subdivided into energy and entropy elastic systems. The free energy per volume G at a deformation γ consists of an energetic and an entropic term (18):

$$(\delta G/\delta\gamma) = (\delta U/\delta\gamma) - T\cdot(\delta S/\delta\gamma) \qquad (9)$$

In most real systems, energy and entropy changes can occur. The elasticity of an ideal network is entropy controlled. In this picture stresses are caused by the chain orientation. From the theory of rubberlike elasticity it can be shown that the shear modulus of an ideal network depends on the number of elastically effective cahins between the crosslinks (19): $G_o = \nu\cdot k\cdot T$ where ν means the number of elastically effective chains in unit volume.

In the light scattering experiments the dependence of the scattered intensity on the scattering vector Q is given by

$$I(Q) = f(c)\cdot S(Q)\cdot P(Q) \qquad (10)$$

where $S(Q)$ is the structure factor of the solution, $P(Q)$ is the form factor of the particles and $f(c)$ depends on the detergent concentration c.

For static light scattering $f(c)$ is given by

$$f(c) = K\cdot n^2\cdot(\delta n/\delta c)^2\cdot(c - cmc)\cdot M_o\cdot M_w \qquad (11)$$

where n is the refractive index, $\delta n/\delta c$ the refractive index increment, M_o the molecular weight of the surfactant, M_w the molecular weight of the scatterers and K an optical constant which has the value of $K = 4,079\cdot10^{-6}$ mol/cm^4 for the used Chromatix KMX-6 apparatus. Hence $I(Q)$ is equal to the Rayleigh ratio R_Θ. In the case of SANS $f(c)$ can be written as

$$f(c) = T\cdot d\cdot((b_M - b_s)/\rho_o^2)\cdot(c - cmc)\cdot M_o\cdot M_w \qquad (12)$$

where T is the transmittance of the sample, d the thickness of the probe, b_M and b_s are the scattering lengths of the micelles and the solvent, respectively, and ρ_o is the density in the interior of the micelles in g/cm^3.

The SANS-data are corrected for background scattering, the scattering due to the solvent D_2O and to the quartz cells. In the case of light scattering the scattering of the solvent H_2O was taken into account. The scattering vector Q is given by $Q = 4\cdot\pi\cdot n\cdot\sin(\Theta/2)/\lambda$ where Θ denotes the scattering angle. The wavelength λ of the neutrons was 10 Å, while in the light scattering experiments λ was 6328 Å. In SANS experiments we have $n = 1$ and $Q_{max} = 0,011$ Å^{-1}. For small Q-values the form factor can always be approximated by

$$p(Q) = e^{-(1/3)\cdot Q^2\cdot R_G^2} \qquad (13)$$

where R_G denotes the radius of gyration which is given for cylinders of the length L and the radius R by

$$R_G^2 = (L^2/12) + (R^2/4) \qquad\qquad (14a)$$

Usually Equation 14a may be approximated by

$$R_G^2 = L^2/12 \qquad\qquad (14b)$$

Especially in this Q-range I(Q) is strongly influenced by the structure factor S(Q) which may oppress the intensity by a factor of ten in the case of strong repulsive interaction. In the case where salt is added S(Q) reaches a value around one, however, and is rather independent of the scattering vector in this Q-range. Therefore we made the assumption $S(\Theta=7°) \simeq S(\Theta=173°)$ when we calculated the R_G-values from static light scattering data in the case when salt was added. Assuming that the radius of the rodlike particles is practically given by the length of the alkyl chains, the headgroup and the counterion and that the density in the interior of the micelles is around $0,9$ g/cm^3 we can calculate both the length of the micelles and $S(\Theta=7°)$ from $R(\Theta=7°)$ and $R(\Theta=173°)$ combining Equations 10, 11, 13 and 14.

Results

Results concerning the rheological properties of the studied systems are given in Figure 1 - 9. Figure 1 shows the magnitude of the complex viscosity for several Alkyltrimethylammoniumsalicylates. Note the rapid increase of $|\eta^*|$ at a particular concentration which is characteristic for each system. Figure 2 shows the storage modulus G' and the loss modulus G" as a function of the angular frequency in a double log plot for a 50 mM concentration of Cetylpyridiniumsalicylate. Of theoretical interest is the plateau value of G' which is reached at high frequencies. The Figures 3a and 3b give the same plots for different concentrations. The plateau values of G' are not reached for all concentrations. Figure 4 gives the $|\eta^*|$-values for different chain length detergents against the angular frequency. The detergent concentration, the ionic strength and the temperature are constant. Note the drastic decrease of $|\eta^*|$ and the shift on the frequency scale. With increasing chain length of the detergent $|\eta^*|$ becomes frequency dependent. Figures 5a and 5b give the corresponding G'- and G"-values for the systems of Figure 4. The Figures 6 and 7 show the influence of added salt on the rheological properties. Figures 6a and 6b show the change of $|\eta^*|$, G' and G" with the added salt concentration for a 20 mM CPySal solution. In Figure 7 the G'-values are given as a function of the frequency for a 20 mM surfactant solution with different salt concentrations. Note that the frequency range of the plateau is first increasing and then decreasing again with the salt concentration. The level of the G'-plateau, however, is little affected by the salt concentration. Figure 8 shows the temperature dependence of the G'-values for a solution of 25 mM CPySal. Figure 9 finally gives the $|\eta^*|$-values for a perfluorosystem against the concentration. The Figures 10 - 12 show data from scattering experiments. Figure 10 gives the R_Θ-values for static light scattering data on RTASal solutions of different chain lengths plotted against the surfactant concentration. All solutions contained

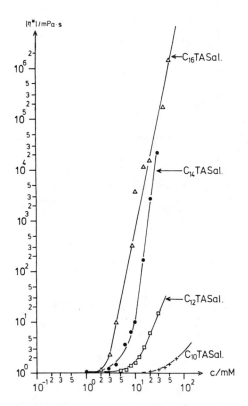

Figure 1. The Magnitude of the Complex Viscosity as a Function of the Detergent Concentration for Different Alkyltrimethylammoniumsalicy-lates (T = 20°C, ω = 0,01 s⁻¹).

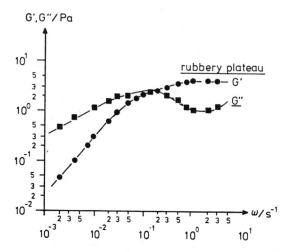

Figure 2. The Storage Modulus G' and the Loss Modulus G" as a Function of the Angular Frequency (CPySal, T = 20°C, c = 50 mM).

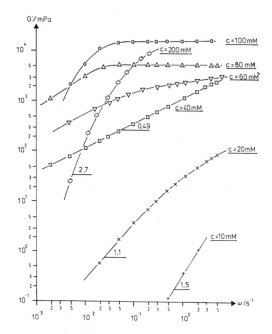

Figure 3a. The Storage Modulus G' as a Function of the Angular Frequency for Different Concentrations of CPySal at T = 20°C.

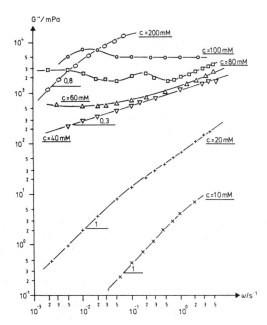

Figure 3b. The Loss Modulus G" as a Function of the Angular Frequency for Different Concentrations of CPySal at T = 20°C.

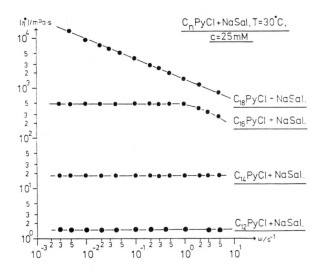

Figure 4. The Complex Viscosity as a Function of the Angular Frequency for Different Chain Homologs of Alkylpyridiniumsalicylates (RPyCl + NaSal, T = 30°C, c = 25 mM).

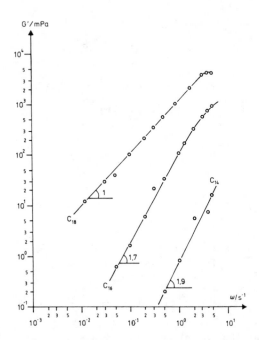

Figure 5a. The Storage Modulus G' as a Function of the Angular Frequency for Different Chain Lengths of Alkylpyridiniumsalicylates
(RPyCl + NaSal, T = 30°C, c = 25 mM).

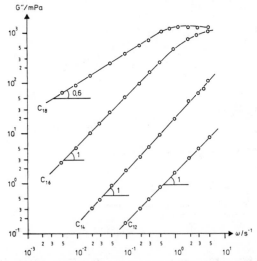

Figure 5b. The Loss Modulus G" as a Function of the Angular Frequency
for Different Chain Homologs of Alkylpyridiniumsalicylates (RPyCl +
+ NaSal, T = 30°C, c = 25 mM).

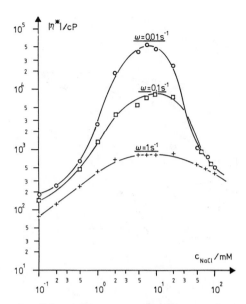

Figure 6a. The Complex Viscosity as a Function of the NaCl-Concentration for Solutions of 20 mM CPySal at T = 20°C .

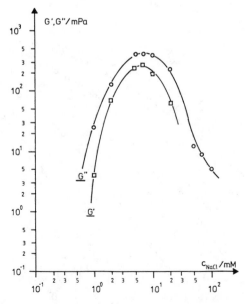

Figure 6b. The storage Modulus G' and the Loss Modulus G" as a Function of the NaCl-Concentration for Solutions of 20 mM CPySal at T = 20°C.

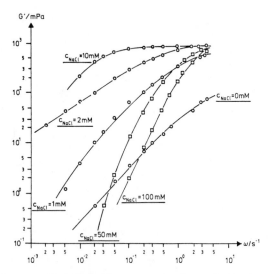

Figure 7. The Storage Modulus G' as a Function of the Angular Frequency for Solutions of 20 mM CPySal with Different NaCl-Concentrations at T = 20°C.

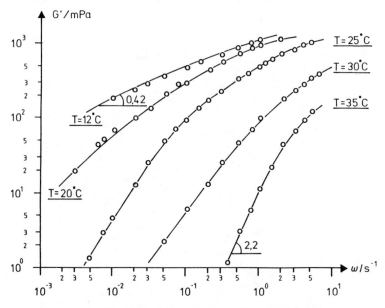

Figure 8. The Storage Modulus G' as a Function of the Angular Frequency for a 25 mM CPySal Solution at Different Temperatures.

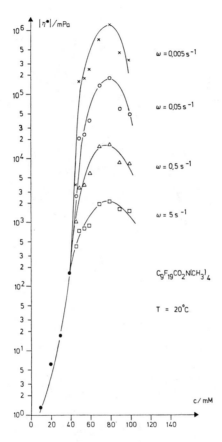

Figure 9. The Complex Viscosity as a Function of the Detergent Concentration for Solutions of $C_9F_{19}CO_2N(CH_3)_4$ at T = 20°C.

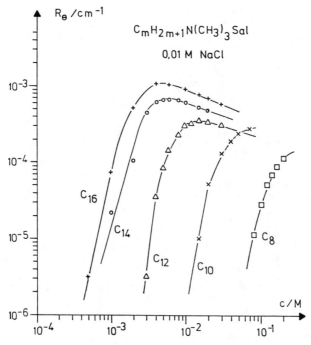

Figure 10. Rayleigh Ratio R_Θ at $\Theta = 7$ as a Function of Detergent Concentration c for Different Chain Lengths at Constant Amount of 10 mM NaCl.

0,01 M NaCl. For each system the forward scattering passes over a maximum with increasing concentration. Figure 11 shows the radius of gyration of the aggregates which were evaluated from the ratio of the forward and backward scattering. Again these values pass over a maximum. Figure 12 gives neutron scattering data for the system CPySal for different concentrations. In this case no salt was added and the influence of the structure factor S(Q) is evident. Furthermore from the scattering maxima the mean distances $\langle r \rangle$ between the micelles can be calculated and as a result we get $\langle r \rangle \sim c^{-(3/2)}$ in this concentration range [20]. This seems to be a general behaviour of detergent solutions. If we assume the radius of the cylinders approximately constant we get [20]

$$\langle r \rangle = k' \cdot (c - cmc)^{(x-1)/3} \tag{15}$$

with the assumption

$$L = k'' \cdot (c - cmc)^{x} \tag{16}$$

where k' and k'' are constants. The dependence of $\langle r \rangle$ on (c - cmc) reflects the growth of the micelles in a certain concentration range. Figure 13 shows a double log plot of $\langle r \rangle$ against (c - cmc) for CTASal with 0,01 M NaCl added. In the dilute range we get x = 0,4 which is close to the theoretical value of x = 0,5 [21] and shows that the micelles grow in length as one would expect. In the semidilute range, however, the exponent x turns to x = -0,5 what means a decreasing length of the micelles. Furthermore the exponent is the same we got from SANS data on the system CPySal in the semidilute range [20]; in this case no salt was added. This shows that in this concentration range the growth of the micelles is controlled by repulsive interaction and further growth of the rods can only be gained by adding more and more salt to overcome the repulsive interaction. Furthermore the S(Q)-data in Table III are decreasing with increasing detergent concentration reflecting the increase of repulsive interaction in this concentration range. As the origin of the repulsive interaction is the same in our system we expect a similar behaviour in the semidilute region which can be seen from Figure 12. The last two plots finally are concerned with NMR-data which were obtained on the CPySal system. Figure 14 shows the change of the linewidths of the CH_2-protons of the alkylchain with time when the solution had been heated to 90°C for a short time period and then was quickly cooled to a reference temperature of 25°C. Note that the curves look very similar for 5 mM and 10 mM solutions. The last Figure 15 shows the linewidths after the solutions were equilibrated plotted against the concentration.

Discussion

General Remarks on the Viscoelastic Properties. Some of the figures in which $|\eta^*|$, G' and G'' are plotted against the angular frequency bear resemblance to rheological data on concentrated polymer solutions. In particular the plateau value of G' reminds of the rubber plateau in melts and concentrated polymer solutions. The rubber plateau in polymer solutions is usually explained on the basis of entanglement networks. These entanglements are produced by coiled polymers in which the end to end distance of the coiled molecules is larger than

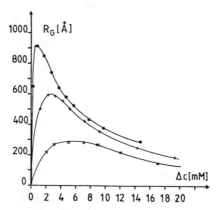

Figure 11. Radius of Gyration at Constant Amount of 10 mM NaCl as a Function of (c - cmc) = Δc for CTASal, C_{14}TASal (+-+-) and C_{12}TASal (x-x-x).

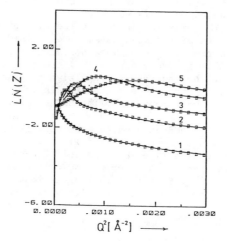

Figure 12. Logarithmic Plot of the Count Rate Z as a Function of the Square of the Scattering Vector Q for Different Concentrations of CPySal (1: 5 mM, 2: 10 mM, 3: 40 mM, 4: 80 mM, 5: 160 mM).

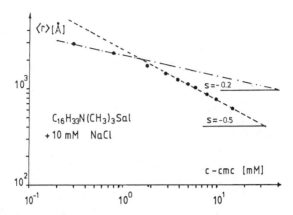

Figure 13. Double Logarithmic Plot of the mean micellar distance against Δc for C_{14}TASal with 10 mM NaCl (The Slope is Given by s = (x - 1)/3 According to Equation 15).

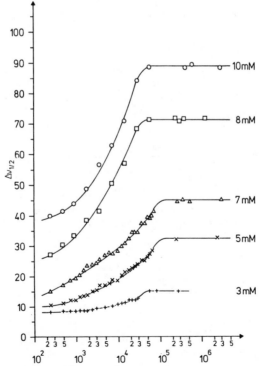

Figure 14. Time Dependence of the Linewidth of the Alkyl Chain Signal for Solutions of CPySal after a Temperature Jump from $90°C$ to $25°C$.

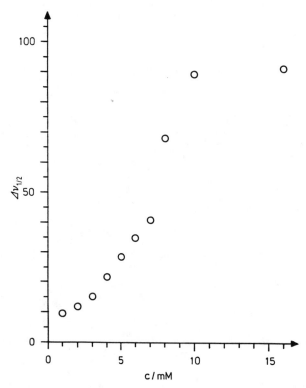

Figure 15. The Concentration Dependence of the Linewidth of the Alkyl Chain Signal for Solutions of CPySal at $T = 25^{\circ}C$.

the mean distance between the polymers. The entanglements therefore
are strictly determined by the dimensions and the concentrations of
the polymer molecules. The interaction energy between the polymers
can be very small or even zero. The situation in the viscoelastic detergent solutions seems to be considerably different. Figures 1 and
3 in combination with Table I show clearly that at the concentration
c* the rods could never form an entanglement network in this classical sence. Their rotational volumes just begin to overlap. On a statistical basis there hardly would be any entanglements at all inspite
of the fact that the solution has elastic properties and behaves like
an entanglement network. The network must therefore be based on interactions between the rods and not on their statistical crowding.

Table I. The Lengths of the Rodlike Micelles as a Function of the Concentration for Solutions of CPySal at Various Temperatures

c/mM	T = 20°C L/Å	T = 25°C L/Å	T = 30°C L/Å	T = 35°C L/Å	T = 40°C L/Å	T = 45°C L/Å	T = 50°C L/Å
1	120	200	70	60	40	40	--
2	210	275	140	110	90	80	60
3	300	350	210	150	145	110	90
4	430	420	270	240	160	170	150
5	550	490	340	275	210	200	160
6	640	570	400	330	280	230	200
7	750	680	460	390	315	275	240
7,5	810	740	500	410	320	300	250

A theory for viscoelastic detergent solutions containing nonspherical particles was recently developped by Hess (22-23) regarding the
coupling between the viscous flow and the molecular alignment. Expressions for the storage modulus $G'(\omega)$ and the loss modulus $G''(\omega)$
are derived (24) for a linear viscoelastic behaviour. On the basis of
this theory a qualitative understanding of our detergent solutions is
possible, but for a quantitative analysis we need more details both
from experiment and theory.

The viscoelastic properties of the solutions depend very much on
the chain lengths of the molecules. In addition to that the rheological properties are strongly influenced by the salt concentration. The
parameters $|\eta^*|$, G' and G'' all pass through a maximum value when the
NaCl concentration is increased. The scattering experiments indicate
that the dimensions of the individual aggregates become always larger
with the salt concentration. On the basis of the entanglement it
would thus be difficult to understand why the viscoelastic properties
should become weaker again.

Dependence of the Shear Modulus on the Concentration.   The experimental results of Figure 3a show that the plateau values increase with
the detergent concentration. Unfortunately, we were not able to reach
the rubber plateau for all concentrations for lack of the frequency
range. From the theory of networks it is possible to calculate the
number of elastically effective chains between the crosslinks from
the shear modulus G_o of the rubber plateau (12). If the network

points are determined by statistics and by intermicellar interaction
the network points should probably be proportional to the number density of the rods. From SANS measurements we could evaluate the number
density of the rods in this concentration range which is given by
$\langle N \rangle = 1/\langle r \rangle^3$. It increases proportional to $c^{3/2}$. This result is at
least in qualitative agreement with the increase of the G_o-values. It
seems to be more difficult to explain the change of the dynamic behaviour of the network with increasing concentration. Up to concentrations of 90 mM the network relaxation times shift to larger values
with increasing detergent concentration. Above this concentration the
relaxation time constants decrease again. The observed changes are
much larger than for the G_o-values. From theoretical points of view
it is possible that this increase of the relaxation time is due to
pretransitional behaviour of the region between the isotropic phase
and a nematic phase (22), and indeed nematic phases were observed at
higher concentrations. A comparison of G_o for 50 mM CPySal and other
systems which were determined by us indicates that the G_o-values do
not vary more than one order of magnitude in completely different systems when the systems are in the concentration range in which the
lengths of the rods are mainly determined by the intermicellar interaction energy. This is clearly evident from the values which are given in Table II for the mentioned systems and several other systems
which have been studied by us.

Table IIa. Values for the Equilibrium Shear Modulus
G_o for Different Aqueous Surfactant Solutions

Detergent System	G_o/Pa	$T/^\circ C$	c/mM
$C_{16}H_{33}N(CH_3)_3C_3F_7CO_2$	16,8	25	50
	90	20	50
CPySal	1,2	20	30
	2,4	20	40
	4,2	20	50
	4,6	20	60
	8,1	20	70
	10,1	20	80
	13,8	20	90
	17	20	100
$C_9F_{19}CO_2N(CH_3)_4$	0,09	20	40
	3,2	20	50
	11,3	20	70
	14,2	20	100
CTASal	0,8	20	15
	1,0	20	20
	2,4	20	30
	3,5	20	40
	4,6	20	50
$C_{14}H_{29}TASal$	0,9	20	30

Table IIb. Values for the Equilibrium Shear
Modulus G_o for Solutions of 20 mM CPySal as a
Function of the Concentration of Added NaCl
at 20 $^{\circ}$C

c_{NaCl}/mM	G_o/Pa
0	0,04
1	0,06
2	0,10
5	0,38
7	0,88
10	0,85
50	0,85
100	0,85

The Dependence of the Viscoelastic Properties on the Chain Length of
the Detergent Ion. Figure 4 shows the magnitude of the complex viscosity for equimolar solutions of C_nPyCl + NaSal as a function of the
angular frequency for different chain lenghts n. In the small frequency region the magnitudes of the complex viscosities are very diffe-
rent. For 2 CH_2-groups the $|\eta^*|$-values change by about one order of
magnitude. The shortening of the chain length leads to a dramatic reduction of the viscous and the elastic properties, as can be seen
from Figures 5a and 5b. We could be tempted to assume that the network becomes weaker with decreasing chain length. The network properties depend on the size and the shape of the micelles. We could argue
that the short-chain detergents do not form rodlike micelles under
the experimental conditions. The light scattering results, however,
seem to rule out this assumption. Typical results are summarized in
Figure 10. The light scattering intensity for forward scattering is
plotted against the concentration of detergent for constant NaCl concentrations. For all four systems the scattering intensity passes
over a maximum with increasing detergent concentration. The same behaviour is shown in Figure 11 by the ratio of the forward to backward
scattering from which it is possible to determine the radius of gyration and hence the lengths of the rods, if we assume that the radii
are practically determined by the alkylchain, the headgroup and the
counterion. We may ask whether other particle shapes are consistent
with our data. From birefringence experiments we know that all these
systems are birefringent; therefore it is only possible to explain
our measurements on the basis of nonspherical particles. Furthermore
at least one dimension of the particles is determined by the lengths
of the monomers. The simplest particle shapes are cylinders and discs.
An interpretation on disclike micelles would lead, however, to very
large aggregates and to very small $S(\theta \to 0)$-values ($\approx 10^{-3}$) and this
seems a rather unlikely result in the presence of 10 mM NaCl. In addition to that we know from SANS-data that the system CPySal which
behaves similar to CTASal forms rodlike micelles which can be seen
from Figure 12 regarding the nonlinear form of $\ln(Z)$ against Q^2 for
higher Q-values which can be seen best from the 5 mM curve as the

others are strongly affected by S(Q) due to intermicellar interaction. The parameters which have been evaluated from these data are given in Table III. These data show that all systems form rods and that in all cases the rods begin rapidly to grow in length with increasing concentration, then with increasing crowding the rods reach a maximum length and finally decrease again in length with increasing detergent concentration.

Table III. Values for the Lengths and the Structure Factors as a Function of the Concentration for Different Surfactant Systems with Constant Amount of 10 mM NaCl Obtained from Static Light Scattering Measurements

CTASal: $R = 22$ Å, cmc = 0,2 mM, $\delta n/\delta c = 0,1844$ ml/g									
c/mM	0,5	1,0	2,0	3,0	4,0	5,0	6,0	8,0	10	15
L/Å	2240	3140	2930	2570	2230	2020	1820	1490	1310	1000
$S(\Theta=7°)$	1,4	3,4	2,3	1,6	1,2	1,0	0,9	0,8	0,7	0,5

C_{14}TASal: $R = 20$ Å, cmc = 0,65 mM, $\delta n/\delta c = 0,1686$ ml/g								
c/mM	2,0	3,0	4,0	5,0	6,0	8,0	10	15	20
L/Å	1750	2060	2030	1910	1750	1480	1240	860	650
$S(\Theta=7°)$	1,6	1,7	1,6	1,5	1,3	1,1	0,97	0,81	0,71

C_{12}TASal: $R = 18$ Å, cmc = 0,285 mM, $\delta n/\delta c = 0,1647$ ml/g							
c/mM	5,0	6,0	8,0	10	12	15	20	30
L/Å	740	960	990	990	930	770	490	330
$S(\Theta=7°)$	1,4	1,3	1,2	1,1	1,0	1,0	1,0	0,91

The light scattering data were carried out at a salt concentration which is somewhat smaller than in the case of the rheological measurements. This, however, has only an effect on the absolute values but not on the relative changes as we have convinced ourselves by experiments. The rheological measurements have been carried out at concentrations at which the size of the rods is determined by the intermicellar interaction. This interaction is very similar for all systems and hence also the sizes of the micelles are similar in this concentration range, too. The large differences in $|\eta^*|$ can therefore not be traced back to different dimensions of the aggregates. The rheological data seem to indicate that the existing structures in the semidilute solutions are probably not too different. With increasing oscillation frequency the $|\eta^*|$-values begin to decrease, first for the longest chain detergent, then for the next longest and so on. Unfortunately we could not carry out experiments with $\omega > 10$ s^{-1}, but from the data we have it looks that the rheological behaviour of all solutions becomes very similar, especially in a frequency range of $\omega \approx 1000$ s^{-1}. This conclusion can be drawn from the G'-values, too. The plots in Figure 5a show that the plateau values of G' seem to be similar for C_{18}- and C_{16}-detergents. For the shorter chain homologs we did not reach high enough frequencies to observe the plateau va-

lues. When we extrapolate such values from the intersection point of
G' and G" they fall in the same order of magnitude. Because these values are proportional to the number of crosslinks of the networks it
is likely that the networks in all systems are very similar. We have
to conclude therefore that the large rheological differences at low
frequencies are not due to different networks but are a consequence
of the different dynamic behaviour of the network. Their characteristic relaxation time constants depend on the chain lengths of the detergent molecules. It is noteworthy to mention that similar plots as
the one in Figure 4 are obtained when the steady state viscosity is
plotted against the shear rate. It can easily be shown that $|\eta^*|$ at a
certain angular frequency coincides with η_∞ at the corresponding
shear rate for most viscoelastic detergent solutions. A limiting condition for this correlation is that the systems do not have a yield
value.

The Influence of Salt on the Viscoelastic Properties. On the application of surfactants it is often necessary to modify the viscosities
of the solutions. As all chemists working with detergents know, this
is very often done by changing the salt concentration. The viscosity
of a solution usually passes over a maximum value when more and more
salt is added. At very high electrolyte concentrations it is sometimes observed that a surfactant solution separates into two liquid
phases. This process is called coacervation. The viscoelastic surfactant solutions also show this kind of behaviour. Typical results are
represented in Figures 6a and 6b. The magnitude of the complex viscosity of a 20 mM CPySal solution has been plotted as a function of the
NaCl-concentration. From these data we can conclude that the added
salt leads to a buildup of structure in the solutions and then to a
breakdown again. The frequency dependent measurements reveal a much
more subtle situation. Results of G' and G" for several salt concentrations are shown in Figure 7. These data show that the plateau values of G' change very little with the salt concentration. It is the
relaxation time of the network which is strongly affected by the salt
concentration. This suggests that the number of crosslinks remains
constant while the dynamic properties of the structures are strongly
influenced by the concentration of the added electrolyte.

The Influence of Temperature on the Viscoelastic Properties. The
viscoelastic properties of the dilute surfactant systems depend on
the temperature of the solutions strongly. Figure 8 shows the values
for the storage modulus G' as a function of the angular frequency at
different temperatures for a 20 mM solution of CPySal. The elastic
properties of the surfactant solutions decrease with increasing temperature. The solution equilibrated at 35°C shows only little elasticity in the frequency range below 1 Hz and at temperatures of 50°C
the solutions behave as Newtonian fluids. The supermolecular structures which are present in these solutions and which are responsible
for the viscoelastic properties seem to be completely destroyed under
these experimental conditions.

The Equilibration Time in Viscoelastic Systems. When working with
viscoelastic detergent systems we should be aware that the systems
may need rather long times to reach equilibrium. The time can be of
the order of days, sometimes longer. In order to obtain reproducible

data it is important to reach equilibrium and the time we have to wait
should therefore be known. In the past we have investigated the structure regeneration process after the system had been heated for a short
time by measuring the change of the viscosity or the buildup of the
electric birefringence. The change in the solutions can also be followed by monitoring the linewidth of the H-NMR-lines. Typical data for
the systems are given in Figure 14. The solutions had been heated for
a short time to a temperature of 80°C and then were quickly cooled
back to 25°C. The measurements show two things which are worth menti-
oning. The equilibration times are not much different for concentrations above or below the concentration c* above which network formation occurs. We can conclude therefore that the formation of the network has nothing to do with the long time constants. It seems more
likely that the slow process is the normal micelle dissolution equilibration which is present in all micellar systems and can vary many orders of magnitude for a given chain length detergent which is combined
with different counterions. This process was shown to be strongly dependent on the chain length.
      The second point of interest is that the linewidth in the equili-
brated state does reflect the transition concentration c*. This is
seen in Figure 15 where the plot of $\Delta\nu$ against c_o shows a break at
around 7,5 mM. This shows that the mobility of the aggregates which
determines the correlation time is further reduced above c*.

Conclusions

All detergent solutions containing rodlike micelles in which the number densities of the rods and their lengths are large enough for their
rotational volumes to overlap have viscoelastic properties. Under
these conditions the rods interact and can build up a supermolecular
structure which has rheological properties of an entanglement network.
The dynamic properties are controlled by the relaxation time constant
τ of this structure and by the number density ν of the contacts. With
little or no interaction between the rods, the relaxation time constant becomes identical with the orientation time of a free rod. Under
these conditions the elastic properties become only apparent at time
scales where the angular frequency reaches the value of the reciprocal
relaxation time. That is about 10^5 s^{-1} for 500 Å long rods. Under
these experimental conditions the storage modulus can be higher than
the loss modulus and reaches a plateau value. The magnitude of this
modulus is determined by the number of crosslinks.The elastic properties for shear rates of the range which are encountered in handling
or swirling solutions in a flask are negligible because the loss modulus G" is much larger than the storage modulus G'. These solutions
furthermore have a low viscosity.
      With increasing attractive interaction between the rods the rela-
xation time constant τ shifts to larger times. As a consequence, the
viscoelastic properties appear in a lower frequency range. The magnitude of the plateau value is little affected by this shift. Usually τ
can be slowed down six orders of magnitude without influencing the
plateau level. Surprisingly G_o is very little dependent on the chain
length of the detergent, the salt concentration and even seems to be
very similar for perfluoro- and hydrocarbon-detergents. It seems to be
only determined by the dimensions and the number density of the rods.
In solutions with overlapping rods these two parameters are controlled

by the intermicellar interaction energy whereby an entropy term seems to play the main role which again does not depend on the chemistry of the system. Therefore for a given concentration of the detergent G_o always lies in the same range. A typical value is 4 Pa for a 50 mM solution.

The results clearly show that the network relaxation time can be varied continuously from 10^{-5} s to many seconds. This means that it is possible that the frequency range in which detergent solutions show elastic properties can be shifted to a desired range.

The results show furthermore that the long time effects which are sometimes observed in viscoelastic detergent solutions are not connected to the network but are determined by kinetic processes in which the rods approach to their equilibrium lengths.

Acknowledgments

We gratefully acknowledge financial support of this work by the Deutsche Forschungsgemeinschaft and the Fonds der Chemischen Industrie.

Literature Cited

1. Missel, P.J.; Mazer, N.A; Benedek, G.B.; Young, C.Y; Carey, M.C. J.Phys.Chem. 1980, 84, 1044
2. Missel, P.J.; Mazer, N.A.; Benedek, G.B.; Young, C.Y.; Carey, M.C. J.Phys.Chem. 1983, 87, 1264
3. Bungenberg de Jong, H.G.; Booij, H.L. "Biocolloids and Their Interaction", Springer Verlag, Wien, 1956
4. Lindman, B.; Wennerström, H. Phys. Rep. 1979, 1, 1
5. Wennerström, H.; Ulmius, J.; Johansson, L.B.A.; Lindblom, G.; Gravsholt, S. J. Phys. Chem. 1979, 83, 2232
6. Gravsholt, S. J. Colloid Interface Sci. 1976, 57, 576
7. Tiddy, G.J.T.; Saul, D.; Wheeler, B.A.; Wheeler, P.A.; Willis, E. J Chem. Soc. Faraday Trans. I 1974, 70, 163
8. Kalus, J.; Hoffmann, H.; Reizlein, K.; Ibel, K.; Ulbricht, W. Colloid Polymer Sci. 1982, 260, 435
9. Hoffmann, H.; Rehage, H.; Platz, G.; Schorr, W.; Thurn, H.; Ulbricht, W. Colloid Polymer Sci. 1982, 260, 1042
10. Hoffmann, H.; Rehage, H. Rheol. Acta 1982, 21, 561
11. Hoffmann, H.; Platz, G.; Rehage, H.; Schorr, W. Adv. Colloid Interface Sci. 1982, 17, 275
12. Hoffmann, H.; Rehage, H.; Schorr, W.; Thurn, H. In "Proceedings of the International Symposium on Surfactants in Solution"; Mittal, K.L., Ed.; in press
13. Nicoli, D.F.; Elias, I.G.; Eden, D. J. Phys. Chem. 1982, 85, 2866
14. Doi, M; Edwards, S.F. J. Chem. Soc. Faraday Trans.II 1978, 74, 918
15. Zimm, B.H.; Crothers, D.M. Proc. Natl. Acad. Sci. US 1962, 48, 905
16. Cox, W.P.; Merz, E.H. J. Polymer Sci. 1958, 28, 619
17. Rehage, H.; Hoffmann, H. Faraday Discuss. Chem. Soc., in press
18. Strenge, K.; Sonntag, H. Colloid Polymer Sci. 1982, 260, 638
19. Flory, P.J. J. Chem. Phys. 1950, 18, 108
20. Kalus, J.; Hoffmann, H.; Ibel, K.; Thurn, H. Ber. Bunsenges. phys. Chem., in press
21. Mukerjee, P. J.Phys. Chem. 1972, 76, 565

22. Hess, S. In " Electro-Optics and Dielectrics of Macromolecules
    and Colloids", Jennings, B.R.; Ed., Plenum Publishing Corp.;
    New York, 1979
23. Hess, S. Z. Naturforsch. 1980, 35a, 915
24. Hess, S. Physica Acta 1983, 118A, 79

RECEIVED January 10, 1985

Enthalpy of Micelle Formation of Mixed Sodium Dodecyl Sulfate and Sodium Deoxycholate Systems in Aqueous Media

K. S. BIRDI

Fysisk-Kemisk Institut, The Technical University of Denmark, Building 206, DK-2800 Lyngby, Denmark

The enthalpy of micelle formation of various mixed sodium dodecylsulfate (NaDDS) and sodium deoxycholate (NaDOC) systems was measured by calorimeter in aqueous systems. The heat of micelle formation, $\Delta H_m^{\ominus}$, showed a maximum around NaDDS:NaDOC molar ratio 1. These data are analyzed in comparison to the aggregation number of mixed micelles and the second virial coefficient, B_2.

The thermodynamic understanding of the aggregation phenomena of surfactant molecules in aqueous media have been investigated by using a wide variety of physico-chemical methods. In recent years, due to the advent of sensitive calorimeters, some enthalpy data on micelle formation have been reported in the literature ([1-11]).

This study is a continuation of our previous investigations, in which the aggregation phenomena of surfactant molecules (amphiphiles) in aqueous media to form micelles above the critical micelle concentration (c.m.c.) has been described based on different physical methods ([11-15]). In the current literature, the number of studies where mixed micelles have been investigated is scarcer than for pure micelles (i.e., mono-component). Further, in this study we report various themodynamic data on the mixed micelle system, e.g., $C_{12}H_{25}SO_4Na$ (NaDDS) and sodium deoxycholate (NaDOC), enthalpy of micelle formation (by calorimetry), and aggregation number and second virial coefficient (by membrane osmometry) ([16]).

Materials and Methods

The microcalorimeter used (LKB, Sweden, bartch 2107) was described in detail in Ref. 11. The mixing procedure was the same as that described in Ref. 11, i.e., the heat of dilution of a surfactant solution (2 mL) was measured on mixing with 2 mL of solvent. In the reference cell the heat of mixing of 2 mL solvent with the

0097–6156/85/0272–0067$06.00/0

same volume of solvent was used, in order to correct for any heat of wetting, etc., inside the cells. The calorimeter was maintained at a constant temperature, 25 ± 0.01 °C.

All chemicals used were of analytical purity grade. NaDDS was used as purchased from B.D.H., U. K. (purity about 99% ([17])). NaDOC was used as supplied by Sigma.

Theoretical Analysis. In these dilution experiments in the calorimeter, the total heat of dilution, q_t, of a surfactant solution will be related to the total concentration and c.m.c. ([1-8,11]):

$$q_t = q_{dil}^m, \quad < c.m.c \qquad\qquad [1]$$

$$= q_{dil}^m + q_{dil}^M + q_{dem}, \qquad > c.m.c. \qquad [2]$$

where q_{dil}^m and q_{dil}^M denote heats of dilution of monomer and micellar species, respectively; and q_{dem} is the heat of demicellization. Because the magnitude of monomeric species, q_{dil}^m, is very small ([1-11]), this quantity can be neglected in the following analyses ([8,11]). Therefore, if we dilute a solution of concentration twice c.m.c. by a factor of two, then we can write ([11]):

$$q_{t,2\ c.m.c.} = q_{dil}^m + q_{dem} \qquad\qquad [3]$$

The heat of demicellization, ΔH_{dem}, can be written ([8,11]):

$$\Delta H_{dem} = q_{t,2\ c.m.c.} / n_M \qquad\qquad [4]$$

$$= q_{dem} / n_M \qquad\qquad [5]$$

where n_M is the concentration of micelles and $\Delta H_{dem} = -\Delta H_{mic}$ (heat of micelle formation). Further, because the experiments are carried out near the c.m.c., it has been argued that the enthalpy measured, ΔH_{mic}, can be assumed equal to the standard micellar enthalpy change, $\Delta H_{mic}^{\ominus}$ ([1-8,11]). All the data measured so far in our laboratory clearly indicate that the heats of dilution show a distinct break, which indicates the difference between the terms q_{dil}^m and q_{dil}^M, as expected from the preceding discussion.

The partial molar enthalpy can be estimated from the slopes of $(q_t/\Delta m)$ ([10]). This analysis will be reported when more data become available. At this stage it is evident that ΔH_m and ΔH_{mic} are constant below and above c.m.c., respectively.

Results

A typical plot of total heat dilution, qt, measured as a function
of concentration (after dilution by as factor of two) of NaDOC:NaDDS
(1:1 molar ratio) is given in Figure 1. The heat of dilution is
endothermic, which means that the heat of micelle formation, when
the concentration is above c.m.c., would be exothermic. These
data also show a break around a region which corresponds to the
c.m.c. as determined by other methods. This observation was
reported from other surfactant systems, e.g., NaDDS and NdeS
(sodium decyl sulfate) (11). Further, it is important that the
micellar equilibria region, i.e., the c.m.c. region, is distinctly
observed in all the systems, regardless of the sizes of micelles
(as discussed later in this chapter). In other words, in the
non-ideal region near c.m.c., the formation of pre-c.m.c.
aggregates (i.e., dimers, trimers, . . . n-mers) is easily
observed from such measurements in all systems with varying
aggregation numbers.

Discussion

In all the measurements carried out in this study, for different
ratios of NaDDS:NaDOC, the dilution curves (Figure 1) of surfactant solution exhibited a clear break that corresponded with the
c.m.c. as determined by other methods. This observation agrees
with literature reports (1-11,18). The present data, however,
show for the first time that mixed micelle systems also behave the
same way as pure micellar systems, as measured by calorimetry.
Further, because the aggregation number, N, of NaDDS is much
larger than that of NaDOC (16), Table I, the variation of enthalpy
around the c.m.c. region is not related to the size of micelles.
At this stage, this analysis cannot be carried out quantitatively.
However, as more data become available such analysis will be
reported.

    The purpose of this study was to analyze the enthalpy data
with the help of data from membrane osmometry on the NaDDS-NaDOC
system (16). This analysis was considered to be necessary, based
on the fact that all current micellar theoretical treatments,
which are based solely on free energy calculations, are empirical;
these theories break down completely for systems at a temperature
different from the one the theory was made to fit (K. S. Birdi,
unpublished). For example, the following relationship between
c.m.c and the aggregation number, N, was given (19):

$$\ln(c.m.c.) = (2\gamma a_0 + g - g')kT$$

$$= ((36\pi v^2/N)^{1/3} 2\gamma + g - g')/kT \qquad [6]$$

$$= k_1/N^{1/3} - k_2$$

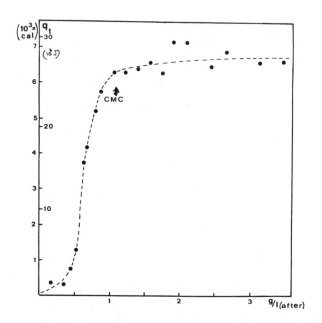

Figure 1. A plot of total heat of dilution (q_r) versus concentration of NaDDS–NaDOC (1:1 molar ratio) (g/L) after dilution by a factor two (at 25 °C, Ionic strength = 0.033, pH = 7.4).

Table I. M_n, N_n, B_2, and ΔH^θ_{mic} Data for Mixed NaDDS–NaDOC Micelles

NaDOC:NaDDS	M_n [a]	N_n [a]	B_2 [a] $(\times 10^5)$ $(\frac{m^3\ mol}{kg})$	ΔH^θ_{mic} (J/mol)
1:0	7354	18	3.0	−920
1:1	13400	38	2.5	−5360
1:2	17730	54	1.9	−4190
0:1	22000	76	0.8	−1590

[a] Data from Ref. 16
Note: Conditions, 25 °C, ionic strength = 0.033, pH = 7.4

where the standard free energy of each monomer in a micelle of aggregation number N is $\mu_N^o = 2ia_o + g$; $N = 4\pi(3v)^2/a_o^3$; a_o is the optimal surface area per amphiphile at the micelle-water interface; the quantity $(g' - g)$ is the hydrophobic energy required to transfer a methylene group (ca. 825 cal/mol or 345 J/mol) from aqueous to micellar phase; v is the volume per monomer; and k_1 and k_2 are constants.

The much-studied NaDDS system was used as a unique example by these investigators to determine the validity of Equation 6. From plots of ln (c.m.c.) versus N^{-3} for NaDDS data in aqueous solutions with varying NaCl solutions (at 21 °C), it was reported that k_1 = 44 and k_2 = 20. From these values, the magnitudes of α = 37 erg dyne/cm (mN/m) and $(g' - g)$ = k_2 kT = 20 kT = 12 kcal/mol (or 49 kJ/mol) were determined. These values were acceptable and of correct magnitudes. We therefore applied the relationship in Equation 6 to another system, DTAB (dodecyltrimethyl ammonium bromide in aqueous solutions with varying concentration of added KBr) at 40 °C. The ln (c.m.c.) versus $N^{-1/3}$ plot was vertical because N remains unaffected by the addition of electrolyte; this result was also reported for NaDDS systems at high temperatures, i.e., ca. 50-60 °C (20, K. S. Birdi, unpublished). We thus find convincing evidence that the exhaustive theories delineated (19,21) are not valid under these circumstances.

Therefore, the present approach was initiated, together with the second virial coefficient, B_2, analyses. It was also shown for the first time (15) that in the case of ionic micelles, the Donnan term of B_2 is proportional to the added salt concentration, m_S^{-1}, as expected from theory. This relationship was valid only in those systems where the aggregation number, N, did not change appreciably with increased m_S. It is thus obvious that ionic micelles must be treated as macro-ions (macromolecules). In the same context, we showed that B_2 goes to zero as the temperature of non-ionic micellar solutions approaches this cloudpoint ("poor solvent") (15).

The mixed NaDDS-NaDOC systems gave the enthalpy of micelle formation, $\Delta H_{mic}^{\ominus}$, which varies with composition as shown in Figure 2.

As the amount NaDOC is increased, the magnitude of H_{mic}^{θ} decreases, but after a minimum (around 1:1 molar ratio) the value increases. In other words, the micellar systems of pure NaDDS and NaDOC exhibit properties that are different with regard to the enthalpic interactions. The value of B_2 varies very little when NaDDS:NaDOC increases from 1:1. These data are in agreement with the H_{mic}^{θ}, where the NaDDS micelles, as formed by the linear alkyl chain, exhibit different energetics than the non-linear alkyl chains of NaDOC. This observation was expected.

Conclusions

The present study reports the variation of enthalpy of micelle formation of mixed NaDDS-NaDOC systems. Our current enthalpy

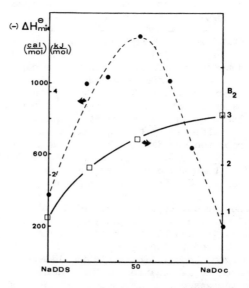

Figure 2. Variation of heat of micellization, ΔH^{Θ}_{mic} (at c.m.c) with the molar ratio NaDDS:NaDOC (at 25 °C, Ionic strength = 0.033, pH = 7.4). Variation of second virial coefficient, B_2 ($\times 10^5 m^3 mol\ kg^{-1}$), with molar ratio of NaDDS:NaDOC.

studies of micellar systems have shown the following (K. S. Birdi, unpublished):

Increase in Alkyl chain	Anionic micelles	$\Delta H_{mic}^{\ominus}$ more exothermic
	Cationic micelles	$\Delta H_{mic}^{\ominus}$ more exothermic
Addition of counter-ions	Anionic micelles	$\Delta H_{mic}^{\ominus}$ more exothermic
	Cationic micelles	$\Delta H_{mic}^{\ominus}$ more endothermic

The enthalpy of a monomer to a micelle of aggregation number, N, can be written as (11):

$$\Delta H_{mic}^{\ominus} = \Delta H_{pho}^{\ominus} + \Delta H_{el}^{\ominus} + \Delta H_{hyd}^{\ominus} \qquad [7]$$

where enthalpies arising from different forces are given: $\Delta H_{pho}^{\ominus}$ is the hydrophobic effect; $\Delta H_{el}^{\ominus}$ arises from the electrostatic interactions; and $\Delta H_{hyd}^{\ominus}$ arises from the hydration of the polar groups. This procedure is analogous to the description used for the free energy of micelle formation, $\Delta G_{mic}^{\ominus}$ (13).

If we compare these $\Delta H_{mic}^{\ominus}$ value variations with the enthalpy of solubility of alkanes in water (21), we find that the latter enthalpy becomes more endothermic with increase in alkyl chain length. The same is valid in the case of n-alcohols solubility data in water (21, K. S. Birdi, unpublished).

Hence, if we argue that the alkyl group of NaDDS is more hydrophobic than NaDOC, then we should have expected endothermic increase with addition of NaDOC. However, from Figure 2 we find the reverse. Thus we conclude that, due to the steric hindrance in the packing of NaDDS and NaDOC alkyl parts, the enthalpy of mixed micelles behaves non-ideally. These conclusions are in agreement with the data of second virial coefficient, B_2 (Figure 2).

At this stage, it is not possible to give a quantitative analysis of these enthalpy data (for each term in the equation). However, work is in progress which is designed to provide the necessary data which would enable us to achieve the former goal.

It is also clear that studies based on c.m.c. and aggregation number data are empirical and cannot provide any quantitative analyses without the enthalpy (and B_2) data.

Acknowledgments

It is a pleasure to thank the Danish Natural Science Research Council for research support of this project. The excellent technical help of J. Klausen is acknowledged.

Literature Cited

1. Pilcher, G.; Jones, M. N.; Espada, L.; Skinner, H. A. J. Chem. Therm. 1969, 1, 381.
2. Goddard, E. D.; Benson, G. C. Trans. Faraday Soc. 1956, 52, 409.
3. Kreschek, G. C.; Hargraves, W. A. J. Colloid Interface Sci. 1974, 48, 481.
4. Espada, L.; Jones, M. N.; Pilcher, G. J. Chem. Therm. 1970, 2, 1.
5. Jones, M. N.; Pilcher, G.; Espada, L. J. Chem. Therm. 1970, 2, 233.
6. Jones, M. N.; Piercy, J. Colloid Polym. Sci 1973, 251, 343.
7. Kishimoto, H.; Sumida, K. Chem. Pharm. Bull. Japan 1974, 22(5), 1108.
8. Pavedes, S.; Tribout, M.; Ferriera, J.; Leonis, J. Colloid Polym. Sci. 1976, 254, 637.
9. De Lisisi, R.; Ostiguy, C.; Perron, G.; Desnoyers, J. E.

   J. Colloid Interface Sci. 1979, 71, 147.
10. Desnoyers, J. E.; Roberts, D; De Lisisi, R.; Perron, G. in "Solution Behaviour of Surfactants"; Mittal, K. L.; Fendler, J. E., Eds.; Plenum: New York; Vol. 1, 1982, p. 343.
11. Birdi, K. S. Colloid Polym. Sci. 1983, 261, 45.
12. Birdi, K. S. in "Colloidal Dispersion and Micellar Behavior"; Mittal, K. L., Ed.; ACS SYMPOSIUM SERIES 9, American Chemical Society: Washington, D.C.; 1975.
13. Birdi, K. S. in "Micellization, Solubilization and Microemulsions"; Mittal, K. L., Ed.; Plenum: New York; 1977.
14. Chattoraj, D. K.; Birdi, K. S. in "Solution Behavior of Surfactants"; Mittal, K. L.; Fendler, J. E., Ed.; Plenum: New York; 1982.
15. Birdi, K. S.; Stenby, E.; Chattoraj, D. K. in "Surfactants in Solution"; Mittal, K. L., Ed.; Plenum: New York; 1983. Proc. Intl. Symp., Lund, Sweden, 1982.
16. Birdi, K. S. Finnish Chem. Lett. 1982, 73(6-8).
17. Birdi, K. S. Anal. Biochem. 1976, 74, 620.
18. Birdi, K. S.; Dalsager, S.; Backlund, S. J. Chem. Soc. Faraday I 1980, 76, 2035.
19. Israelachvili, J. N.; Mitchell, D. J.; Ninham, B. W. Faraday Trans. II 1976, 72, 1525.
20. Mazer, N.; Benedek, G.; Carey, M. C. J. Phys. Chem. 1976, 80, 1075.
21. Tanford, C. "The Hydrophobic Effect"; Wiley: New York; 1973.

RECEIVED January 8, 1985

Role of the Fluidity of the Interfacial Region in Water-in-Oil Microemulsions

A. M. CAZABAT, D. LANGEVIN, J. MEUNIER, O. ABILLON, and D. CHATENAY

Laboratoire de Spectroscopie Hertzienne de l'Ecole Nationale Supérieure, 24 Rue Lhomond, 75231 Paris Cedex 05, France

The behavior of water in oil microemulsions has been studied using different techniques : light scattering, electrical conductivity, viscosity, transient electrical birefringence, ultrasonic absorption. All these experiments lead us to propose a picture of the microemulsions structure which assigns an important role to the fluidity of the interfacial region.

Microemulsions are transparent fluid mixtures of water, oil, surfactant and cosurfactant (alcohols). At small water fraction, w/o microemulsions are dispersions of water droplets surrounded by a surfactant layer in a continuous oil phase. The microemulsion structure at larger water fraction has been studied with different techniques and some results are presented subsequently. A qualitative microemulsion picture is proposed to explain   the data.

Light Scattering Data and Critical Consolute Point

It was proved by scattering studies (1-9) that water in oil microemulsions at low water content are dispersions of identical spherical droplets in a continuous oil phase. Dilution procedure and light scattering measurements (both the intensity and the correlation function of the scattered light) allow to measure the osmotic compressibility and the diffusion coefficient of the droplets :

$$\frac{\partial \pi}{\partial \phi} = \frac{kT}{v} (1 + \beta\phi) \text{ where } v = \frac{4}{3} \pi R^3$$

$$D = \frac{kT}{6\pi\eta_0 R_H} (1 + \alpha\phi)$$

ϕ is the volume fraction of the droplets, R is the droplet radius and R_H the hydrodynamic radius; η_0 is the continuous phase viscosity; α and β are virial coefficient ($\beta=8$, $\alpha=1,5$ for hard-sphere-like systems).

A lot of microemulsions, located on demixing surfaces (At <u>larger</u> alcohol content, <u>inside</u> the 1-phase domain, definite droplets may no longer exist ($\overline{10}$) ($\overline{11}$)) in the phase diagram (Figure 1) have been studied. For this purpose, the constituents (oil and alcohol), the amount of surfactant or the water salinity have been varied. The surfactant was SDS (sodium dodecyl sulphate). The composition for several microemulsions series is indicated in Table I.

The dilution procedure, first proposed by Schulman ($\underline{7}$) has been improved by Graciaa ($\underline{12}$). First, oil is added to a transparent microemulsion. The sample becomes milky and transparency is obtained again by adding alcohol (with a certain amount of water). This is repeated several times. The added alcohol is plotted versus the added oil. A linear plot indicates a constant composition of the added substances which constitute the microemulsion continuous phase. The validity criteria for dilution procedure is ($\underline{13}$) :

$$\phi \lesssim 0.15 \text{ for } -20 \lesssim \beta \lesssim 0$$

$$\phi \lesssim 0.30 \text{ for } \beta \gtrsim 0$$

At larger concentration ϕ , the system cannot be described as a droplets dispersion : a more complicated structure, possibly bi-continuous, appears and deviation from dilution line is observed. (The same occurs for $\beta < -20$ even at low ϕ values ($\underline{13}$)). The measured values of β and R are reported for different microemulsions in Table I. The β values smaller than 8 (hard-sphere-like systems) can be accounted for by introducing a supplementary attractive potential. The osmotic pressure can be written as the sum of two terms : a hard-sphere term and an attractive perturbation. Such a perturbative treatment leads to introduce a hard-sphere radius R_{HS} ($\underline{2}$),($\underline{6}$).

The strength of the attractive potential is found to increase when :
                    -the water to soap ratio is increased (larger droplets)
                    -the alcohol chain length is decreased
                    -the amount of salt is decreased.
The attractive interaction increase can be associated with an increasing fluidity of the interfacial region as evidenced from an increasing difference between R, R_H, R_{HS} (increasing penetration of the continuous phase into the surfactant layer and increasing interpenetration of droplets during a collision). As the droplets interpenetrate during collisions, the Van der Waals forces become very large. The Van der Waals contribution has beeen calculated ($\underline{14}$) and found in good agreement with experimental data (($\underline{14}$) ($\underline{15}$)).

However, the static and dynamic behaviors are well correlated except for small discrepancies which may be accounted for by the transient character of the droplets ($\underline{6}$) (they can exchange constituents during sticky collisions and during very short times : <1μs, much shorter than the droplet diffusion time, longer than 10μs). No satisfactory description of the "averaged" droplet motion is available at present time.

As the attractive potential V between droplets increases, β decreases. The normalized osmotic compressibility curves $v/kT.\partial\pi/\partial\phi$ versus ϕ pass closer and closer to the ϕ axis and become tangent to

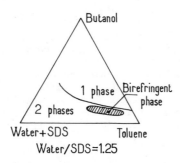

Figure 1.   Phase diagram for a mixture of water, toluene, SDS, butanol (water/SDS ratio:1.25).

Table I. Composition, droplets radius (R) and osmotic virial coefficients for several microemulsions series.

Micro-emulsion	oil	alcohol	water/SDS	salinity	R(Å)	β	Ref
ACP	cyclohexane	pentanol	1.25	0	49	7	8
BCP	cyclohexane	pentanol	2.5	0	75	0	8
ATP	toluene	pentanol	1.25	0	44	-7	6
αTB	toluene	butanol	1	0	34	-12	6
ATB	toluene	butanol	1.25	0	42	-20	6
βTB *	toluene	butanol	1.75	0	60	-30	13
γTB *	toluene	butanol	3	0	90	$\underset{\sim}{<}$-40	13
TB10	toluene	butanol	3.3	10%	150	6	18
TB8	toluene	butanol	5	8%	200	2	18

* *dilution not possible above* ϕ *>0.02*

the ϕ axis for the critical value V_c of the attractive potential.
The contact point $\phi = \phi_c$ is a critical consolute point. The calculated critical values of the virial coefficient and of the droplet
volume fraction ($\beta = -21$ and $\phi_c \approx 0.13$) for a hard-sphere model with
an attractive potential are in qualitative agreement with the experimental observations (Figure 2). Around those critical values, a very
large turbidity is observed. If the temperature is varied, the microemulsion separates into two turbid microemulsions. Angular variations
of the scattered intensity and of the diffusion coefficient are observed (16) but the correlation function remains exponential. All
these features are characteristic of the vicinity of a critical
consolute point. The data can be fitted with theoretical predictions
(17) :

$$I(q) = I_0 / (1 + q^2 \xi^2) \;\; ; \;\; I_\rho \sim \xi^2$$

$$D(q) = \frac{kT}{6\pi\eta\xi} \;\;\; K(q\xi)$$

with $K(X) = \dfrac{3}{4} \left[X^2 + 1 + (X^3 - \dfrac{1}{X}) \; \text{Arctg} \; X \right] \;\; ; \;\; K(0) = 1$

q is the scattering wave vector, η the microemulsion viscosity and
ξ the correlation length of the concentration fluctuations.
        Three determinations of ξ are available (from $I(q)$, $D(q)$ and
$D(0)$)and are all in satisfactory agreement. Values as large as 800 Å
have been measured.
        Interfacial tension and interfacial thickness measurements on
several quasi critical systems with salt are in agreement with this
picture (18).
        The universal character of the scaling laws does not allow to
know whether we observe critical concentration fluctuations of individual molecular constituents or of droplets assemblies. The fact
that measured critical values of β and ϕ are close to the calculated critical values for droplets assemblies ($\beta = -21$ and $\phi_c \sim 0.1$)
indicates that the second hypothesis seems more probable.

Electrical Conductivity

The electrical conductivity of hard-sphere-like microemulsions
increases smoothly as the volume fraction ϕ is increased. On the
contrary, the conductivity of microemulsions with attractive interaction between droplets increases steeply around $\phi_p \sim 0.08-0.14$ (Figure
3). The behavior of the conductivity may be accounted for by percolation theories and ϕ_p is identified to the percolation threshold.
However, in such systems one must distinguish between geometrical
percolation and conductivity percolation.
        For the former case, the theoretical value of ϕ_p is found to be
$\phi_p \sim 0.15$ for hard-sphere-like systems. This value corresponds to
the volume fraction at which an infinite path of droplets in contact
appears.
        However in microemulsions, to ensure ions transport through the
sample, it is not sufficient to suppose that the droplets are merely
in contact.Thus one is led to suppose that during a collision between
two droplets there is an opening of pores in their interfa-

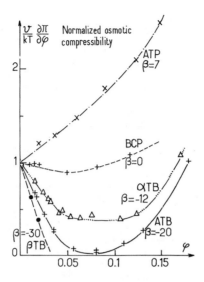

Figure 2. Normalized osmotic compressibility versus droplets
volume fraction for various microemulsions.

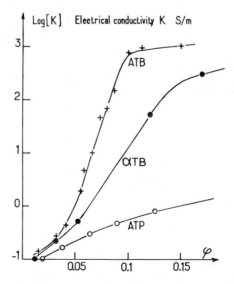

Figure 3. Electrical conductivity versus droplets volume fraction
for three series of microemulsions.

cial layer : the water cores of the droplets may then exchange and
ensure charge transport. This process is possible only in the case
of a fluid interfacial layer which means that the electrical percolation is observed only in the case of systems with sufficiently
attractive interactions between droplets.

Usually geometrical connectivity and concentration fluctuations
are not related. However, in our case, the electrical percolation
is not a simple geometrical connectivity. This fact can explain that
electrical percolation and critical points seem to be associated and
that $\phi_p \sim \phi_c$ in ATB microemulsions (19) (13).

Ultrasonic Absorption and Viscosity Measurements (20),(21)

Ultrasonic absorption and viscosity measurements are reported on
Figures 4 and 5 for the ATP and ATB systems. The viscosity and ultrasonic absorption variations for the hard-sphere-like microemulsion
ATP versus ϕ are progressive while for ATB microemulsion, viscosity
and ultrasonic absorption data show a large anomaly around $\phi_a \sim 0.1$.

Far from this volume fraction, viscosity and ultrasonic absorption of the ATP and ATB microemulsions are identical. In this case,
ultrasonic absorption can be attributed to alcohol exchange trough
the droplets interfacial layers.

Around the ϕ_a concentration, the ultrasonic absorption and viscosity anomalies for ATB microemulsion can be extracted from the
difference between ATB and ATP data. These anomalies cannot be explained by a critical phenomenon : the measured critical exponents
are not the theoretical ones (20),(21). The best explanation is the
opening of pores, needing and opening energy, through the interface
with a cooperative effect when the droplets concentration ϕ is
large enough. This cooperative effect and the clusters formation can explain the viscosities and ultrasonic absorption anomalies
at $\phi = \phi_a$: around this value, the number of pores steeply increases
and is very sensitive to a perturbation like ultrasonic wave or
shearing (22). The transient character of the connections explains
that the observed viscosity anomalies are not as high as in polymer
gelation.

Transient Electrical Birefringence Data (23)

The Kerr constant B is defined as $B = \Delta n/\lambda E^2$ where Δn is the steady
state birefringence induced by the applied electric field E and λ
is the wavelength of the light beam used for detection. Experimental
results for various microemulsions are reported in Figures 6,7,8. In
Figure 6, the contribution of the continuous phases has been substracted ($B_{\phi_c} \sim 0 - 45 \times 10^{-16}$ mV^{-2}).
Let us recall the most important experimental facts :
-the variation of B versus ϕ is never linear even at low volume fraction
-the characteristic time constant of the decay of the induced
birefringence is at least of the order of 1µs
-the decay of the induced birefringence is non exponential in
most cases.
The second fact allows to eliminate, as a possible source of
the induced birefringence, the coupling between the electric field

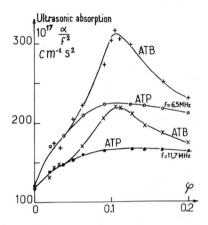

Figure 4. Ultrasonic absorption versus droplets volume fraction in ATP and ATB microemulsions at 6.5 and 11.7 MHZ.

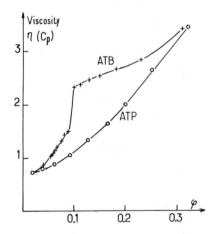

Figure 5. Viscosity versus droplets volume fraction in ATP and ATB microemulsions.

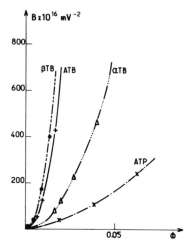

Figure 6. Kerr constant of several series of microemulsions
at low droplets volume fraction ϕ. The contribution of the
continuous phase has been substracted.

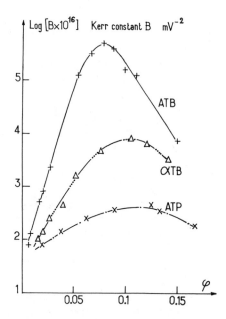

Figure 7. Kerr constant versus droplets volume fraction for three
series of microemulsions.

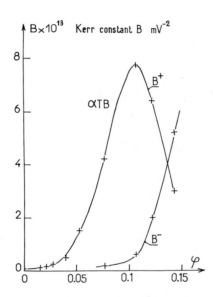

Figure 8. Kerr constant versus droplets volume fraction ϕ for
αTB microemulsions. Key: +, positive contribution; -, negative
contribution.

and the individual molecules of the samples (the characteristic
times would be much smaller).

The non linear behavior of B versus ϕ leads to eliminate
some other mechanisms:

-orientation of elongated droplets, or deformation of
droplets in the electric field

-local coupling of the interfacial layer with the electric
field (this process can be considered only in the case of microemulsions in which the persistence length ξ_p of the interfacial layer
is smaller than the spontaneous radius of curvature of the interface).

We shall then account for the observed phenomena by the orientation of transient aggregates (postulated few years ago (24),
recently observed in attractive systems by neutron scattering (9))
formed during sticky collisions between droplets. Such an explanation agrees with the first experimental fact (non linear behavior
of B versus ϕ) and with the measured characteristic times at low
volume fraction (1 to 10 µs). The characteristic rotation time for
a single droplet with R=50 Å would be 70ns which is much shorter
than the measured times.

B is very small for hard-sphere-like microemulsions ATP and
increases very much for attractive systems (Figure 7).

The Kerr constant around $\phi \sim 0.1$ becomes very large for ATB
microemulsions and the decay curve becomes exponential.The measured
decay time τ is in fairly good agreement with the lifetime τ_ξ of a
critical density fluctuation of size ξ. ξ is the correlation length
of the density fluctuations previously measured in ATB microemulsion
by light scattering and

$$\tau_\xi = \frac{6\pi\eta\xi^3}{kT}$$

Calculated τ_ξ and measured τ values (23) are given in Table II
for ATB microemulsions.

At larger ϕ , a surprising phenomenon was observed : a fast, negative contribution B^- to the birefringence signal arises and
grows rapidly with ϕ(Figure 8). In ATB microemulsion which cannot
be studied at $\phi > 0.15$ because of its large electrical conductivity,
this process could not be observed.

The short decay times associated to B^- ($\sim 1\mu s$) suggest that this
process reflects local reorganizations of the interfacial film
possibly prefigurating the inversion.

Conclusion

The preceding analysis assigns an important role to the interaction between droplets or equivalently the fluidity of the interfacial region in the microemulsions structure.

In attractive microemulsions, the formation of transient aggregates close to ϕp by opening of pores in the interfacial region
during sticky collisions qualitatively explains the experimental
data.

The role of the observed local reorganization in the interfacial
film at a lower scale than droplets aggregates (electrical birefringence data) remains to be understood.

Table II. Calculated (τ_ξ) and measured (τ) values of the birefringence decay time for ATB microemulsions.

ϕ	$\xi(\overset{\circ}{A})$	$\tau_\xi(\mu s)$	$\tau(\mu s)$
0.07	240	120	150
0.08	280	190	230
0.09	300	230	200

Literature Cited

1. Schulman, J. H. and Friend, J. A.   J. Colloid Interface Sci. 1949,
    4, 497.
2. Calje, A. A.; Agterof, W. G. M. and Vrij, A. in "Micellization
    Solubilitzation and Microemulsions"; Mittal, K. L., ed.; Plenum:
    New York, 1977; Vol. 2.
3. Graciaa, A.; Lachaise, J.; Chabrat, P.; Letamendia, L.; Rouch, J.;
    Vaucamps, C.; Bourrel, M.; Chambu, C. J. Phys Lett. 1977, 38;
    253, 1978, 39, 235. .
4. Zulauf, M. and Eicke, H. F. J. Phys. Chem. 1979, 83, 480.
5. Gulari, E.; Bedwell, B.; Alkhafaji, S. J. Colloid Interface Sci.
    1980, 77, 202.
6. Cazabat, A. M.; Langevin, D. J. Chem. Phys. 1981, 74, 3148.
7. Schulman, J. H.; Stockenius, W.; Prince, L. M. J. Phys. Chem.
    J. Phys. Chem. 1959, 63, 1677.
8. Dvolaitzky, M.; Guyot, M.; Lagues, M.; Le Pesant, J. P.;
    Ober, R.; Sauterey, C.; Taupin, C. J. Chem Phys 1978, 69, 3279.
9. Ober, R.; Taupin, C. J. Phys. Chem. 1980, 84, 2418.
10. Stilbs, P.; Lindman, B. J. Phys. Chem. 1981, 85, 2587.
11. Sjoblom, E.; Henriksson, U. in "Surfactants in Solutions";
    Mittal, K. L., ed.; Plenum, New York, 1983, to be published.
12. Graciaa, A. Thesis, Universite de Pau, France, 1978.
13. Cazabat, A. M. J. Phys. Lettres 1983, 44 L 593.
14. Lemaire, B.; Bothorel, P.; Roux, D. J. Phys. Chem. 1983, 87, 1023
    and 1028.
15. Roux, D.; Bellocq, A. M.; Bothorel, P. in "Surfactant in Solutions";
    Mittal, K. L., ed.; Plenum, New York, to be published.
16. Cazabat, A. M.; Langevin, D.; Meunier, J.; Pouchelon, A.
    J. Phys. Lettres 1980, 41, 441.
17. See for instance Swinney, H. L.; Henry, D. L. Phys. Rev. A 1983,
    8, 2586.
18. Cazabat, A. M.; Langevin, D.; Meunier, J.; Pouchelon, A.
    Adv. Coll. Int. Sci. 1982, 16, 175.
19. Sorba, O. These de 3eme cycle, universite Pais VI, 1983.
20. Cazabat, A. M.; Langevin, D.; Sorba, O. J. Phys. Lettres 1982,
    43, L-505.
21. Zana, R.; Lang, J.; Sorba, O.; Cazabat, A. M.; Langevin, D.
    J. Phys. Lettres 1982, 43, L-829.
22. Bennet, K. E.; Hatfield, J. C.; Davis, H. T.; Macosko, C. W.;
    Scriven, L. E. in "Microemulsions"; Robb, I. D., ed.; Plenum:
    New York, 1982.
23. Guering, P.; Cazabat, A. M. J. Phys. Lettres 1983, 44, L601.
24. Eicke, H. F.; Shepherd, J. C. W.; Steineman, A. J. Colloid
    Interface Sci. 1976, 56, 168.

RECEIVED January 8, 1985

Effect of Chain Length Compatibility on Monolayers, Foams, and Macro- and Microemulsions

M. K. SHARMA, S. Y. SHIAO, V. K. BANSAL, and D. O. SHAH

Departments of Chemical Engineering and Anesthesiology, University of Florida, Gainesville, FL 32611

The effect of chain length compatibility of various surfactants on molecular packing, foams, macro- and microemulsion structures, solubilization and oil displacement efficiency in porous media has been studied using different techniques. Moreover, the solubilization of water in microemulsions was studied in detail as a function of alkyl chain length of oils and cosurfactants. The solubilization behavior is discussed in terms of partitioning of alcohol among oil, water and the interface depending upon the chain length of alcohol and oil, as well as in terms of molecular packing at the interface in relation to the disorder produced by the chain length compatibility effects. It is proposed that the chain length compatibility strikingly affects the properties of interfacial film, which in turn influences emulsion stability, foam stability, solubilization capacity, molecular packing at the interface, fluid displacement efficiency and effective gas mobility in oil recovery processes.

The formation and stability of foams and emulsions depend on the structure of surfactants and cosurfactants employed in these systems. It has been reported that the structure of alcohol strongly influences the properties of microemulsions (1-5). Moreover, the interfacial composition and alcohol partitioning between aqueous phase and oil are influenced by the alkyl chain length of oil and alcohol (6,7). The structural aspects of microemulsions using various techniques such as X-ray diffraction, viscometry, light scattering, ultracentrifugation, electron microscopy and electrical conductivity have been reported by previous investigators (8-10). From these studies, it is proposed that microemulsions are isotropic, clear or translucent and thermodynamically stable dispersions of oil, water and emulsifiers with the droplet diameter ranging from 100 - 1000 Å.

0097–6156/85/0272–0087$06.00/0
© 1985 American Chemical Society

Schick and Fowkes (11) studied the effect of alkyl chain length of surfactants on critical micelle concentration (CMC). The maximum lowering of CMC occurred when both the anionic and nonionic surfactants had the same chain length. It was also reported that the coefficient of friction between polymeric surfaces reaches a minimum as the chain length of paraffinic oils approached that of stearic acid (12). In order to delineate the effect of chain length of fatty acids on lubrication, the scuff load was measured by Cameron and Crouch (13). The maximum scuff load was observed when both hydrocarbon oil and fatty acid had the same chain length. Similar results of the effect of chain length compatibility on dielectric absorption, surface viscosity and rust prevention have been reported in the literature (14-16).

The gas/liquid and liquid/liquid systems are relevant to biomedical and engineering applications. The large interfacial area in foams, macro- and microemulsions is suitable for rapid mass transfer from gas to liquid or liquid to gas in foams and from one liquid to another or vice versa in macro- and microemulsions. The formation and stability of these systems may be influenced by the chain length compatibility which may also influence the flow through porous media behavior of these systems. Therefore, the present communication deals with the effect of chain length compatibility on the properties of monolayers, foams, macro- and microemulsions. An attempt is made to correlate the chain length compatibility effects with surface properties of mixed surfactants and their flow behavior in porous media in relation to enhanced oil recovery.

Experimental

Materials. Sodium dodecyl sulfate was supplied by Aldrich Chemical Company, Milwaukee, WI, and various alkyl alcohols were obtained from Supelco, Inc., Bellefonte, PA, with purity greater than 99%. For microemulsion preparation, sodium stearate (>> 99% pure) was supplied by Matheson, Coleman and Bell, Inc. and sodium myristate was supplied by K & K Laboratories, Inc. All the oils (>> 99% pure) were obtained from Chemical Samples Company. Double-distilled water was used in all experiments. For monolayer studies, the 4.0 mM solutions of all pure alkyl alcohols were prepared in a mixture of methanol, chloroform and n-hexane in a volume ratio 1:1:3.

The sand used as porous media was purchased from AGSCO Corp., Peterson, NJ. The transducer used for the measurements of pressure drop across the porous medium was supplied by (Validyne DP-15), Validyne Engineering Corporation, Northridge, CA. The recorder was obtained from (Heath/Schlumberger 225), Heath Company, Benton Harbor, MI. The water was pumped using Cheminert Metering Pump (Model EMP-2), Laboratory Data Control, Riviera Beach, FL.

Methods. Using an Agla microsyringe, alkyl alcohol solution (0.025 ml) was spread on the subsolution of 0.01 M HCl. A time interval of 5 minutes was allowed for spreading solvents to evaporate or diffuse in the subsolution from the monolayers. The monolayer was compressed at a constant rate by an electrically operated motor. The surface pressure area curve for a monolayer was recorded automatically by x-y recorder. Three to five monolayers of each mixture were studied and the results reported are average values. The reproducibility of data was $\pm$ 0.15 Å^2/molecule. The detailed discussion of the apparatus is given elsewhere (17).

The macroemulsions were prepared by mixing an aqueous surfactant solution and oil in a volume ratio 3:1. Oil phase also contained an equimolar alkyl alcohol. The emulsions were produced by using an ultrasonic device (Model Wl85) for two minutes. The volumes of the emulsions were recorded at different time intervals. The microemulsions were prepared by mixing surfactant (1 g), alcohol (4 or 8 ml), oil (10 ml) and water (1 ml) to get a clear solution. The water solubilization capacity was determined by adding more water slowly from a graduated 1 ml pipette to the microemulsion, until turbidity was observed and two-phase formation occurred upon standing. The surface tension of freshly prepared aqueous solutions was measured by Wilhelmy plate method (18) and surface viscosity was measured by a single knife-edge rotational viscometer (19).

For flow through porous media studies, the sandpacks used as porous media were flushed vertically with carbon dioxide for an hour to replace interstitial air. Distilled water was pumped and the pore volume (PV) of the porous medium was determined. By this procedure, the trapped gas bubbles in the porous media can be easily eliminated because carbon dioxide is soluble in water. For determining the absolute permeability of the porous medium, the water was pumped at various flow rates and the pressure drop across the sandpack as a function of flow rate was recorded. After the porous medium was characterized, the mixed surfactant solutions of known surface properties were injected. This was followed by air injection to determine the effect of chain length compatibility on fluid displacement efficiency, breakthrough time and air mobility in porous media.

Results and Discussion

Mixed Monolayers. Figure 1 shows the excess area/molecule when C_{18} alkyl alcohol was mixed with alkyl alcohols of various chain lengths. In this figure, A_{20} represents the molecular area at surface pressure of 20 dynes/cm, whereas A_c and A_e represent the area/molecule at zero surface pressure in the condensed and expanded states of the mixed monolayers. The excess area/molecule is the difference between the experimentally measured area/molecule in the mixed monolayers and that expected from the simple additivity rule (17). The comparison of the results of area/molecule in pure and mixed monolayers indicates that the mixed alkyl alcohols of different chain lengths form a

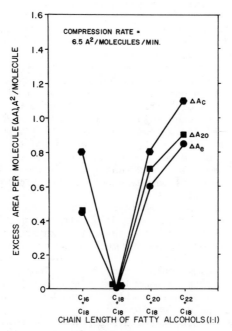

Figure 1.   Excess Area per Molecule   for   Various   Alkyl   Alcohol
            Mixtures   (1:1)   with $C_{18}H_{37}OH$ as the Common Component
            at pH 2.0.

considerably expanded mixed monolayer. It is also evident that as the difference between the alkyl chain length of the components increased, the molecular area in the mixed monolayers also increased. Among excess area/molecule measured at different states of the monolayers, the excess area/molecule in the condensed state is most strikingly influenced by the differences in chain length. Similar results were also observed when the common components in the mixed monolayers were C_{16}, C_{20} and C_{22} alkyl alcohols (17).

It was reported by previous investigators (20-22) that a change of 0.3 to 1.5 Å in the intermolecular spacing between lipid molecules in the monolayer strikingly influences the rate of the enzymic hydrolysis and interaction of metal ions in the monolayers. By assuming a molecular area to be a circle, one can calculate the intermolecular spacing in the monolayers. The distance between the centers of adjacent molecules (2R) can be viewed as the intermolecular spacing in the monolayers. For example, at surface pressure of 20 dynes/cm, the excess area/molecule of 0.90 Å² was observed in the mixed monolayers. Moreover, the experimentally observed molecular area is 21.0 Å², whereas the average area per molecule according to simple additivity rule is 20.1 Å² at 20 dynes/cm in the mixed monolayers of C_{18} + C_{22} alkyl alcohols. The corresponding diameter for the circles (2R) or intermolecular spacing for these two areas are calculated to be 5.18 Å and 5.06 Å, respectively. The increase in the intermolecular spacing due to expansion of the monolayer would be 0.12 Å (5.18 - 5.06 = 0.12 Å). It is very likely that a change in the intermolecular spacing of the order of 0.12 Å may be of significant importance in determining the properties of mixed surfactant systems such as foams and macroemulsions.

Our explanation for these chain length compatibility effects is shown schematically in Figure 2. The equal chain length surfactant molecules form a condensed mixed monolayer as compared to the dissimilar chain length molecules. However, in mixed monolayers of different chain lengths, it is very likely that the portion of molecule above the height of the adjacent molecules exhibits thermal motion such as oscillational, vibrational and rotational modes. Moreover, if these thermal disturbances were limited to the portion above the height of the adjacent molecules, it would not expand the mixed monolayer and molecular area would remain the same. However, the thermal motion most probably propagates along the chain towards the polar group of the molecule which in turn causes expansion in the mixed monolayers and exhibits greater molecular area.

Figure 3 represents the effect of compression rate on excess area per molecule in mixed monolayers. As the rate of compression increases, the excess area per molecule in mixed monolayers increases. These results suggest that the disordered segments of alkyl chain may begin to orient themselves in a more ordered state with decreasing rate of compression, resulting in smaller excess area per molecule. Moreover, it is interesting to note that the rate of compression did not influence significantly the molecular area in the monolayers of pure components. Based on the

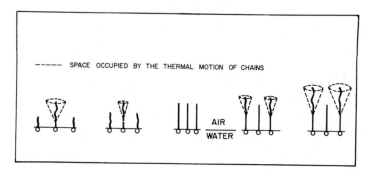

Figure 2. Schematic Diagram of the Chain Length Compatibility
and the Thermal Motion of the Terminal Segments of
Molecules.

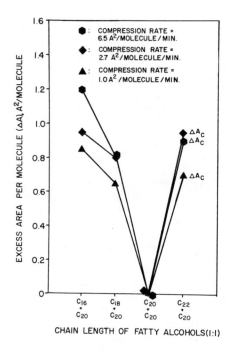

Figure 3. Effect of Compression Rate on the Excess Area/Molecule
for Various Alkyl Alcohols with $C_{20}H_{41}OH$ Alcohol as a
common Component at pH 2.0.

concept of chain length compatibility in monolayers, this study was further extended to investigate the surface chemical aspects of several colloidal systems such as foams, macro- and microemulsions.

Foams. In order to correlate the chain length compatibility with surface properties of foaming solutions and bubble size in foams, sodium lauryl sulfate ($C_{12}H_{25}SO_4Na$) and various alkyl alcohols ($C_8H_{17}OH$-$C_{16}H_{33}OH$) were used in a molar ratio 10:1 as mixed foaming agents. Figure 4 shows the photomicrographs of various foams containing sodium lauryl sulfate (5.0 m M) and different alkyl alcohols (0.5 m M) at 15 minutes after foams were produced. The mixed surfactants of equal alkyl chain length produced the smallest bubbles as compared to the mixed surfactants of unequal chain length (e.g. $C_{12}H_{25}SO_4Na + C_nH_{(2n+1)}OH$, where n = 8, 10, 14 and 16). It is evident that as the difference in chain length increases, the bubble size also increases.

Table I shows various surface and microscopic properties such as surface tension, surface viscosity, foaminess (i.e. foam volume generated in a given time) and bubble size in foams of the surfactant solutions as a function of chain length compatibility. The results indicate that a minimum in surface tension, a maximum in surface viscosity, a maximum in foaminess and a minimum in bubble size were observed when both the components of the mixed surfactant system have the same chain length. These results clearly show that the molecular packing at air-water interface influences surface properties of the surfactant solutions, which can influence microscopic characteristics of foams. The effect of chain length compatibility on microscopic and surface properties of surfactant solutions can be explained as reported in the previous section.

Macroemulsions. Figures 5 and 6 show the emulsion stability for various emulsions containing sodium alkyl sulfates (C_{12}, C_{14}) and sodium alkyl soaps (C_{12}, C_{14}) as the common hydrophilic emulsifiers, and various alkyl alcohols as the hydrophobic emulsifiers. The stability of various macroemulsions was measured by recording the volume of the unseparated emulsion in a given period of time. The volume of the emulsion was maximum when both mixed emulsifiers had the equal chain length (Figures 5 and 6). As the difference in alkyl chain length increases, the volume of the macroemulsion decreases. It clearly shows that the surfactant molecules with similar alkyl chain length pack tightly at liquid/liquid interface similar to that observed at gas/liquid interface. Moreover, the tight packing at the liquid/liquid interface seems to reduce the rate of water or oil separation from the macroemulsion phase. These results indicate that the gas/liquid (foams) and liquid/liquid (macroemulsions) systems behave in the same manner in the presence of mixed surfactants.

Table II shows the effect of chain length compatibility on oil recovery, fluid displacement efficiency, breakthrough time and effective gas mobility in porous media. For gas/liquid systems (e.g. Foams), a maximum in fluid displacement efficiency, a

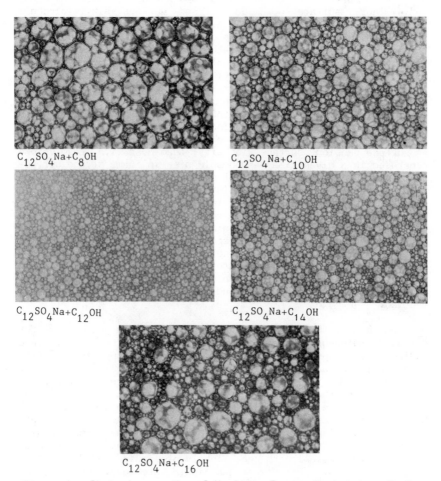

$C_{12}SO_4Na+C_8OH$ $C_{12}SO_4Na+C_{10}OH$

$C_{12}SO_4Na+C_{12}OH$ $C_{12}SO_4Na+C_{14}OH$

$C_{12}SO_4Na+C_{16}OH$

Figure 4. Photomicrographs of Various Foams Containing Sodium Lauryl Sulfate (5 mM) and Different Alkyl Alcohols (0.5 mM).

Table I. Effect of Chain Length Compatibility on Surface Properties of Mixed Surfactant Solutions

Systems Surface Properties	$C_{12}H_{25}SO_4Na$ $+$ $C_8H_{17}OH$	$C_{12}H_{25}SO_4Na$ $+$ $C_{10}H_{21}OH$	$C_{12}H_{25}SO_4Na$ $+$ $C_{12}H_{25}OH$	$C_{12}H_{25}SO_4Na$ $+$ $C_{14}H_{29}OH$	$C_{12}H_{25}SO_4Na$ $+$ $C_{16}H_{33}OH$
Surface Tension (dynes/cm)	25.8	24.3	22.9	25.6	30.3
Surface Viscosity (s.p.)	9.8×10^{-3}	14.5×10^{-3}	32.0×10^{-3}	26.2×10^{-3}	21.0×10^{-3}
Foam Volume in 1st Min. (ml)	284	310	480	298	260
Bubble Size Diameter (cm)	0.21	0.13	0.05	0.16	0.17

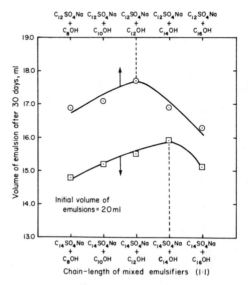

Figure 5. Emulsion Stability for the Mixed Systems of
$C_{12}H_{25}SO_4Na$ (5 mM) and Various Alkyl Alcohols (5 mM)
with the Water-Tetradecane Ratio 3:1.

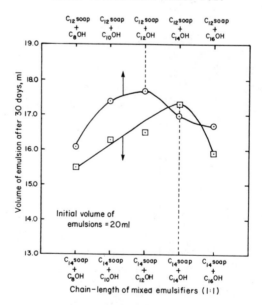

Figure 6. Emulsion Stability for the Mixed Systems of $C_{12}H_{25}Na$
(4.5 mM) and Various Alkyl Alcohols (4.5 mM) with the
Water-Tetradecane Ratio 3:1.

Table II. Effect of Chain Length Compatibility on Flow Through Porous Media Behavior of Foams and Macroemulsion Systems

Systems / Measured Parameters	$C_{12}H_{25}SO_4Na$ + $C_8H_{17}OH$	$C_{12}H_{25}SO_4Na$ + $C_{10}H_{21}OH$	$C_{12}H_{25}SO_4Na$ + $C_{12}H_{25}OH$	$C_{12}H_{25}SO_4Na$ + $C_{14}H_{29}OH$	$C_{12}H_{25}SO_4Na$ + $C_{16}H_{33}OH$
Foams					
Fluid Displacement Efficiency, %	77.8	78.1	83.0	69.6	68.9
Breakthrough Time (min.)	25.1	41.3	46.8	30.6	27.2
Oil Recovery, %					
Air Foam	36.5	38.7	43.3	39.1	35.9
Steam Foam	49.8	51.5	55.0	49.5	47.6
Surfactant Flooding					
Oil Recovery, %	33.2	35.8	41.1	36.8	35.0

maximum in breakthrough time, a minimum in effective gas mobility
and a maximum in oil recovery were observed when both the components of mixed surfactant system had the same chain length.
Similar results were also observed by displacing oil with solutions of surfactants of equal chain length (Table II). These
results can be utilized to design the surfactant formulations for
optimum performance in enhanced oil recovery processes.

Microemulsions. The effect of chain length compatibility on
microemulsion formation was studied by measuring the amount of
water solubilized in the system. The amount of water solubilized
as a function of oil chain length for sodium stearate system in
the presence of various alcohols is shown in Figure 7. It was
observed that for n-butanol containing microemulsions, the maximum
amount of water solubilized decreased continuously with increase
in oil chain length, whereas for n-heptanol it increased continuously. For n-pentanol and n-hexanol containing microemulsions,
water solubilization reached a maximum value for tridecane and
dodecane systems. Similar results were also observed for sodium
myristate system (Figure 8). A maximum in water solubilization
was observed when microemulsion contains n-hexanol and decane.
From the observed findings, one can conclude that when the oil
chain length is increased, the solubilization capacity of
microemulsions can decrease, increase or exhibit a maximum,
depending upon the structure and chain length of alcohol used.
From the results of sodium stearate and sodium myristate systems,
it is inferred that the water solubilization reached a maximum
when alcohol chain length (l_a) plus that of the oil (l_o) is equal
to that of the surfactant (l_s).

Our proposed explanation for the maximum water solubilization
and chain length compatibility effect observed for pentanol and
hexanol containing systems is schematically shown in Figure 9.
The size of microemulsion droplets is expected to increase with
increasing water content which in turn increases the interfacial
area. Up to a certain extent, the alcohol from the oil phase can
partition into the interface to stabilize the additional interfacial area. As the alcohol in the oil phase is depleted, further
growth of water droplets would result in an increase of interfacial tension at the oil/water interface due to an increase in the
area per molecule. This will destabilize the microemulsion as
well as prevent further solubilization of water. It has been
shown theoretically (23) that for microemulsion formation, the
interfacial tension at the oil/water interface should be about
10^{-3} dynes/cm. When the chain length of oil plus alcohol is not
equal to that of the surfactant $(l_a + l_o < l_s)$, there will be a
region of disordered hydrocarbon chains near the terminal methyl
groups around the globules (Figure 9) due to thermal motion which
produces a disruptive effect on the packing of surfactant
molecules and increases area per molecule at the interface. It
was suggested (17) that the thermal motion of alkyl chains which
causes disruptive effect can increase the intermolecular distance
between surfactant molecules by about 0.05 A. This change influences various properties of the systems such as evaporation of

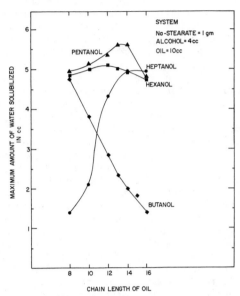

Figure 7. Effect of Alcohol and Oil Chain Length on Water Solubilization Capacity of Sodium Stearate Containing Microemulsions.

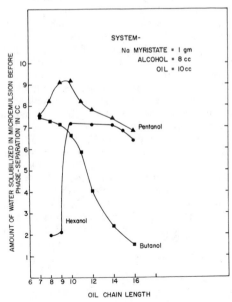

Figure 8. Effect of Alcohol and Oil Chain Length on Water Solubilization Capacity of Sodium Myristate Containing Microemulsions.

water through monolayers, contact angle, surface tension, boundary
lubrication, surface viscosity, bubble size in foams and emulsion
stability of mixed surfactant systems. Therefore, our proposed
explanation based on chain lengths of oil, alcohol and surfactant,
is that the maximum water solubilization in microemulsion occurred
at $l_o + l_a = l_s$, which is due to the maximum cohesive interaction
between hydrocarbon chains (or due to minimum disruptive effect in
the interfacial region) under these conditions (Figure 9).

Further studies indicated that not only the molecular packing, but also the total amount of alcohol and surfactant at the
interface influence the water solubilization capacity of the
microemulsions (25). As the amount of alcohol and surfactant at
the interface increases, the water solubilization capacity in
microemulsions also increases. Figure 10 schematically illustrates the proposed explanation for the solubilization behavior of
heptanol and butanol containing microemulsions. Previous study
(24) on the surfactant partitioning in various oil/water systems
indicates that the surfactant partitioning in oil phase decreases
with increasing alkyl chain length of oil from n-hexane to n-
hexadecane. Similar trend is also expected for n-heptanol. Since
the solubility of n-heptanol in water is negligible, the decrease
in n-heptanol partitioning in oil phase as the chain length of oil
is increased from C_6 to C_{16} would result in an increase in the
heptanol concentration at the interface (e.g. increase in the
alcohol/soap molar ratio at the interface, Figure 10). This
indeed was confirmed by the titration method (Figure 11) as
described by Bowcott and Schulman (24). As the chain length of
oil is increased, the intercept, which represents the molar ratio
of alcohol to soap at the interface, increases.

For microemulsion systems containing butanol, the solubilization capacity decreased with increasing oil chain length (Figures
7 and 8). Our proposed explanation for this observation is
schematically shown in Figure 10. As the oil chain length is
increased, a decrease in butanol partitioning occurs in the oil
phase. Since the total amount of butanol is constant in each system, a concomitant increase in the butanol concentration in the
water phase and at the interface occurs. Unlike heptanol containing microemulsions, the water pool in butanol systems gradually
becomes water plus butanol pool as the butanol has substantial
solubility in water. Therefore, the water plus butanol pool as a
solvent will solubilize some of the soap molecules in the pool and
remove them gradually from the interface, which results in a
decrease in the total interfacial area and hence the concomitant
decrease in the capacity of water solubilization. This study suggests that the maximum water solubilization occurs when the soap
and alcohol molecules remain predominantly adsorbed at the interface as well as when the chain length of oil plus alcohol equals
to that of the surfactant.

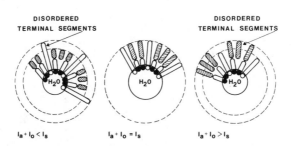

Figure 9. Schematic Diagram of the Disordering of the Terminal Segment of Alkyl Chain Protruding out of the Interfacial Film and the Solubilization Capacity of Microemulsions l_a, l_o and l_s are the Chain Lengths of Alcohol, Oil and Surfactant Molecules, respectively.

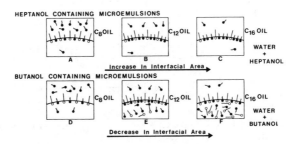

Figure 10. Schematic Illustration of the Effect of Increasing Oil Chain Length on the Partitioning of Alcohol in the Water, Oil and Interfacial Region.

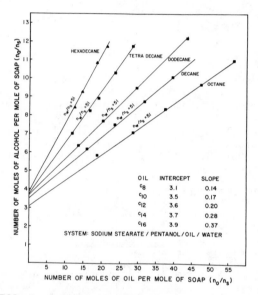

Figure 11.   Effect of Oil Chain Length on the   Oil/Alcohol   Titration Plots for Microemulsions.

Literature Cited

1.   Cooke, C. E.; Schulman, J. H. in "Surface Chemistry"; Ekwall, P; Groth, K.; Runnstrom-Reio; Eds.; Academic Press, New York, 1965, pp. 231-251.
2.   Frank, S. G.; Zografi, G. J. Colloid Interface Sci. 1965; 29, 28.
3.   Sjoblom, E.; Friberg, S. J. Colloid Interface Sci. 1978, 67, 16.
4.   Higuchi, W. I.; Misra, J. J. Pharm. Sci. 1962, 51, 455.
5.   Shah, D. O.; Walker, R. D. Jr.; Hsieh, W. C.; Shah, H. J.; Dwivedi, S.; Nelander, J.; Pepinsky, R.; Dermer, D. W., SPE Paper 5815 presented at Improved Oil Recovery Symposium of SPE of AIME, Tulsa, OK. 1976, March 22-24.
6.   Bansal, V. K.; Chinnaswamy, K; Ramachandran, C.; Shah, D. O., J. Colloid Interface Sci. 1979, 72, 524-537.
7.   Chan, K. S.; Shah, D. O. J. Dispersion Sci. Technol. 1980, 1, 55-95.
8.   Stoeckenius, W.; Schulman, J. H.; Prince, L. M. Kolloid-Z. 1960, 169, 170.
9.   Schulman, J. H.; Riley, D. P. J. Colloid Sci. 1948, 3, 383.
10. Schulman, J. H.; Stoeckenius, W.; Prince, L. M. J. Phys. Chem. 1959, 63, 1677.
11. Schick, M. J.; Fowkes, F. M. J. Phys. Chem. 1957, 61, 1062.
12. Fort, T., Jr., J. Phys. Chem. 1962, 66, 1136.
13. Cameron, A.; Crouch, R. F., Nature 1963, 198, 475.
14. Meakins, R. J., Chem. Ind. 1968, 1768.
15. Ries, H. E., Jr.; Gabor, J., Chem. Ind. 1967, 1561.
16. Sharma, M. K.; Shah, D. O., Proc. 18th Intersociety Energy Conversion Engineering Conference, 1983, August 21-26, 527-534.
17. Shiao, S. Y., Ph.D. Thesis, University of Florida, Gainesville, FL., 1976.
18. Wilhelmy, L., Ann. Physik 1863, 119, 117.
19. Brown, A. G.; Thuman, W. C.; McBain, J. W., J. Colloid Sci. 1953, 8, 491.
20. Shah, D. O.; Schulman, J. H., J. Lipid Res. 1967, 8, 215.
21. Shah, D. O.; Schulman, J. H., J. Lipid Res. 1967, 8, 227.
22. Shah, D. O.; Schulman, J. H., J. Colloid Interface Sci. 1967, 25, 107.
23. Ruckenstein, E. in "Micellization, Solubilization and Micro-emulsions," K. L. Mittal, ed.; Plenum Press, New York, 1977, Volume 2, pp. 755-778.
24. Bowcott, J. E. L.; Schulman, J. H., Z. Electrochem. 1955, 58, 283.
25. Pithapurwala, Y. K., Ph.D. Thesis, University of Florida, Gainesville, FL, 1984.

RECEIVED December 20, 1984

Phase Diagrams and Interactions in Oil-Rich Microemulsions

D. ROUX and A. M. BELLOCQ

Centre de Recherche Paul Pascal, Domaine Universitaire, 33405 Talence Cedex, France

Pseudoternary phase diagrams of the water-dodecane-SDS-pentanol and water-dodecane-SDS-hexanol systems have been investigated in detail. A great variety of new domains has been evidenced in the oil rich part of these diagrams including, one-, two-, three- and four-phase liquid regions. An interpretation of these diagrams is proposed : it is shown that interactions between water domains play an important role in microemulsion stability.

In several theoretical models of microemulsion stability (1-4), the free energy includes an entropic term mainly due to dispersion of water and oil domains and two types of enthalpic contributions : a term of interfacial tension and terms of interaction between domains of same nature (water-water or oil-oil). These interactions result from coulombic and Van der Waals forces. Recently it has been shown that a balance between entropy and interfacial tension could also interpret phase transition (5-6). However these models only generate two-phase equilibria in distinction to the experiments which show the existence of multiphase equilibria. In order to find three-phase equilibria where a microemulsion coexist with a nearly pure water phase and a nearly pure oil phase, Talmon and Prager have taken into account curvature effects. They found a three-phase region but they had to assume a very special form for the curvature energy. In a recent publication De Gennes et al. (6) suggest that the curvature term is not sufficient to reproduce three-phase equilibria and conclude that more complex effects involving strong attractive interactions are requested to explain the three-phase equilibria. But up to now, interaction terms are neglected in the models proposed by Talmon and Prager and De Gennes et al.

Several structural studies have shown that in the oil rich part of the phase diagram, microemulsions consist of a dispersion of monodisperse water droplets in interaction (7-8). In recent papers (9-10), we have investigated by light scattering the effect of the micellar

0097-6156/85/0272-0105$06.00/0
© 1985 American Chemical Society

size and of the molecular structure of the components on the intermicellar interactions. Our results evidence that interactions are
depending on the size of the droplets and on the length of alcohol,
the molecular volume of oil and the polar head area of the surfactant.
However, it is shown that one of the most important molecular parameter is the alcohol chain length. The intermicellar interactions are
all the more attractive as the alcohol is shorter. When interactions
become strongly attractive a critical behavior is evidenced (11).
Then for example in the series of systems : water–dodecane–SDS–pentanol or hexanol or heptanol a critical behavior is only observed with
pentanol. It appears therefore of particular interest to investigate
the phase diagrams of the water–dodecane–SDS–pentanol (A) and water-
dodecane–SDS–hexanol (B) systems.

One of the main objectives of this paper is to examine the importance of the interactions between water in oil micelles on the
stability of oil rich microemulsions. For this purpose, we have investigated the phase diagrams of the systems A and B made with pentanol and hexanol. We examine not only the micellar one–phase region
but also the polyphasic regions around it. This study allows us to
locate critical points in system A and also to evidence in both systems a great variety of new domains including one-, two-, three- and
four-phase regions. In the first part of this paper, we describe in
detail several pseudoternary diagrams. In the second part, we show
that a model of droplets in interaction allows us to interpret the
existence of a line of critical points in the oil rich region of system A (12).

Experimental study : phase diagrams of the water–dodecane–pentanol (or hexanol)–sodium dodecyl sulfate (SDS) systems

The phase diagrams of quaternary mixtures have to be represented in a
three dimensional space. It is useful to present pseudoternary sections of this diagram. The W/S representation (which keeps constant
the water over surfactant ratio) gives a good description of the oil
rich side of the phase diagram. In figure 1 are shown the experimental pseudoternary phase diagrams obtained with hexanol and pentanol.
In a first step, the W/S volumic ratio has been fixed at 1.8. This
value corresponds to the microemulsions previously studied by light
scattering (9-10). Both diagrams present a great variety of one-,
two- or three-phase domains. The sequence of phases observed along
the paths AA' and BB' is described. The variable along AA' is the
alcohol concentration. The path BB' follows the demixing line. Two
points can be noticed :

i) Both systems exhibit similar phase diagrams ; specially two
one–phase regions are observed below the demixing line which bounds
the microemulsion region : in region II the microemulsions are birefringent and in region III, which is poor in alcohol, microemulsions
are isotropic.

ii) The main difference between the two diagrams is the existence, in the case of pentanol, of a critical point (P_c) and of two
liquid–liquid isotropic regions V and XII around P_c. These latter are
separated by the three–phase region XI where the three microemulsion
phases in equilibrium are isotropic. Near P_c, the two phases in equilibrium have very close compositions. The LL region V is separated

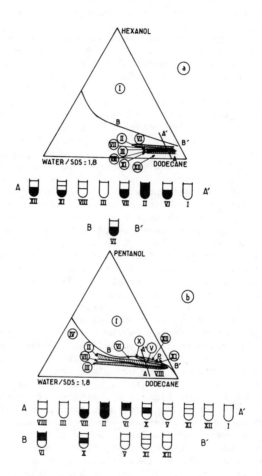

Figure 1 . W/S = 1.8 pseudoternary diagrams at T = 21.5°C (expressed
in volume) of the :
a) water-dodecane-SDS-hexanol system
b) water-dodecane-SDS-pentanol system.
The dark regions correspond to liquid birefringent phases.

from the two-phase liquid-liquid birefringent region VI by a narrow
three-phase zone X. In these equilibria the middle phase is birefringent and the two others are isotropic. The four regions X, V, XI
and XIII do not exist in the phase diagram with hexanol. An enlarged
drawing of the oil rich part of the pentanol diagram is given in figure 3c.

For each system several pseudoternary diagrams corresponding to
different values of the W/S ratio have been investigated. In the
hexanol system, small changes of the phase diagram are observed as
the W/S ratio is varied between 1.8 to 5. (figure 2). The effect of
an increase of this W/S ratio is the shift of the isotropic one-phase
region (I) towards the high alcohol concentrations. But qualitatively
the phase diagram remains unchanged. In the three pseudoternary phase
diagrams studied one can notice that there is no critical point. In
all the cases, the inverted micellar domain is bounded by a two-phase
region where this isotropic phase is in equilibrium with either the
lamellar phase (II) or the other isotropic phase (III). In this system one does not observe a region where two inverted micellar phases
are in equilibrium. We emphasize that in this system which does not
present a critical point in the oil rich region, the micellar phase
located along BB' separates with an organized birefringent phase. On
the contrary, in the pentanol system which presents a critical point,
the system separates in the L-L region XII around P_c into two isotropic microemulsions in equilibrium. This remark seems to be general.
Indeed, a critical behavior has been observed in the water-SDS-buta-
nol-toluene system by light scattering (13). We have observed that
the mixtures which are far in composition from the critical point separate with an organized phase whereas those which are close of this
point separate into two isotropic phases.

In the pentanol system as the W/S ratio is changed, a new critical point is evidenced ; the set of these points constitutes within
the three-dimensional phase diagram a line of critical points. One
can ask the question to know where are the ends of this line. Several
possibilities can be considered : i) the line crosses the whole diagram and goes from one face of the diagram to another one, ii) the
line abuts on a three-phase region either on a critical end point or
on a tricritical point. We have examined in detail the pseudoternary
diagrams defined by the following W/S ratios : 1, 1.2, 1.4, 1.8 and
2.2. The oil rich part of some of these diagrams is shown in figures
3a-3d. Analysis of these pseudoternary diagrams allows one to eliminate the first proposal. Indeed, the pseudoternary diagram defined by
W/S = 1 is similar to those observed with hexanol. The zone XII and
its critical point as well as the zones X, V and XI are no longer
observed in this plane. Moreover, this study shows that the extent of
these regions decreases with the W/S ratio. They disappear for a W/S
ratio included between 1 and 1.1.

This study allows us to evidence a new one-phase region noticed
L^* on figures 3. The microemulsions located in this region scatter
light and exhibit a flow birefringence when their oil content is larger than 85 %. In the planes W/S = 1.2 and 1.4 we have also observed
a four-phase liquid region : LL^*L_BL where one of the phases is birefringent (L_B) and the three other ones are isotropic however one of
them is flow birefringent (L^*). Three of the four three-phase regions
which surround the four-phase region have been observed in the plane

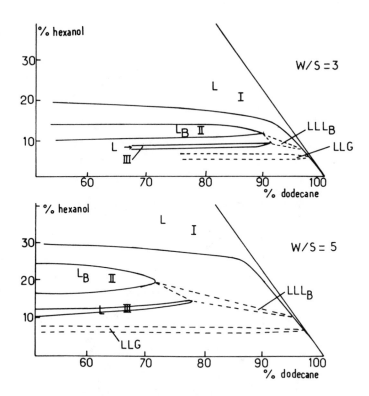

Figure 2 . W/S pseudoternary diagrams at T = 21.5°C (in volume) of
the water–dodecane–SDS–hexanol system.
L, L$_B$ designate respectively an isotropic phase and a birefringent phase.

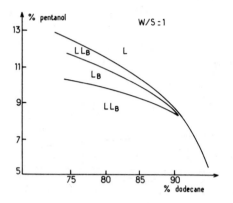

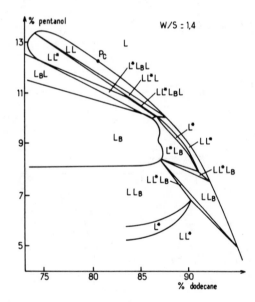

Figure 3. W/S pseudoternary diagrams at T = 21.5 °C (in volume) of the water–dodecane–SDS–pentanol system. Top: W/S = 1; bottom: W/S = 1.4.   L, L_B, and L^* designate respectively an isotropic phase, a birefringent phase and a flow birefringent liquid phase.

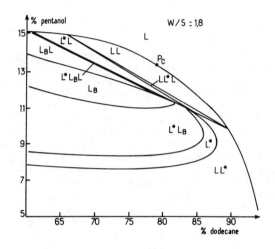

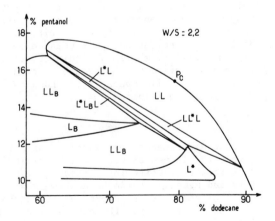

Figure 3. Continued. Top: W/S = 1.8; bottom: W/S = 2.2.

defined by W/S = 1.2 ; these equilibria are the following LL^*L, L^*L_BL
and LL^*L_B. Systematic progression from three phases to four phases to
three phases with changing either oil or alcohol concentration is reported for the first time. As the W/S ratio increases, the two following effects are observed : the disappearance of the four-phase region, and the shifts of the birefringent (L_B) and flow birefringent
(L^*) regions towards the water and surfactant rich region.

Thus the study of several W/S pseudoternary diagrams of system A
leads us to observe a great variety of one-phase regions and of phase
equilibria. In addition this study allows us to evidence a line of
critical points which limits are within the tetrahedron of representation of the states of the system. On the contrary in the system B,
any critical point is evidenced as the W/S ratio is changed up to 5.

The phase behavior observed in the quaternary systems A and B is
also evidenced in ternary systems. Figure 4 shows the phase diagrams
for systems made of AOT-water and two different oils. The phase diagram with decane was established by Assih (14) and that with isooctane has been established in our laboratory. At 25°C the isooctane system does not present a critical point and the inverse micellar phase
is bounded by a two-phase domain where the inverse micellar phase is
in equilibrium with a liquid crystalline phase, as for system B or
system A when the W/S ratio is below 1.1. In the case of decane, a
critical point has been evidenced by light scattering (15). Assih and
al. have observed around the critical point a two-phase region where
two microemulsions are in equilibrium. A three-phase equilibrium con-
nects the liquid crystalline phase and this last region.

In conclusion, the same phase behavior is evidenced when we
change the alcohol or the W/S ratio in a quaternary system, and the
oil in a ternary system. This behavior can be characterized by two
types of phase diagrams. In the first type no critical point occurs.
In this case, the inverse micellar phase is bounded by a two-phase
region where it is in equilibrium with a liquid crystalline phase.
The second type is characterized by the occurence of a critical point.
In this case, the inverse micellar phase is bounded by a region where
two micellar phases are in equilibrium.

Discussion

Some features of these diagrams can be explained with the help of
light scattering results. Indeed existence of a critical point is always related with strong attractive interactions. Light scattering
experiments in inverted micellar phase have shown that interactions
increase as the alcohol chain length decreases or as the W/S ratio
increases (i.e. as the micellar size increases). Therefore one can
assume that the line of critical points observed in the pentanol system above a limit value of the W/S ratio is due to a strengthening of
the interactions. At W/S equal to 1.1 interactions between droplets
are sufficient to induce a liquid-gas type phase separation. On the
same manner, interactions can explain the differences observed in the
phase diagrams of the ternary systems containing AOT and water as
isooctane is replaced by decane. Indeed it has been shown that interactions increase with the molecular volume of oil (10).

In the following we present a quantitative interpretation of the
phase diagram based on a model of interacting particles. We show that

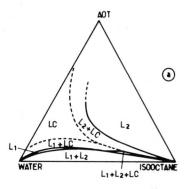

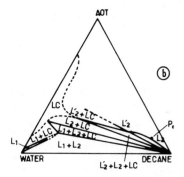

Figure 4 . Phase diagrams of the ternary systems. Water, AOT, iso-
octane or decane. The phase diagram with decane has been
established by Assih et al.

the LL zone XII observed in the diagram of the pentanol system where two microemulsions coexist can be interpreted as a liquid-gas transition due to interactions between inverse micelles. The liquid phase has a high micellar concentration and the gas phase has a low concentration of micelles.

Using the hard sphere adhesive state equation proposed by Baxter (16), it is possible to calculate the demixing line due to interactions. This state equation corresponds to the exact solution of the Percus-Yevick equation in the case of an hard sphere potential with an infinitively thin attractive square well. In our calculation we assum that the range of the potential is short in comparison to the size of the particles (in fact less than 10 %).

The Baxter's state equation includes only one parapeter τ which is related to the attractive interactions.

$$\frac{P}{\rho k T} = \frac{1 + \eta + \eta^2}{(1-\eta)^3} - \eta \, \lambda \, \frac{18 \, (2 + \eta) - \eta \, \lambda^2}{36 \, (1-\eta)^3}$$

with : $$\lambda = 6 - \tau + \frac{\tau}{\eta} - \left[\left(6 - \tau + \frac{\tau}{\eta} \right)^2 - 6 \left(1 - \frac{2}{\eta} \right) \right]^{1/2}$$

η is the volumic fraction of particles and ρ the density number.

Baxter has shown that below a certain value of τ ($\tau = \tau_c$) a phase separation of liquid-gas type occurs. τ is related to the second virial coefficient B of the osmotic pressure which value can be experimentally determined from light scattering experiments.

$$\tau = \frac{12}{8 - B}$$

In previous papers, the experimental values of B for several w/o microemulsions have been measured. As already pointed out, these values are function of both the radius of the micelles and the alcohol chain length (9-10). The attractive interactions between micelles increase as the micellar radius increases and as the alcohol chain length is shorter. We have proposed an interaction potential between w/o micelles which allows to account for the scattering results (10). This potential V(r) results from the possibility of penetration of the micelles. V(r) is proportional to the volume of interpenetration of micelles. The penetration is limited by the molecules of alcohol located inside the interfacial film. r is the distance between two micelles.

$$\begin{cases} \dfrac{V(r)}{kT} = 0 & r > 2R \\[2em] \dfrac{V(r)}{kT} = -\dfrac{\Pi}{6} \, \Delta\rho \, (2R - r)^2 \left(2R + \dfrac{r}{2} \right) & 2R - \ell < r < 2R \\[2em] \dfrac{V(r)}{kT} = +\infty & r < 2R - \ell \end{cases}$$

where R is the radius of micelles and ℓ is the difference between the lengths of surfactant and alcohol. $\Delta\rho$ is a parameter depending only on the oil and surfactant. Its value was determined from light scattering experiments (10). In the case studied $\Delta\rho$ = 0.00071 $\overset{\circ}{A}^{-3}$ and ℓ = 8.82 $\overset{\circ}{A}$ for pentanol and ℓ = 7.56 $\overset{\circ}{A}$ for hexanol.

For the calculation of B, numerical integration is requested

$$B = 8 + \frac{24}{(2R)^3} \int_{2R-\ell}^{2R} \left[1 - e^{-\frac{V(r)}{kT}} \right] r^2 \, dr$$

Following Barboy (17) the coexistence curve is obtained by numerical resolution of the two following equations :

$$\left. \begin{array}{l} P(\tau,\eta_1) = P(\tau,\eta_2) \\ \mu(\tau,\eta_1) = \mu(\tau,\eta_2) \end{array} \right\}$$

where η_1 and η_2 are the micellar volumic fractions of the low and high density micellar phases in equilibrium. μ is the chemical potential, it is related to P by :

$$\frac{\eta}{v_o} \frac{\partial\mu}{\partial\eta} = \frac{\partial P}{\partial\eta}$$

where v_o is the micellar volume ; an analytical expression for μ is given by Barboy (17).

The demixing curves in the W/S pseudoternary diagrams for the hexanol and pentanol systems have been calculated according to the above theoretical treatment. These lines have been determined in the following way. The calculation of the state equation is applied to a dilution line ; along such a line the inverse micelles have a constant radius R. The micelles contain the whole water (volume V_w), the surfactant (volume V_s) and a part of the alcohol V_A^m. The rest of alcohol V_A^o is in the oil continuous phase. We suppose that the alcohol-oil ratio in the continuous phase is constant and is equal to k. Besides, in the calculation of the micellar radius R one assumes that the surfactant and the alcohol molecules which are situated at the interface have a constant area per chain s. In most of the previous studies s has been found constant and equal to 25 $\overset{\circ}{A}^2$. This value is taken equal for the alcohol and surfactant chains. Consequently :

$$R = \frac{3v_s}{s} \frac{\left[(k + 1)(V_{ws} + V_A) - k \right]}{k \frac{v_s}{v_a}(V_{ws} - 1) + \frac{V_{ws}}{x + 1} + (k + 1)\frac{v_s}{v_a} V_A}$$

- v_s is the molecular volume of surfactant (412 $\overset{\circ}{A}^3$ for SDS)
- v_a is the molecular volume of the alcohol (180 $\overset{\circ}{A}^3$ for pentanol and 209 $\overset{\circ}{A}^3$ for hexanol)
- k is the ratio of the alcohol (V_A^o) and oil volumes in the oil continuous phase (k = 0.133)
- $x = V_w/V_s$; $V_{ws} = V_w + V_s$; $V_A = V_A^o + V_A^m$

All the parameters involved in the calculation of the phase separation curve are deduced from experimental values. Application of

this calculation to the W/S = 1.8 plane leads to a calculated demixing line very close to the experimental one in the case of pentanol, particularly in the region XII (figure 5). Besides a critical
point P_c^T is found near the experimental one P_c^E. On the contrary in
the case of hexanol the theoretical demixing line is found largely
below the experimental one. These results corroborate our hypothesis
of liquid-gas transition for the region XII of the phase diagram with
pentanol where two microemulsions are in equilibrium. In the hexanol
system, the phase separation is not due to micellar interactions but
results from other factors such as interfacial tension or curvature
and leads to the formation of a lamellar phase. The coexistence curve
corresponding to such effects is termed "lamellar" demixing line. In
the case of the pentanol system, it exists a competition between
these last effects and interactions. This competition gives rise to
the three-phase equilibria observed in region X.

In our model a critical radius R_c appears ; its value is 52 Å in
the pentanol system. Then in each W/S plane a critical dilution line
corresponding to such a radius is obtained. Its location depends on
the W/S ratio. As this ratio decreases, the critical line lies at a
lower alcohol content. In the case, where the critical dilution line
is below the "lamellar" demixing line, the phase diagram is expected
to be similar to that observed with the hexanol system for which the
interactions are not predominant. This is the case of the pseudoternary sections defined by a W/S ratio less than 1.1. Besides this provides an explanation for the simultaneous disappearance of the regions X, V, XI, XII.

A similar interpretation of phase diagrams has been recently
proposed by Safran and Turkevich (18). These authors have considered
the effects of interaction and curvature on the stability of microemulsions. They suggest that unstabilities of spherical microemulsion
droplets lead the system to separate with water in order to prevent
micellar growth above a limit radius R_ℓ. Taking into account interactions with a phenomenologic treatment, they show that phase separation due to interaction is also possible and they found a critical
radius R_c. If R_c is greater than R_ℓ a water phase is formed and if R_c
is lower than R_ℓ phase separation gives rise to two micellar phases
with a critical point. This theoretical treatment reflects very well
the behavior we observe but it is not in full accordance with our
experimental results. The main difference is that when interactions
are not preponderant the phase separation does not occur with a water
phase but with a lamellar phase.

Our theoretical treatment is based on the same physical idea.
But in our interpretation the comparison with the experimental behavior is quantitative and the calculation of the demixing line is
based on the experimental dependence of the interactions upon micellar radius or alcohol chain length. We can notice that calculations
made by Safran et al. lead the authors to suppose interactions increasing with the radius what is experimentally well established.

For the water, AOT and decane system, light scattering experiments are in progress in view to apply the proposed interaction potential to ternary systems. As expected the first results clearly
indicate that interactions increase as the system approaches the critical point. Besides preliminary calculations confirm that a liquid-
gas type transition must occur very close to the experimental demixing line.

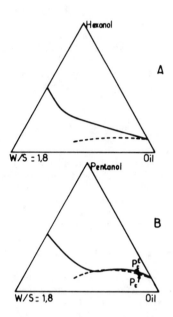

Figure 5 . Comparison between experimental (full line) and calculated (dashed line) demixing curves in the oil rich region
of the hexanol and pentanol systems. P_C^E is the experimental critical point. P_C^T is the theoretical critical point.

Conclusion

In conclusion the study of the phase diagram of the two following
quaternary mixtures A : H_2O - $C_{12}H_{26}$ - SDS - pentanol and B : H_2O -
$C_{12}H_{26}$ - SDS - hexanol leads us to observe several one-phase regions
in the oil rich part of the phase diagram. These one-phase regions
are connected the ones to the others by a great variety of multiple
coexisting phases equilibria (two, three or four). In the case of
the pentanol system, as the W/S ratio is greater than 1.1 a line of
critical point is evidenced.
     Light scattering measurements and theoretical treatment strongly
support the idea that attractive interactions between inverse micelles play an important role in the stability of oil rich microemulsions. In the system containing pentanol, attractions between ω/o
micelles can be sufficient to give rise to a phase separation between
two microemulsion phases.

Acknowledgments

The authors are grateful to P. BOTHOREL, J. PROST, P. BAROIS,
C. COULON, for many stimulating discussions. They wish to thank
O. BABAGBETO and M. MAUGEY for their technical assistance.

Literature Cited

1. Ruckenstein, E., and J.C. Chi, J. Chem. Soc. Faraday Trans II,
    71, 1960.
2. Miller, C.A., R. Hwan, W.J. Benton, and T. Fort, J. Coll. Int.
    Sci., 61, 554 (1977).
3. Huh, C., J. Coll. Int. Sci., 71, 408 (1979).
4. Cantor, R., Macromolecules, 14, 1186 (1981).
5. Talmon, Y., and S. Prager, J. Chem. Phys. 69, 2984 (1978).
6. Jouffroy, J., P. Levinson, and P.G. De Gennes, J. Phys., 43,
    1241 (1982).
7. Dvolaitzky, M., M. Guyot, M. Lagues, J.P. Le Pesant, R. Ober,
    C. Sauterey, and C. Taupin, J. Chem. Phys., 69, 3279 (1978).
8. Cazabat, A.M., D. Langevin, J. Chem. Phys., 74, 3148 (1981).
9. Brunetti, S., D. Roux, A.M. Bellocq, G. Fourche, and P. Bothorel,
    J. Phys. Chem., 87, 1028 (1983).
10. Roux, D., A.M. Bellocq, and P. Bothorel, Proceedings of the In-
    ternational Symposium on Surfactants in Solution, Ed. by K.L.
    Mittal and B. Lindman (1984).
11. Fourche, G., A.M. Bellocq, and S. Brunetti, J. Coll. Int. Sci.,
    28, 302 (1982).
12. Roux, D., A.M. Bellocq, and M.S. Leblanc, Chem. Phys. Lett., 94,
    156 (1983).
13. Cazabat, A.M., D. Langevin, and O. Sorba, J. Phys. Lett., L 505
    (1982).
14. Assih, T., P. Delord, and F. Larché, Proceedings of the Inter-
    national Symposium on Surfactants in Solution, Ed. by K.L.
    Mittal and B. Lindman (1983).
15. Huang, J.S., and M.W. Kim, Phys. Rev. Lett., 47, 1462 (1981).
16. Baxter, R.J., J. Chem. Phys., 49, 2770 (1968).
17. Barboy, B., J. Chem. Phys., 61, 3194 (1974).
18. Safran,S.A., and L.A. Turkevich, Phys. Rev. Lett.,50,1930 (1983).

RECEIVED June 8, 1984

Critical Points in Microemulsions
Role of van der Waals and Entropic Forces

D. CHATENAY, O. ABILLON, J. MEUNIER, D. LANGEVIN, and A. M. CAZABAT

Laboratoire de Spectroscopie Hertzienne de l'Ecole Nationale Supérieure, 24 Rue Lhomond, 75231 Paris Cedex 05, France

Interactions in microemulsions have been studied using light scattering techniques. In water in oil systems, hard sphere interactions are dominant. The remaining interactions are usually attractive and are of the Van der Waals type. The case of oil in water microemulsion is less known. In those systems, interactions seem to be of a quite different kind. Entropic forces are thought to be important in those media, as will be shown by the study of a simplified system.
Both interfacial and bulk properties have been investigated.

Interactions in microemulsions are not perfectly known. We will present in the following an experimental study which will contribute to clarify this problem. We shall mainly emphasize the differences between water in oil and oil in water microemulsions.

Description of the system

We have studied a mixture of oil, brine, surfactant and co-surfactant of following composition :
brine.......................... 46.8% (wt)
Oil (toluene)................... 46.25%(wt)
surfactant (SDS)............... 1.99%(wt)
co-surfactant (butanol)........ 3.96%(wt)

By increasing the brine salinity, we observe the sequence of equilibria
$$S < S_1 \quad (S_1 = 5.4 \text{ wt%})$$
-Winsor I (organic phase coexisting with an oil in water microemulsion)
$$S_2 > S > S_1 \quad (S_2 = 7.5 \text{ wt%})$$
-Winsor III (middle phase microemulsion coexisting with aqueous and organic phases)

0097–6156/85/0272–0119$06.00/0

$$S > S_2$$

-Winsor II (water in oil microemulsion coexisting with an aqueous phase).

In the two phase domains ($S < S_1$ and $S > S_2$), we were able to dilute the microemulsion phases. This led us to picture the microemulsions phases as a dispersion of droplets of oil (resp. water) in water (resp. oil). These droplets are surrounded by a layer of surfactant and co-surfactant. This dilution procedure allows us to define the composition of the continuous phase (1) (2). No dilution procedure was found in the three phase domain $S_1 < S < S_2$, thus supporting the idea that the structure of the middle-phase microemulsion is more complicated than a mere dispersion of droplets.

Interfacial properties

Experimental results:
The interfacial tensions between coexisting phases were measured by surface light scattering (3). The results are presented on figure 1-a. The two curves intersect at the optimal salinity $S^* = 6.3$ and their common value is $\gamma^* = 4.5 \ 10^{-3}$ dyn/cm.

In the two phase domains, we measured the interfacial tensions ($\gamma_{om} \cdot \gamma_{wm}$) between successive dilutions of the microemulsion phases and the excess organic or aqueous phases. The measured values did not deviate significantly from the initial values. We then measured the interfacial tensions between the microemulsion continuous phase and the excess phase. The results of these measurements are given in figure 1-b. The values of the measured interfacial tensions in that case are equal within 30% to those obtained with the original samples.

In the three phase domain, we have measured the interfacial tensions γ_{ow} between the aqueous and organic phases (see fig 1-b). in this domain, we found that $\gamma_{ow} = \text{Sup} |\gamma_{om} , \gamma_{wm} |$.

Origin of the interfacial tension in the case $\gamma > \gamma^*$:
The experimental results presented above suggest that the interfacial properties are not fully dependent on the bulk properties of the coexisting phase. The origin of the low interfacial tensions is then the presence of a surfactant and co-surfactant layer at the interface.

In order to evaluate the order of magnitude of the interfacial tensions, one may write the free energy of a microemulsion in the form :

$$F = F_c + F_d + F_i$$

F_c is a curvature energy

F_d is the entropy of mixing of the constituents

F_i is the energy of interaction between the elements of the structure.

These different terms may be evaluated for different possible structures of the microemulsion phase. Then the interfacial tensions can be deduced from the knowledge of F_c, F_d and F_i.

Different evaluations of F_d (4-5-6-7) (1981,1983) and F_i (6-7-8-9-10) have been proposed . In all cases, the contribution of these two terms is found negligible (which is consistent with our

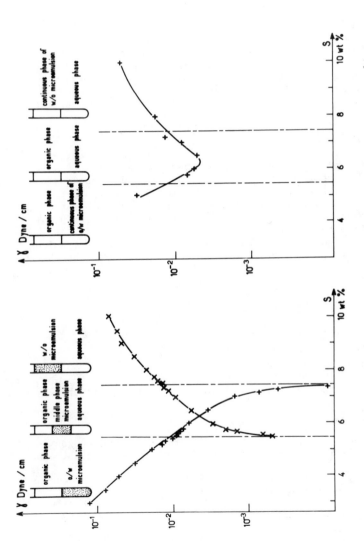

Figure 1. a) Interfacial tension between the different phases versus salinity: +, interfacial tension between microemulsions and top phases; x, interfacial tension between microemulsions and bottom phases. b) Interfacial tensions between organic and aqueous phases containing no micelles: top and bottom phases in the three phase domain, top phase and continuous phase of the microemulsion in the left, bottom phase and continuous phase of the microemulsion on the right.

experimental observations). For instance, in the case of a dispersion of droplets which may be viewed as hard spheres, the contribution of these two terms is found to be (6-7-11)

$$\gamma_{d+i} \simeq \frac{-kT}{4\pi R^2} \ln \phi$$

which gives $\gamma_{d+i} \simeq 7.10^{-3}$ for $\phi = 0.1$, $R = 100\overset{o}{A}$(typical experimental values in the two-phase domain). This value is significantly lower than the measured ones in the two-phase domains far from the phase boundaries, where the hard sphere picture is valid. In that case, the values of the interfacial tension are essentially due to curvature effects. An estimation of the contribution of the curvature terms (12) lead to the value $\gamma_c = 10^{-1}$ dyn/cm in reasonable agreement with experimental results.

Origin of low interfacial tensions in the case $\gamma > \gamma^*$:
Close to the boundaries S_1 and S_2 in the three phase domain, the interfacial tensions were found to be very low. In that case, the theoretical model presented above is no longer valid, first of all because the middle phase microemulsion structure is not simply a droplet dispersion. Furthermore the interaction term F_i becomes evidently dominant and is difficult to evaluate since the nature of the forces is not perfectly known. However, such low interfacial tensions are characteristic of critical consolute points. It was then tempting to check that the behavior of the interfacial tensions was compatible with the universal scaling laws obtained in the theory of critical phenomena. In these theories the relevant parameter is the distance ε to the critical point defined by :

$$\varepsilon = \frac{\mu_i - \mu_i^c}{\mu_i}$$

where
μ_i is the chemical potential of component i.

μ_i^c is the value of the chemical potential of component i at the critical point.

Scaling laws predict that :

$\Delta\rho = \Delta\rho_\theta . \varepsilon^\beta$ where $\Delta\rho$ is the density difference of the two coexisting phases. $\Delta\rho_0$ is a non-universal prefactor, β an universal exponent

$$\gamma = \gamma^0 \varepsilon^\phi .$$

Eliminating ε between these two relationslead to :

$$\gamma = \gamma^0 \left| \frac{\Delta\rho}{\Delta\rho_0} \right|^{\phi/\beta}$$

The universal exponent (ϕ/β) depends on the type of theory. Renormalization group theory predicts $\phi/\beta = 4$ while mean-field theory predicts $\phi/\beta = 3$.

On figure 2, the measured values of the interfacial tensions are plotted against the density difference on a log-log graph. For $\gamma < \gamma^*$, the results are in good agreement with renormalization group theories (Cazabat 1982).

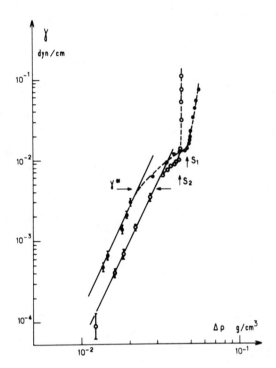

Figure 2 .
  Interfacial tensions versus density differences between phases:
  o, middle phase microemulsion, top phase : o, middle phase
  microemulsion, bottom phase. The horizontal arrows indicate
  the common value $\gamma^{\star}$ of the two interfacial tensions at the
  optimal salinity : the vertical arrows indicate the density
  difference relative to the transition between three phase and
  two phase domains. The line represents the expected scaling
  law : their slope is 4.

ϕ/β = 4.2 ± 0.3 for the interfacial tension between the middle phase
microemulsion and the excess oil.
ϕ/β = 4.4 ± 0.6 for the interfacial tension between the middle phase
microemulsion and the excess water.
     Above γ^*, a deviation from scaling laws is found, which is in
agreement with our preceding interpretation of the origin of the
interfacial tensions.
Relation between the different interfacial tensions
In the three phase domain, the measured values of the three interfacial tensions satisfy the Antonov inequality :

$$\gamma_{ow} < \gamma_{om} + \gamma_{wm}$$

this inequality is characteristic of a non-wetting situation (13). This
was checked by showing that a drop of the middle phase microemulsion
did not spread at the interface between the oil and water phases
around the salinity S^* (fig 3). Approaching the two boundaries S_1 and
S_2 it should be possible to observe a wetting-non wetting transition.
Experiments are currently under way to investigate this point.

Bulk properties
Introduction :
Bulk properties of the microemulsion phases were investigated by various techniques (light scattering, viscosity, conductivity, ultrasonic absorption).
     These techniques allowed us to test some structural models of
microemulsions and to study forces, exchanges and connectivity phenomena in these media.
Light scattering data in the two-phase domains for the boundaries S_1
and S_2 :
     In that case, all light scattering data are well accounted for
by the model of mono-disperse droplets submitted to Brownian motion.
The dilution procedure (2) associated to the measurements of the intensity of the scattered light allows to study the osmotic compressibility of the samples as a function of the volume fraction of the
dispersed phase. In that case the scattered intensity I is given by :
(9-10):

$$I \sim \left| \frac{\pi}{\phi} \right|^{-1} \quad \text{and} \quad \frac{\pi}{\phi} = \frac{3kT}{4\pi R^3} \ (1 + B\phi)$$

where
     π is the osmotic pressure
     ϕ is the volume fraction of the droplets
     k is the Boltzmann constant
     T is the absolute temperature
     R is the droplet radius
     B is the first virial coefficient

The value of B is characteristic of the type of interaction between
droplets :
     $B = 8$ hard-sphere interaction
     $B > 8$ repulsive interaction
     $B < 8$ attractive interaction

Quasi elastic light scattering techniques allow to measure the diffusion coefficient D of the droplets, the variation of which with the

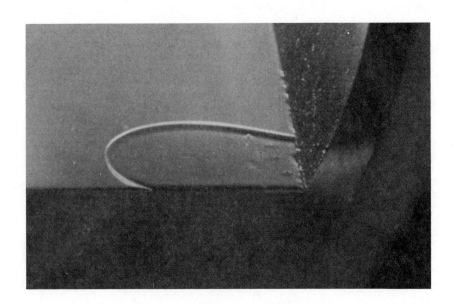

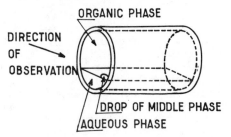

Figure 3 .
  A drop of middle phase at the interface between aqueous phase and oil phase. The aqueous phase, the oil phase and the middle phase are obtained by phase separation at S=6.5. The drop of middle phase does not spread at the interface showing that

$$\gamma_{w,o} < \gamma_{m,o} + \gamma_{m,w} .$$

volume fraction is given by (10):

$$D = \frac{kT}{6\pi\eta_0 R_H} (1 + \alpha\phi)$$

where η_0 is the continuous phase viscosity
R_H is the hydrodynamic radius of the droplet
α is a virial coefficient

The virial coefficient takes into account the direct interactions
between droplets (like B) and the hydrodynamic interactions. Light
scattering data in the two-phase domains are gathered in table I.
    From these data some essential features can be retained. First
of all the droplets become bigger and bigger and the interactions
between them become more and more attractive when one approaches the
boundaries S_1 and S_2. For $S < S_1$, we found an increasing difference
between R and R_H when S increased. For $S < S_2$, the differences
between R and R_H and the attractive interactions are well taken into
account by a model which allows an interpenetration of the interfacial
layers (15). In this model, the forces between droplets originate from
Van der Waals forces between aliphatic chains.
Light scattering data close to the boundaries S_1 and S_2 :
For $S \sim S_2$, the variations of I and D with the scattering wave-vectors
are in good agreement with the predictions of the theories of critical
phenomena. These theories lead to three independent determinations
of the correlation length ξ of the critical concentration fluctuations
(14).
ξ is extracted from the variations of I with the scattering wave-
vector.
ξ_0 is determined from the extrapolation of the variation of the diffusion coefficient at zero scattering wave-vector.
ξ_D is extracted from the variation of the diffusion coefficient with
the scattering wave-vector.
    These three determinations lead to the same value (within experi-
mental accuracy) of the correlation length (cf. table II). These
results are in agreement with our preceding interpretation of the
origin of low interfacial tension close to the boundary S_2.
    For $S \sim S_1$, we observed variations of I with the scattering
wave-vector but this was not the case for the diffusion coefficient.
Furthermore, the autocorrelation function of the scattered intensity
showed significant deviations from an exponential form. This indicates an increasing polydispersity of the droplets when S increases.
The interpretation of the bulk properties of the microemulsions
phases, close to S_1, in terms of critical phenomena, is then less
satisfying. Near this boundary, the samples are further from a critical consolute point than in the case of the boundary S_2. As far as
bulk properties are concerned, light scattering experiments are
rather sensitive to droplets elongation as it will be observed in
viscosity measurements.

Complementary results
Introduction :
In order to precise the different phenomena which occur at the two
phase boundaries S_1 and S_2, the system has been studied with complementary techniques : viscosity, connectivity experiments (electrical
conductivity, forced Rayleigh scattering). These experiments will

Table I. Droplet Sizes and Interaction Parameters of the Microemulsion in the Two Phase-Domain.

Salinity	R	R_H	B	α	
3.5 %	85 Å	95 Å	5	−1	HARD SPHERE
5	130	190	3	−5	
5.2	155	240	0	−6	
S = 5.4					
S = 7.4					
8	180	200	2	−7	
10 %	225	150	6	−2	HARD SPHERE
HARD SPHERE			8	1.5	

Table II. Values of the Correlation Lengths in the Microemulsions Close to the Boundaries. Three determinations of these Values Are Given as Described in the Text.

DATA

	Salinity	ξ_0 (D_0)	$\xi(I)$ | $I(\theta)$ |	$\xi(D)$ | $D(\theta)$ |
	5.2 %	100 Å	90 Å	< 100
	5.3	120	125	−
S_1	5.4	190	170	−
	5.5	110	130	−
	5.6	80	100	−
	7.2	110	150	−
	7.3	220	250	280
S_2	7.4	610	600	750
	7.5	300	320	350
	7.6	230	280	280

lead to a differentiation between the two boundaries S_1 and S_2.
Viscosity :
When dilution is possible $(S < S_1$ or $S > S_2$), we have studied the
viscosity of the samples as a function of ϕ, the droplet volume
fraction. In each case, we have determined the quantity :

$$= \lim_{\phi \to 0} \frac{\eta - \eta_0}{\eta_0 \phi}$$

where η is the viscosity at the volume fraction ϕ, η_0 is the continuous phase viscosity. For $S > S_2$ (oil rich system), we found $\lambda = 2.5$
which is characteristic of a sphere suspension. For $S \leqslant S_1$ (water
rich system), we found that λ increased rapidly approaching the
phase boundary, for instance $\lambda = 3.7$ for $S=3.5$ and $\lambda = 4.3$ for
$S=5$. In that case, viscosity results together with light scattering
ones may be taken into account by describing the microemulsion phase
as a polydisperse dispersion of ellipsoids (3).
Connectivity experiments :
In the case of oil rich systems, the connectivity of water core may
be studied by conductivity experiments (16). When dilution was possible, we have studied the conductivity of the microemulsion phases
as a function of the droplets volume fraction (3).

   In those systems, as the attractive interactions increase (i.e.S
decreases), a steep increase of the conductivity at a volume fraction
around 0.15 is more and more visible (fig.4).

   The percolation phenomenon is visible only in systems with sufficiently attractive interactions, and shows that there are pore ope-
nings in the overlapping region of the interfacial layers. A collective phenomenon must be supposed to explain such a strong anomaly
of a transport coefficient.

   In the case of water rich systems, the connectivity of the system cannot be evidenced in a so-direct way. Nevertheless, forced
Rayleigh scattering and fluorescence photo-bleaching experiments are
currently under way to elucidate this point (with L. Léger , Collège
de France).

Study of an oil free system
Introduction :
In this case of water rich systems, there is no evident model of interaction and Van der Waals forces between droplets are too weak to
lead to a critical point which seems to be of ot quite different
kind from the one near S_2 . The phase diagram of the system is possibly too complicated to study this particular point but it seems
that the critical point near S_1 is a lower critical point. In that
case, one may think that entropic forces are important in the medium.
In order to confirm this point, we have studied simpler systems :
oil free systems.

   These oil free systems have been studied for three reasons :
-if entropic forces (water structuration around the droplets) are at
the origin of the critical point near S_1 , one may expect that these
forces will not be affected if one removes the oil-core of the droplets. In that case, the main structural elements are the interfacial
layer (made of surfactant and alcohol) and the surrounding brine.
-we have experimentally evidenced that for about the same relative
concentration of brine, alcohol and SDS, the oil-free system was

very close to a lower critical point at the brine salinity S_1 .
-in oil free systems, we have suppressed one component : the phase
diagram of the system is then simpler and may be studied with not too
much difficulty.

For instance, we have studied a system of following composition :
brine (salinity 6,6%)......   50 cc
butanol...................   5.75 cc
SDS......................   2.55 g

This system had a lower critical point at T_c = 24.85 °C.

Preliminary results

As above, the interfacial and bulk properties have been studied by
light scattering (17). All the results that we obtained are in fairly
good agreement with the renormalization group theory of critical phenomena and the experimental variations of the interfacial tensions and
correlation lengths are given in table III.

Typical experimental results for the anisotropy of the scattered
intensity and the diffusion coefficient are given in figs. 5-a and 5-b.
Both results are in good agreement with the theoretical predictions
(Ornstein-Zernike law for the intensity. Kawasaki formula for the
diffusion coefficient)(14).

Some questions remain unanswered about the values of the prefactor measured in our experiments (table III). The value of the prefactor γ_0 of the interfacial tension variations is very small :
$\gamma_0 \sim 0.2$ dyn/cm, compared to the values obtained in the case of critical points for pure fluids or binary mixtures (γ_0 a few tens of dyn/cm).
This small value remains to be understood. The values of the prefactors of the correlation lengths are of the order of magnitude of a
molecular size. Experiments are currently under way to study
thoroughly those systems.

Conclusion

All these experimental results show a strong asymmetry between the
two boundaries S_1 and S_2 . This has been confirmed by ultrasonic
experiments made by Hirsch et al.   ·

In oil-rich systems, the interpenetration of the interfacial
layers allows to explain the origin of the attractive Van der Waals
interactions and to describe the behavior of the conductivity (opening of pores). This picture is corroborated by ultrasonic adsorption
experiments which show a strong adsorption near S_2 . This maximum of
the ultrasonic adsorption is well accounted for by the exchanges of
surfactants in the microemulsion phase, these exchange becoming
more and more important as the interactions increase leading to a
critical point.

In the case of the water-rich systems, the approach to the critical point seems to be quite different. Ultrasonic adsorption experiments show no evidence of surfactant exchange even close to the
boundary S_i(18). We rather observe an elongation of the droplets. In
those systems, entropic forces must be dominant as evidenced by the
study of a simpler system.

More experimental and theoretical work is needed to fully understand all these phenomena in micellar systems.

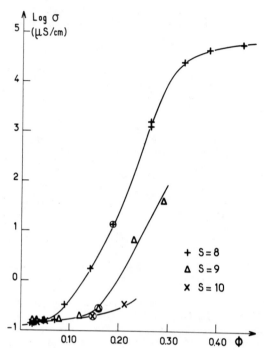

Figure 4 .
    Electrical conductivity versus volume fraction for three series
    of water in oil microemulsions versus salinity. Circles cor-
    respond to the microemulsions obtained from phase separation.

Table III. Critical Exponents and Prefactors for the Interfacial
Tensions and the Correlations Lengths in the Case of the Oil-Free
System.

$T > T_c$ (2 phases)	$T < T_c$ (1 phase)
$\gamma \sim .2 \; \epsilon^{1.2}$	
$\xi_{lower} \quad 7 \; \epsilon^{.64} \; (\overset{\circ}{A})$	$\xi \sim 15 \; \xi^{.6}$
$\xi_{upper} \quad 6 \; \epsilon^{.63}$	$13 \; \xi^{.62}$

$$\left| \; \xi = \Delta T/T_c \; \right|$$

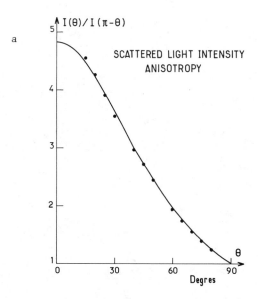

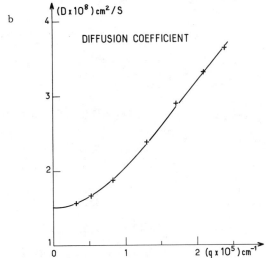

Figure 5 .
Anisotropy of the Scattered intensity and of the diffusion coefficient for a system without oil in the monophasic domain and for $\varepsilon = (T-T_c)/T_c = 1.8 \ 10^{-4}$.
a) Variation of the ratio of the light scattered at the θ angle and the light scattered at the $(\pi-\theta)$ angle versus the scattering angle θ. The full line represents the best fit with Ornstein-Zernike formula. It corresponds to $\xi(I) = 593$ Å .
b) Variation of the diffusion coefficient versus scattering wave vector. The full line represents the best fit with Kawasaki formula. It corresponds to $\xi(D) = 613$ Å . The value of ξ deduced from D(0) is $\xi(D_0) = 600$ Å .

Acknowledgement
We are very undebted to Professor Widom for his suggestion concerning
the wetting experiments.

Literature Cited

1.   Graciaa A., J. Lachaise, A. Martinez, M. Bourrel, C. Chambu,
      C.R. Acad. Sci. Série B, 282, 547 (1976)

2.   Cazabat A.M.,   J. Phys. Lettres, 44, L-593 (1983)
      De Gennes P.G., C. Taupin. J. Phys. Chem 86, 2294 (1982)

3.   Cazabat A.M., D. Langevin, J. Meunier, A. Pouchelon,
      Adv. Coll. Int. Sci. 16, 175 (1982)

4.   Talmon Y., Prager S.,   J. Chem. Phys. 69, 2984 (1978)

5.   Ruckenstein E.,   Chi. J.,   J. Chem. Soc. Far. Trans. 2, 71
      1960, (1975).

6.   Ruckenstein E., Soc. Pet. Eng. Journal 21, 393, (1981).

7.   Ruckenstein E., Proceedings of the Lund Symposium. July
      1982 to be published.

8.   Overbeek J., Th., Far Disc. Chem. Soc. 65, 7 (1978)

9.   Calje A.A., Agterof W.G.M. and Vrij A.,
      In Micellization , Solubilization and Microemulsions
      ed. K.L. Mittal (Plenum Press, N.Y. 1977), Vol 2.

10.   Cazabat A.M., Langevin D.,   J. Chem. Phys. 74, 3148 (1981)

11.   Israelachvili J.,
      Proceedings of the Enrico Fermi Summer School on Amphiphiles
      July 1983, to be published, (see also D. Langevin, same
      reference).

12.   Michell D.J., Ninham B.W., to be published.

13.   Widom B., J. Chem. Phys. 62, 1332 (1975)

14.   Swinney H.L., Henry, D.L., Phys. Rev. A 8, 2586 (1973)

15.   Lemaire B., Bothorel P., Roux D., J. Phys. Chem 1983,
      87, 1023 and 1028.

16.   Lagües M., Ober R., Taupin C., J. Phys. Lettres 39 , L487
      1978.

17.   Abillon O., to be published.

18.   Hirsh E., Debeauvais F., Candau F,. Lang J., Zana R.,
      to be published.

RECEIVED December 26, 1984

Thermal and Dielectric Behavior of Free and Interfacial Water in Water-in-Oil Microemulsions

D. SENATRA[1], G. G. T. GUARINI[2], G. GABRIELLI[2], and M. ZOPPI[1]

[1] Department of Physics, University of Florence, Largo E. Fermi, 2 Arcetri, 50125 Florence, Italy
[2] Department of Chemistry, University of Florence, Via G. Capponi, 9, 50121 Florence, Italy

The low temperature properties of a dodecane-hexanol-K.oleate w/o microemulsion from 20°C to -190°C were studied vs. increasing water content (C,mass fraction) in the interval 0.024-0.4, by Differential Scanning Calorimetry and dielectric analysis (5 Hz-100 MHz). A differentiation between w/o dispersions is obtained depending on whether they possess a "free water" fraction. Polydispersity is evidenced by means of dielectric loss analysis. Hydration processes occurring, at constant surface tension, on the hydrophilic groups of the amphiphiles, at the expenses of the free water fraction of the droplets, are shown to develop "on ageing" of samples exhibiting a time dependent behavior. An energy balance between endothermic and exothermic processes of the system is presented and a model proposed based upon the expansion of the interphase region by means of the formation of 4-H_2O-molecule struc_tures on the polar groups of the surfactants.

It is well known that in recent years microemulsions have attracted relevant interest from both theoretical and practical points of view.

However there is still a great discussion about the definition of their representative properties and even about the proper meaning of the word "microemulsion". Indeed such a name would indicate a dispersed system while microemulsions show the appearence of true solutions,i.e.,of homogeneous systems. Since an essential requisite for the existence of a microemulsion is the presence of water (1), we think that a study of the fundamental properties of such systems should require the use of experimental approaches specifically apt to reveal "in primis" the behavior of water. Moreover,besides of course the hydrocarbon,being the other components necessary for the existence of a microemulsion,amphiphilic compounds acting as surface active agents, also techniques suitable for the study of systems with a high surface-to-volume ratio, are requested.

0097–6156/85/0272–0133$06.00/0

The aim of the present work is twofold: to reach a deeper knowledge of the rôle of water in determining the properties of microemulsions and to identify to the greatest possible extent, the condition
of organization and stability of the interphasal region.

Low frequency dielectric analysis in parallel with differential
scanning calorimetry (DSC) were used to study a water-in-dodecane
microemulsion by only changing its water content, without modifying
the initial proportions of the other components.
Dielectric analysis was chosen to study:

i) the change of the w/o microemulsion as a function of increasing water content by considering "water" both a component and a highly specific dielectric probe, because of the large difference between
ϵ - water (80.37) and ϵ - dodecane (2.014) at 20°C; (2)

ii) the modification of the system's properties upon solidification, at several different concentrations (3);

iii) the frequency dependence of the dielectric loss of liquid
samples in order to use interfacial polarization phenomena (Maxwell-
Wagner absorption) (4-5) to obtain information about the degree of
dispersity of the system and, consequently, of the extension of the
interface.
Thermal analysis was employed to acquire:

i) general thermodynamic information about a multicomponent system;

ii) semiquantitative evaluation of the heats associated with experimentally identified transitions with particular regard to those
of water. This is thought possible because of the small and defined
water content of w/o microemulsions;

iii) a description of the present system based on properties
which do not depend on whether the structure is known (5-6).

Both experimental approaches are supposed suitable to: a) characterize the microemulsion in the two different states: liquid and solid; b) identify the presence of a "free water" fraction in microemulsion systems and the concentration at which the latter becomes detect
able; c) distinguish between different types of w/o dispersions depending on whether they possess a free water content.

Materials and Methods

Microemulsions. Components of the initial mixture of the actual system are: dodecane,n-hexanol and potassium oleate whose % weights are,
respectively, 58.6%, 25.6% and 15.8% (7-8) with K-oleate/dodecane=0.4
(g/ml) and hexanol/dodecane=0.2 (ml/ml). In order to solubilize the
K-oleate,2.4% of water by weight was added to the former mixture
which was therefore kept at 20°C,sealed into a quartz bottle,for a pe
riod of one year without using it (5-6) (9). Water-in-oil microemulsion samples were produced at 20°C, by adding to the above mixture
very small amounts of double distilled water from a Super-Q-Millipore
System with a 0.2 μm Milli Stack filter. In order to speed up the for
mation of isotropic liquid specimens, the samples were gently stirred
for about 15 minutes. Particular care was devoted to preserve the integrity of the samples by maintaining them sealed into neutral glass
containers at a constant temperature of 20°C. The sample water con -
tent was expressed by the mass fraction C=water/(water+oil). The concentration interval investigated estends from 0.024 up to 0.4. The
corresponding range with C in volume fraction, is $0.019 \leq C \leq 0.34$.

The average uncertainty on the sample water content was evaluated to be less than $1°/_{oo}$.

Previous findings on the actual microemulsion are given in references ($9-17$) where the system's phase map vs.concentration in the temperature interval ($-20°C + 80°C$), viscosity measurements,dielectric analysis of liquid samples against both concentration and frequency, the thermally stimulated dielectric polarization release (TSD), electro-optical phenomena, light scattering, Raman spectroscopy and sound propagation investigations are reported.

Dielectric analysis.   The real (ε') and the imaginary (ε'') part of the relative complex permittivity were determined,at a given fixed frequency, as a continuous function of decreasing temperature in the interval from $20°C$ down to $-190°C$ and, at a constant temperature of $20°C$, as a function of frequency in the range (5 Hz- 100 MHz). The dielectric analysis was applied to several samples within the aforementioned concentration interval following the procedure described in references (5) ($11-12$). The low temperature measurements were performed by applying a rate of 0.18 °C/s (10.8 °C/min).

The temperature dependence of the microemulsion was described by reporting the behavior of both ε' and $\varepsilon''/\varepsilon'$ (loss tangent) against decreasing temperature. The latter parameters were determined with an average uncertainty of 5% and 10%, respectively.

The sample temperature,measured with calibrated thermocouples directly inserted into the specimen, was known with an error of ±0.05 °C throughout the whole temperature interval investigated.

The dielectric loss (ε''_{diel}) of liquid microemulsion samples at $20°C$, was obtained by subtracting from the experimentally observed loss (ε''_{ob}), the dc conductance contribution assessed from the lowest frequency measurements where, ε''_{dc} follows the relation $\varepsilon''_{dc} = \lambda/\omega$, λ being the specific conductivity. (18)

The dielectric loss study was performed by means of a cell allowing an electrode distance up to 10 mm (6). The ε''_{diel} values were determined with an average uncertainty of 8%. For the procedures followed to calibrate the sample holders and to define the low temperature calibration values, see references ($5-6$) ($11-12$).

Enthalpy Measurements.   A Perkin Elmer Differential Scanning Calorimeter Model DSC 1-B was used throughout. The instrument measures the heat flow rate (cal/s) by maintaining the sample and the reference isothermal to each other while they are heated or cooled with a linear known temperature rate (Scan speed, °C/min). (19)

The DSC curve obtained is the recording of the heat flow rate (dH/dt) as a function of temperature.quantitative data were obtained by evaluating the total heat absorbed or released during endothermic ($\Delta H > 0$) or exothermic ($\Delta H < 0$) processes that take place, at constant pressure, in the known weight of the sample, within a temperature interval ΔT. The area of each recorded DSC peak in the ΔT interval, is a measure of the enthalpic change (ΔH) associated with the given thermal process occurring in the microemulsion samples. However, in order to define the enthalpy change of a process it is necessary to define the temperature at which the process takes place. The latter essential condition was fulfilled by measuring the transition points and the enthalpies of fusion of both standard materials and the individual components of the microemulsion, with particular regard

to those used to formulate the initial mixture of the system. (20-21)
By means of a careful calibration, we assumed as the temperature
at which a given thermal process takes place,the value assessed for
the temperature at which the half width of the DSC peak occurs on the
starting side of any given peak. The reference values obtained and,
therefore used throughout the present work are reported in "Table I".
Four different scan speeds were tested, namely: 2°C/min, 4°C/min ,
8°C/min and 16°C/min. For each concentration the thermograms were
made at both decreasing and increasing temperatures.
Both the dielectric and the calorimetric results were found to
be quite well reproducible in the whole concentration range studied
but for two concentration values, C=0.270 and C=0.353, where time dependent phenomena were observed upon ageing of the samples. Because
of the latter effect, the two analyses were performed by taking the
measurements respectively, 3-7-10-15-20 and 30 days after the preparation of each concentration tested. The latter study had the aim to
follow the spontaneous evolution of the system until time independent
both dielectric and calorimetric behaviors were obtained. The dielectric study of the time dependent phenomena was carried out at the
fixed frequency of 10 KHz.

Results

Low Temperature Dielectric measurements.   Depending on the sample wa
ter content, three dielectric trends were observed for both ϵ' and
ϵ''/ϵ' as a function of decreasing temperature.
In Figure 1-a, curves "F" (Flat), show the behavior in the concentration range $0.024 \leq C \leq 0.222$ while, curves "SPS" (Single-Peak-Shaped),
refer to the interval $0.222 < C < 0.31$. In Figure 1-b, the trend that
characterizes the samples in the range $0.31 < C < 0.4$ is reported (DPS-
Double-Peak-Shaped). The sample temperature upon solidification with
a gradient of 10.8°C/min is also plotted in both pictures.
At C=0.270 and C=0.353 the dielectric properties of the system
were found to change upon ageing of the samples. Because of the "ageing effect",the SPS dielectric curves of C=0.270 as well as the DPS
ones of C=0.353, flatten and a trend like that observed on the samples with wery low water content results (Figure 1-a,curve F_{aged}).

Dielectric Loss.   The frequency dependence of the dielectric loss
(ϵ''_{diel}) of liquid microemulsions (T=20°C) is reported in Figure 2.

Table I   Reference Values (+) (†)

Component	$T_m(°C)$	Enthalpy (cal/g)	Symbol used
n-dodecane	- 9.60	51.74	ΔH_d^o
n-hexanol	-51.62	36.00	ΔH_h^o
water	0.00	79.70	ΔH_w^o
heavy water	+ 3.90	74.96	ΔH_D^o
water (T=-10°C)	...	74.66	$(\Delta H)_{-10°C}$

(+) For literature values See references (20-21);(÷) Each figure
represents tne mean of three independent measurements.

Each curve shows the typical behavior of a sample in the aforemention ed C-intervals, with: C_1=0.194; C_2=0.304; C_3=0.324. The curve 4-F_{ag} corresponds to the $\varepsilon"_{diel}$ of aged samples with C=0.270 and C=0.353, measured 30 days after their preparation. The dielectric loss observ- ed on the corresponding samples 3 days after their preparation, was found to be of the C_2 and C_3 type, respectively. In Figure 2 the ab- breviations "F-SPS-DPS-F_{ag}", are used to indicate the low temperature dielectric behavior that corresponds to the above samples. Besides the "aged", the data refer to measurements taken 3 days after the preparation of the samples. On stable samples, no change was observed in the loss values as a function of time.

DSC Analysis.   Samples with concentration in the interval $0.024 \leq C \leq 0.222$ are characterized by the thermal curves shown in Figure 3-a and, in the range $0.222 < C < 0.4$, by the DSC recording plotted in Figure 3-b (full line). The two families of curves differ only in the 273 K peak. The latter process was ascribed to the melting of a "free water" fraction of the sample water content. The same kind of exper- iment performed on a D_2O /O microemulsion (dashed line), from 180 K to room temperature,confirmed the above interpretation. The DSC ther- mograms were found to be temperature rate independent and showed a reversible behavior at both decreasing and increasing temperatures.
    In Figure 4, a synthetic representation is given of all the en- thalpic changes recorded, expressed in calories per gram of sample. In Figure 4: ΔH_1 = 1st DSC-peak at T= -60°C; ΔH_2 is comprehensive of the -40°C broad band and the small endotherm at the melting tempera- ture of the n-hexanol ($\simeq$ -52°C); ΔH_d = dodecane enthalpy change ($\simeq$-10 °C); ΔH_w = free water enthalpy change (0°C). Numbered points refer to the "ageing effect": ΔH_w and ΔH_d vs. time at C=0.270 and C=0.353 (time= 3-7-15-20-30 days after the sample preparation). Bars are standard errors of the mean of 6 independent experimental runs.
    The ageing effect is characterized by a decrease of the amount of "free water" until ΔH_w =0; correspondingly, the enthalpy change associated with the fusion of the dodecane (ΔH_d) increases. The sat uration trend followed by both ΔH_w and ΔH_d as the concentration ex- ceeds C=0.3 is quite well observable.
    The enthalpy changes associated with well identified thermal events -melting of water and of dodecane- are plotted vs.concentra- tion in Figure 5 where ΔH_w and ΔH_d are expressed in calories per gram of component.

Paralleling Dielectric and Calorimetric Results.   The concentration interval investigated can be divided into three main regions each characterized by definite and well reproducible both dielectric and calorimetric behaviors, namely:
1.- $0.024 \leq C \leq 0.222$ where: ΔH_w =0 (Figure 3-a) and $\Delta H_d > \Delta H_d^o$ (Figure 5); liquid samples exhibit at 20°C a "Maxwell-Wagner" absorption $\varepsilon"_{diel}$ centered at a frequency higher than 100 MHz, (Figure 2,curve 1-F); the permittivity low temperature dependence follows a flat trend (Figure 1-a, curves F).
2. - $0.222 < C < 0.31$ where: $\Delta H_w \neq 0$ (Figure 3-b) and $\Delta H_d < \Delta H_d^o$ (Figure 5) ; the dielectric loss shows the appearance of an additional peak in the first MHz range (Figure 2,curve 2-SPS); the behavior of ε' and $\varepsilon"/\varepsilon'$ vs. temperature shows that both quantities increase with de-

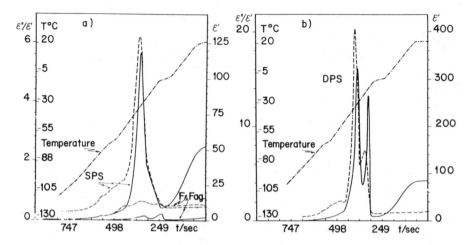

Figure 1. Low temperature dielectric behavior of the w/o micro-
emulsion. Dashed line (ε'); full line ($\varepsilon''/\varepsilon'$). (a): C=0.293 (SPS);
C=0.195 (F). (b): C=0.134 (DPS). "Reproduced with permission from
Ref. 5. Copyright 1984,'Italian Physical Society'."

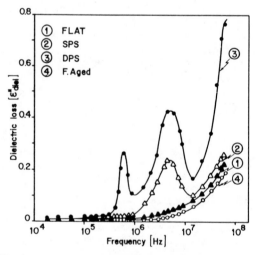

Figure 2. Frequency dependence of the dielectric loss of liquid
microemulsion samples at T=20°C.

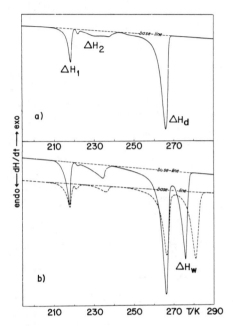

Figure 3. Typical DSC curves of w/o microemulsion samples with concentration in the intervals: (a) $0.024 \leq C \leq 0.222$ with $\Delta H_w = 0$; (b) $0.222 < C < 0.4$ with $\Delta H_w \neq 0$. Dashed curve: DSC recording of a D_2O/o microemulsion. "Reproduced with permission from Ref. 5. Copyright 1984, 'Italian Physical Society'."

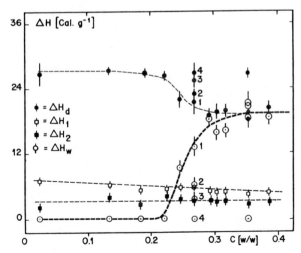

Figure 4. Enthalpy changes of all the thermal processes recorded by DSC. $\Delta H_w \rightarrow$ "free water"; $\Delta H_d \rightarrow$ "dodecane". ΔH is expressed in cal/g of the sample. "Reproduced with permission from Ref. 5. Copyright 1984, 'Italian Physical Society'."

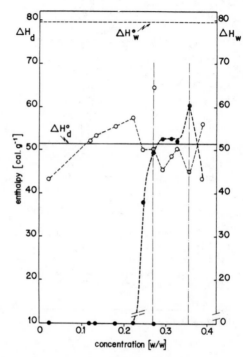

Figure 5. Behavior of the enthalpy change associated with the melting of the sample free water and dodecane content against concentration. ΔH is expressed in cal/g of component.

creasing temperature going through a maximum around $-35°C$, a temperature that does not correspond to the sample transition temperature,
(Figure 1-a, SPS curves).

3. $- 0.31 < C < 0.4$ where: $\Delta H_w \neq 0$ (Figure 3-b), ΔH_d oscillates around
ΔH_d^o (Figure 5); the frequency dependence of $\varepsilon"_{diel}$ is distinguished
by the appearance of an additional loss peak at the end of the KHz
region (Figure 2, curve 3-DPS); the dielectric properties of the system vs. temperature are characterized by Double-Peak-Shaped curves
with ε' and $\varepsilon"/\varepsilon'$ maxima at $-15°C$ and $-33.4°C$ respectively (Figure
1-b, DPS curves).

Within the 2nd and 3rd region, time dependent phenomena were
found to occur upon ageing of the samples with $C=0.270$ and $C=0.353$,
as a result: i)- the "free water" fraction evaluated from the experimentally measured ΔH_w (or $\Delta H_w^o - \Delta H_w$) tends to zero accompanied with
a parallel increase of the heat of melting of the dodecane up to val-
ues higher than ΔH_d^o. (See Table II, "Ageing Effect"); ii)- the low
temperature dielectric peak of both SPS and DPS types flatten (Figure
1-a, curves F and F_{ag}); iii)- the complex dielectric loss spectra exhibited by the samples 3 days after their preparation, evolve into a
simpler behavior with only one main absorption whose maximum develops
beyond the 100 MHz upper limit of the frequency interval investigated
(Figure 2, curves 2-SPS,3-DPS and curve $4-F_{aged}$).

The analysis of interfacial polarization phenomena was used as
an experimental means of verifying whether the results obtained on
temperature processed samples, as in dielectric and calorimetric measurements, are in accordance with those gathered on the corresponding
samples in the liquid state (temperature-unprocessed). The agreement
between the results obtained by means of different experimental approaches confirms that : (a)- we are dealing with basic and general
properties of the microemulsion which appear to manifest themselves
independently of the particular measuring technique employed; (b)-
the temperature change does not affect the main properties of the system. The latter result is in accordance with previous findings on the
"electret" properties of the actual microemulsion (3) (5) (22).

Hydration Processes.   From the experimentally measured ΔH_w and the
known water content of the sample, the amount of water that is "not
free" (hidden water) was estimated and the ratio, $R = N_w /(N_S + N_C)$,
between the number of hidden water molecules (N_w) and that of all the
hydrophilic groups available in the sample (N-surfact. + N-cosurf.)
calculated. The results reported in Table II show, first of all, that
R is a small number and, secondly that, as $\Delta H_w \gtreqless 0$, the R ratio increases by $\simeq 4$. Then, by measuring the total heat absorbed upon melting by a given sample during a whole DSC run, (ΔH_{tot}^S, in cal/g of sample), and calculating the difference between this quantity, measured
at a time $t_o=3$ days and at $t > t_o$ after the preparation of the same
concentration, it is possible to ascertain that the ageing effect implies some energy saving process, the difference given in "equation 1",
being the "energy saved":

$$\Delta H_{t_o-t}^S = (\Delta H_{tot}^S)_{t_o} - (\Delta H_{tot}^S)_{t> t_o} \qquad (1)$$

Such a trend does, on average, occur upon water addition also
in stable samples where time dependent phenomena could not be detected at least within the macroscopic time scale used. As it is shown

Table II. Ageing Effect, Free Water and Interfacial Water (+)

C	Days	ΔH_d {1}	ΔH_w	Free H$_2$O (%)	Hidden H$_2$O (%)	R (++)	$|\Delta|$ {2}	% Inter-{3} facial water	ΔH_{tot} {4}	$\Delta H_{\Delta t}$ {5}	ΔH_δ {6}
0.134	3	52.70	00.00	---	---	2.88	0.96	1.28	35.93		
0.195	3	54.67	00.00	---	---	4.82	2.93	3.92	37.58		
0.222	3	57.24	00.00	---	---	5.29	5.50	7.36	35.99		
0.247	3	50.09	38.03	47.73	52.27	3.19	1.65		41.00		
0.270 (o)	3	51.64	49.27	61.82	38.18	2.71	0.10		43.24		
"	7	53.56	23.67	27.70	70.30	4.82	1.82	2.43	38.72		
"	10	62.11	13.42	16.84	83.16	5.69	10.37	13.88	38.58	4.78	4.82
"	15	58.40	13.10	16.44	83.56	5.72	6.66	8.94	38.46		
"	20	59.19	00.00	---		6.92	7.45	9.97	33.92	4.65	4.82
"	30	63.10	00.00	---		6.92	11.36	15.20	34.07		
0.293	3	46.21	62.48	78.39	21.61	1.66	5.51		44.91		
0.304	3	50.69	54.52	68.41	31.59	2.56	1.05		44.71		
0.315	3	52.47	52.72	66.15	33.85	2.88	0.73		46.73		

(+) Each figure represents a measure during one "DSC" experimental run. (o) Ageing effect.

{1} in: (cal/g) of component; {2} $\Delta = \Delta H_d - \Delta H_d^o$ gives an estimate of the endothermic contribution due to the melting of "interfacial water" at $T \approx -10°C$. {3} Estimate of the % water that by disappearing from the bulk phase of the droplets becomes adsorbed on the hydrophilic groups of the surfactants molecules. The calculation is made with respect to $(\Delta H_w)_{-10°C}$ given in Table I. {4} (ΔH) : in (cal/g) of sample. {5} From "equation 3". {6} From "equation 4". (++) "R" is the ratio (N "hidden water molecules")/(total number hydrophil. groups

in Figure 4, this completely general behavior of the system can be easily evidenced by considering that, ΔH_w and ΔH_d saturate as $C \simeq 0.3$ and, correspondingly, the other thermal processes (ΔH_1 and ΔH_2) give decreasing enthalpic contributions proving that the system, as a func tion of increasing water content, succeeds in decreasing the exten- sion of the interphase region.

The increase of the R ratio by $\simeq 4$, the vanishing of ΔH_w and the gradual decrease of ΔH_{tot}^S shown in Table II, suggest that a hydration process may develop through a migration of water molecules from the droplet core toward the interphase region. Such a process occurs at the expenses of the free water fraction of the sample upon adsorption of H_2O molecules on the hydrophilic groups of the amphiphiles and ends, upon ageing of the sample, with a total change of 1-to-4 water molecules per hydrophilic group of the surfactant molecules.

Recently, Steinbach and Sucker (23) reported about the forma- tion of 4-H_2O-molecule structures that may develop on the hydrophilic groups of surface active compounds upon dilatation of a 1-H_2O-molecu- le structure, by adsorbing 3-water molecules from the subphase at a water-air interface. In the case of the water-oil interphase of the microemulsion, the dispersed droplet consits of an interphasal choro- na that surrounds an inner water core; the free water fraction of the latter (bulk-H_2O)is the subphase that, acting as a reservoir, sup- plies H_2O molecules to the interphase region. Since the formation of hydrated structures takes place at constant surface tension (23), the above mechanism allows the water-oil interface to expand without af- fecting the surface pressure necessary to maintain the system's equi- librium. In this way while the area of every polar head of the amphi- phile remains constant, the interphase area stabilized by a single polar head increases up to the amount corresponding to the definite area requirement of the 4-H_2O-molecule structure (23) (5-6).

Depending on the sample water content, it follows that if the free water fraction is larger than that required for the formation of the "surface complexes", there will be a portion of free water left within the inner core of the dispersed phase. This interpretation ,put forwardto explain that hydration structures may form also in samples that do not exhibit a time dependent behavior, is however consistent with the previously described saturation trend followed by ΔH_w a- gainst increasing concentration. Such an experimental result proves, unambiguously,that the free water fraction of the samples does not become any larger, despite the increase in the system's water content, (Figures 4 and 5).

Hydration processes may therefore "solve" the apparent paradox that the droplets size increases, as shown by light scattering meas- urements, while, thermodynamically, no increase in the sample free water content is observable. In fact, upon formation of hydrated structures, the diameter of the dispersed phase, as a whole, does indeed increase, however the dimension of the droplet's core is kept within some constant values. The dimensions of the inner water core of the droplets as well as the number of H_2O molecules that can be adsorbed on the hydrophilic groups to form hydrated structures should reasonably, depend on the type of components constituting a microemul sion (23) .

Surface Enthalpy Increment and Enthalpy Decrement.   In a previous paragraph we reported that in the ageing effect (Table II-Figure 4), as $\Delta H_w \rightarrow 0$, ΔH_d increases up to values higher than ΔH_d^o (Table I). The greater amount of heat absorbed by the dodecane "more endothermic" transition, was ascribed to the presence of another endothermic contribution occurring in the same temperature interval ($\simeq -10^\circ C$). This overimposed thermal process could arise from the fusion of that fraction of the free water content that, by disappearing from the water bulk phase, becomes adsorbed on the hydrophilic groups to form the 4-H$_2$0-molecule structures (interfacial water).

The enthalpy of association of the hydrated structure is 4.1 kcal/mole (23); the process is exothermic in nature. This energy is thus released when the change in the structural configuration from 1- to-4 water molecules takes place. Therefore, in the particular case of the samples exhibiting a time dependent behavior, the formation of such structures could be verified to a certain extent. We calculated:

1. The total heat (ΔH_s^o) required for the fusion of "each H$_2$0 molecule" of the 4-water-molecule structure, as follows:

$$\Delta H_s^o = (\Delta H_w)_{-10^\circ C} - (\Delta H_l^o)_{ass} = 17.77 \text{ cal/g} \qquad (2)$$

where: (ΔH_w) at $-10^\circ C$ is the enthalpy of fusion of "interfacial water" (Table I), in accordance with our previous assumption; while ($\Delta H_l^o)_{ass} = 4.1/4 \times 18.02 = 56.89$ cal/g is the heat released by each associated water molecule forming the hydrated structure. The quantity ΔH_s^o is called the "surface enthalpy increment".

2. By means of "equation 1.", the difference between the experimentally measured (ΔH_{tot}) at t=t$_o$ and that at the time t=t' at which the R ratio was increased by 3 ,

$$(\Delta H_{tot}^S)_{t_o-t'} = (\Delta H_{tot}^S)_{t_o} - (\Delta H_{tot}^S)_{t'} = \Delta H_{\Delta t} \qquad (3)$$

Then we tried to find the relation between the above quantities. It was found that, under the conditions given in point (2.), the experimentally measured difference of "equation 3." follows the empyrical law:

$$\Delta H_\delta = \Delta H_{\Delta t} = \Delta H_s^o \times \alpha \text{ (cal/g of sample)} \qquad (4)$$

α being a coefficient that gives the fraction of water available in the given weight of the sample under test (6). The quantity ΔH_δ is called "the enthalpy decrement" and interpreted as the total heat released in a sample upon formation of the 4-H$_2$0-molecule structures and used to partly or fully compensate the endothermic heat of melting when the R ratio is increased by 3.

From the observation that $\Delta H_d > \Delta H_d^o$ occurs not only with the disappearance (ageing effect) but also with the absence of the sample free water content (ΔH_w =0), it follows that in the very first C-range $(0.024 \leq C \leq 0.222)$ where no free water could be detected, the "more endothermic" behavior of ΔH_d may be ascribed to the fusion of interfacial water. Thus the endothermic processes dominating the energetics of the system are the formation of the w/o dispersion with the connected development of the interphase region and, the melting of the water around $-10^\circ C$. We recall that the assumption of a melting temperature of $-10^\circ C$ for bound water, is not in contrast with the findings reported in literature for "interfacial" water (24).

By extending the energy balance analysis to samples with C > 0.3, focusing the attention on the "less endothermic" behavior followed by the dodecane enthalpy change in the range $0.3 < C < 0.4$, further support to the development of hydration processes is found. With respect to the ΔH_d^o reference value, among the possible both negative and positive contributions to the measured enthalpic changes, it follows that, within the above concentration interval, two main processes seem to prevail which both lead to a "resultant" exothermic type of contribution: the formation of the hydrated structures and the decrease of the system's interphasal region. Both events may compensate the endothermic contributions due to the melting.

Polydispersity.   In previous papers (3) (5-6) (9) we reported that permanently polarized microemulsion specimens can be produced by rapid freezing isotropic liquid samples, poled at 20°C, with a low level electric field. Moreover it was shown that these samples behave like pyroelectric bodies with the polar axis in the direction of the field. Such a behavior implies that, under the action of the field, apart from electrostriction phenomena that affect the liquid system as a whole, the dispersed phase undergoes a deformation that can be either a real shape change, as in coacervate drops, or a "generalized" deformation due to a migration of charges over microscopic distances, as in electrets. The introduction of a sort of "shape polarity" to explain the mechanism of polarization of non-rigid spherical droplets dispersed in a continuous medium, is however consistent with the frequency dependence of the dielectric loss plotted in Figure 2. In fact, by introducing a "shape factor" that accounts for the degree of elongation of the droplet, according to Sillars' theory of interfacial polarization (25) (5-6), it follows:

i)   the loss produced by a dispersion of spherically shaped droplets is much less than that caused by the same amount of material dis tributed in the form of elongated spheroids;

ii)   the dielectric loss will shift its maximum to a lower frequency upon a shape change of the dispersed phase toward a more elongat ed configuration.

As a consequence, a system containing particles of different shape will exhibit a sequence of loss maxima, centered at frequencies lower than that at which occurs the maximum loss for spheres.

Therefore, in the case of droplets differing in size -under the assumption that larger droplets will deformate more than the smaller ones- the dielectric loss analysis may help in distinguishing among the losses caused by the different size-distributions of the droplets.

The behavior of ε''_{diel} vs. frequency plotted in Figure 2 suggests that in the intervals $(0.222 < C < 0.31)$ and $(0.31 < C < 0.4)$, the samples are polydispersed and that two types of polydispersity develop upon water addition which are distinguished by the loss curves 2-SPS and 3-DPS, respectively.

According to Sillars' theory, the aged samples showing the behavior of curve $4-F_{ag}$, should consist of a homogeneous distribution of small spherical droplets like the one obtained in the interval $(0.024 \leq C \leq 0.222)$, upon minimal addition of water (curve 1-F). Despite the increased sample water content, the experimental evidence supports again an interpretation in terms of spherical-undeformed droplets, which however, being larger, need to be more rigid to account for the observed trend. (5-6). Such a condition is achieved by the

formation of the $4-H_2O-$ molecule structures accompanied by an entropy
decrease and by the disappearance of the free water ($\Delta H_w \neq 0$). The
above entropy decrease is due to the diminution of the freedom of the
water molecules and possibly also to the fact that, being the formation of hydrated structures enthalpically favoured, such event may
take place also with a relative entropy decrease.

Discussion

Two different types of w/o dispersions were found to develop in the
intervals (0.024$\leq$ C$\leq$ 0.222) and (0.222< C< 0.4). The former, is distinguished by ($\Delta H_w = 0$) and rather high values of the R ratio ($\simeq 3$-5).
The latter is characterized by the presence of a free water fraction
as a well detectable component of the system ($\Delta H_w \neq 0$), with R ranging around a value of 2. (See Table II)
     In the "transition" region (0.222< C< 0.31) where a $\Delta H_w \neq 0$ con-
tribution starts being detectable with $\Delta H_w \simeq \Delta H_d$ and $\Delta H_d \simeq \Delta H_d^o$, (see
Table I and Figure 5), a concentration was found at which R$\simeq 1$ (Table
II). This result suggests that at the concentration C=0.293 the inter
phase region has reached the proper extension to allow the formation
of droplets consisting of a continuous monolayer of surfactants with
$1-H_2O$-molecule per hydrophilic group, anchored to a shell of interfacial water (melting at -10°C), that encloses an inner core containing a free water fraction. The difference between the total sample
water content and the measured ΔH_w confirms in fact that there is a
portion of water which is neither "free" nor "interfacial".
     The properties that characterize the w/o dispersions recognized
in the above concentration intervals are in good agreement with the
definition usually adopted to distinguish micellar solutions from w/o
microemulsions (26-30). However, by following the evolution of the
system against water addition , it appears that a continuous change
occurs in the structural configuration of the dispersed phase which
mostly concerns the interphasal region.
     The conclusions drawn on the basis of the dielectric loss anal-
ysis of liquid samples, support the interpretation that a very gradual confluence of the different types of dispersions takes place. Such
an interpretation could explain the instauration of polydispersed
samples in terms of the coexistence, at equilibrium, first, of micellar aggregates with w/o microemulsion droplets and, successively, of
a microemulsion with $1-H_2O$-per hydrophilic group monolayer, in equilibrium with a hydrated type of microemulsion (4-water molecule per
polar head of the surfactant hydrophilic groups monolayer). The latter interpretation is in accordance with Steinbach and Sucker findings that the two types of structures ($1-H_2O$ and $4-H_2O$ molecule),
may coexist at equilibrium (23).
     As shown in Figure 5 by the two dashed vertical lines, the con-
centrations C=0.270 and C=0.353, at which time dependent phenomena
were observed, may be regarded as the upper limits of the two types
of polydispersed samples. Beyond C=0.353 the onset of the bicountin-
uous structure is thought to develop even if macroscopically, the
samples appear isotropic and homogeneous besides being stable.

Conclusion

From the ensemble of the results collected it is possible to draw
some final considerations:

1) really complementary and converging information was obtained by
    parallel study of calorimetric ans dielectric properties of the
    w/o microemulsion; in particular, according to our working hypo-
    thesis, deeper knowledge of the rôle of both water and interphase
    has been gained and it was thus possible to evidence the exist-
    ence of two types of water ("free" and "interfacial") due to the
    presence of interphasal structures characterized by different sur
    face water complexes;
2) the above leads to the possibility of distinguishing among some
    main types of dispersion which can be traced to, say, micelles,
    microemulsions and bicontinuous structures;
3) the observed continuity of behavior of the system's calorimetric
    properties,a part from ΔH_W , can be accounted for by the presence
    of an equilibrium between any two contiguous structures developed
    upon water addition;
4) the reversibility of the DSC experiments and the agreement be-
    tween high and low temperature dielectric measurements indicate
    that the microemulsion freezes unaltered just like a tempered
    multicomponent system.

Acknowledgments

Finantial support of this work by the "Ministero della Pubblica Istru
zionz" (MPI) and the "Gruppo Nazionale di Struttura della Materia"
(GNSM) of the C.N.R., is gratefully acknowledged.
     The authors express their gratitude to Mr. Paolo Parri of the
Chemistry Department for his kind assistance in the preparation of
the drawings.

Literature Cited

1. Prince, L. M. "Microemulsions Theory and Practice"; L.M. Prince
    Ed.; Academic Press:New York, 1977; Chap. 1-3.
2. "Tables of Dielectric Constants of Pure Liquids", N.B.S. Circu-
    lar 514, U.S. Department of Commerce, 1951.
3. Senatra, D. and Gambi, C. M. C., in "Surfactants in Solutions"
    K. Mittal Ed.; Plenum: New York, Vol.3, 1983, p. 1709.
4. Davies, M. In "Dielectric Properties and Molecular Behavior"; N.
    E. Hill et al. Ed.; Van Nostrand Reinhold: London,1969,Chap. 5.
5. Senatra, D., Guarini, G.G.T., Gabrielli, G. and Zoppi, M. In "
    Physics of Amphiphiles. Micelles,Vesicles and Microemulsions";
    V. De Giorgio and M. Corti Eds.;Italian Physical Soc., Int.
    School of Physics "E. Fermi",Course 90th, North Holland:Amster-
    dam, 1984, in press.
6. Senatra, D., Guarini, G.G.T., Gabrielli, G. and Zoppi, M. J. de
    Physique, 1984, in press.
7. Shah, D. O., Science, 1971, 171, 483.
8. Prince, L. M., J. Colloid Interface Sci., 1975, 52, 182.
9. Senatra,D., Gambi, C.M.C. and Neri, A.P., J. Colloid Interface
    Sci., 1981, 79, 443.
10. Ballarò, S., Mallamace, F., Wanderlingh, F., Senatra, D. and Giu
    bilaro, G., J. Phys. C, 1979, 12, 4729.
11. Senatra, D. and Giubilaro, G., J.Colloid Interface Sci., 1978,
    67, 448.

12. Senatra, D. and Giubilaro, G., J.Colloid Interface Sci., 1978,
     67, 457.
13. Senatra, D., J. of Electrostatics, 1982, 12, 383.
14. Senatra, D., Il Nuovo Cimento B, 1981, 64, 151.
15. Ballarò, S., Mallamace, F. and Wanderlingh, F., Phys. Lett. 1979
     70A, 497.
16. Mallamace, F., Migliardo, P., Vasi, C. and Wanderlingh, F., Phys.
     Chem. Liq., 1981, 2, 47.
17. Ballarò, S., Mallamace, F. and Wanderlingh, F., Phys. Lett.1983,
     77A, 2, 3, 198.
18. Coster, W. C., Magat, M., Schneider, W. C. and Smith, C. P.,
     Trans. Faraday Soc., 1946, 42A, 213.
19. Wendlandt, W. W. in "Thermal Methods of Analysis", J. Wiley and
     Sons: New York, 2nd Edition, 1974, Chap. 5.
20. Landolt-Boernstein in "Zahlenwerte und Funktionen", 1961, 6, II
     Band, 4 Teil, Springer Verlag: Berlin.
21. Handbook of Chemistry and Physics, C. D. Hogman, R. C. Weast,
     and S.M. Selby, Edrs.,Chemical Rubber Publish.: Cleveland, Ohio,
     1959.
22. Van Turnhaut, J., In "Electrets", G. M. Sessler Ed.;Topics in
     Applied Physics, Vol. 33, Springer-Verlag: Berlin, New York,
     1980, Chap. 3.
23. Steinbach, H. and Sucker, Chr., Advances in Colloid Interface
     Sci., 1980, 14, 43.
24. Bassetti, V., Burlamacchi, L. and Martini, G., J. Am. Chem. Soc.
     1979, 101, 5471.
25. Sillars, R. W., J. Instr. Engrs.: (London), 1937, 80, 378.
26. Sjöblom, E. and Friberg, S., J. Colloid Interface Sci., 1978,67,
     16.
27. Friberg, S. and Burasczanska, I., Progr. Colloid and Polymer.
     Sci., 1978, 63, 1.
28. Jouffroy, J., Levinson, P. and de Gennes, P. G., J.de Physique,
     1982, 43, 124.
29. Zulauf, M. and Eicke, H. F., J. Phys. Chem., 1978, 83, 480.
30. Robb, I. D., In " Microemulsions ", Plenum: New York, 1982.

RECEIVED June 8, 1984

Macro- and Microemulsions in Enhanced Oil Recovery

M. K. SHARMA and D. O. SHAH

Departments of Chemical Engineering and Anesthesiology, University of Florida, Gainesville, FL 32611

The physicochemical aspects of micro- and macroemulsions have been discussed in relation to enhanced oil
recovery processes. The interfacial parameters (e.g.
interfacial tension, interfacial viscosity, interfacial charge, contact angle, etc.) responsible for
enhanced oil recovery by chemical flooding are described. In oil/brine/surfactant/alcohol systems, a
middle phase microemulsion in equilibrium with excess
oil and brine forms in a narrow salinity range. The
salinity at which equal volumes of brine and oil are
solubilized in the middel phase microemulsion is
termed as the optimal salinity. The optimal salinity
of the system can be shifted to a desired value by
varying the concentration and structure of alcohol.
It was observed that the formulations consisting of
ethoxylated sulfonates and petroleum sulfonates are
relatively insensitive to divalent cations. The results show that a minimum in coalescence rate, interfacial tension, surfactant loss, apparent viscosity
and a maximum in oil recovery are observed at the
optimal salinity of the system. The flattening rate
of an oil drop in a surfactant formulation increases
strikingly in the presence of alcohol. It appears
that the addition of alcohol promotes the mass transfer of surfactant from the aqueous phase to the interface. The addition of alcohol also promotes the
coalescence of oil drops, presumably due to a decrease in the interfacial viscosity. Some novel con-
cepts such as surfactant-polymer incompatibility,
injection of an oil bank and demulsification to promote oil recovery have been discussed for surfactant
flooding processes.

During the past two decades, much attention has been focused on enhanced oil recovery by chemical flooding processes in order to increase the world-wide energy supply. It is well recognized that the

0097-6156/85/0272-0149$06.75/0
© 1985 American Chemical Society

macro- and microemulsion systems play an important role in the oil
recovery processes. Recently, several major research findings in
this area have been reported in the review articles and books (1-4).
Wagner and Leach (5), Taber (6) and Melrose and Bradner (7)
have suggested that capillary forces are responsible for entrapping
a large amount of oil in the form of oil ganglia within the porous
rocks of petroleum reservoirs. Foster (8) has also shown that interfacial tension at crude oil/brine interface, which plays a dominant role in controlling capillary forces, should be reduced by a
factor 10,000 times (e.g. 10^{-3} to 10^{-4} dynes/cm) to achieve an
efficient displacement of the crude oil. Such low interfacial tension can be achieved by appropriate surfactant formulations. The
oil droplets can be deformed easily in the presence of low interfacial tension. Figure 1 schematically illustrates a two dimensional view of the surfactant-polymer flooding process. After injecting
a surfactant slug into the reservoir, a polymer slug is injected for
mobility control. During this process, the displaced oil droplets
coalesce and form an oil bank (Figure 2).
        Once an oil bank is formed in the porous medium, it has to be
propagated through the porous medium with minimum entrapment of oil
at the trailing edge of the oil bank. The maintenance of ultralow
interfacial tension at the oil bank/surfactant interface is necessary for minimizing the entrapment of the oil in the porous medium.
The leading edge of the oil bank coalesces with additional oil
ganglia. Moreover, besides interfacial tension and interfacial
viscosity, another interfacial parameter which influences the oil
recovery is the surface charge at the oil/brine and rock/brine interfaces (9,10). It has been shown that a high surface charge density leads to a lower interfacial tension, lower interfacial viscosity and higher oil recovery as shown in Figure 3.
        In 1959, Wagner and Leach (11) suggested that increased oil
recovery could be obtained by changing wettability of rock material
from oil-wet to water-wet. Melrose and Bradner (7) and Morrow (12)
also suggested that for optimal recovery of residual oil by a low
interfacial tension flood, the rock structure should be water-wet.
Previous investigators (13,14) have used sodium hydroxide to make
the reservoir rock water-wet. Slattery and Oh (15) have shown that
intermediate wettability may be less desirable than either oil-wet
or water-wet rocks. Since, chemical floods satisfy many of these
conditions, they have been considered promising for enhanced recovery of oil. The mechanism of oil displacement in porous media
has been reviewed by Bansal and Shah (16) and more recently by
Taber (17).

Enhanced Oil Recovery By Microemulsion Flooding

In this section, several important aspects of microemulsions in
relation to enhanced oil recovery will be discussed. It is well
recognized that the success of the microemulsion flooding process
for improving oil recovery depends on the proper selection of
chemicals in formulating the surfactant slug.
        During the past decade, it has been reported that many surfac-
tant formulations for enhanced oil recovery generally form multi-
phase microemulsions (18-20). From these studies, it is evident

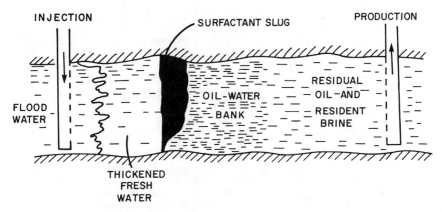

Figure 1.   Schematic illustration of the surfactant-polymer flooding
            process.

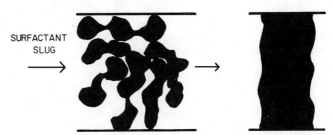

DISPLACED OIL GANGLIA MUST COALESCE TO FORM
A CONTINUOUS OIL BANK; FOR THIS A VERY LOW
INTERFACIAL VISCOSITY IS DESIRABLE

Figure 2.   Schematic presentation of the role of coalescence of
            oil ganglia in the formation of the oil bank.

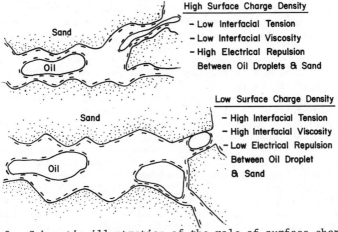

Figure 3.   Schematic illustration of the role of surface charge
            in the oil displacement process.

that a variety of phases can exist in equilibrium with each other.
Figure 4 shows the effect of salinity on the phase behavior of oil/
brine/surfactant/alcohol systems. The microemulsion slug partitions
into three phases (Figure 4), namely, a surfactant-rich middle phase
and a surfactant-lean brine and oil phase (21-23) in the intermediate salinity range. This surfactant-rich phase was termed as the
middle phase microemulsion (23). The middle phase microemulsion
consists of solubilized oil, brine, surfactant and alcohol. The
l → m → u transition of the microemulsion phase can be obtained by
varying any of the eight variables listed in Figure 4.

Interfacial Tension. It is well established that ultralow interfacial tension plays an important role in oil displacement process
(18,20). Figure 5 schematically illustrates the factors affecting
the magnitude of the interfacial tension. Using this conceptual
approach, one can broaden and lower the magnitude of interfacial
tension as well as increase the salt tolerance limit of the surfactant formulation. Experimentally, Shah et al. (24) demonstrated a
direct correlation between interfacial tension and interfacial
charge in various oil-water systems. It was established that the
interfacial charge density plays a dominant role in lowering the
interfacial tension. Figure 6 shows the interfacial tension and
partition coefficient of surfactant as a function of the salinity.
The minimum interfacial tension occurs at the same salinity where
the partition coefficient is near unity. The same correlation between interfacial tension and partition coefficient was also observed by Baviere (25) for paraffin oil/sodium alkylbenzene sulfonate (average MW 350)/isopropyl alcohol/brine system.

Chan and Shah (26) proposed a unified theory to explain the
ultralow interfacial tension minimum observed in dilute petroleum
sulfonate solution/oil systems encountered in tertiary oil recovery
processes. For several variables such as the salinity, the oil
chain length and the surfactant concentration, the minimum in interfacial tension was found to occur when the equilibrated aqueous
phase was at CMC. This interfacial minimum also corresponded to the
partition coefficient near unity for surfactant distribution in oil
and brine. It was observed that the minimum in ultralow interfacial
tension occurs when the concentration of the surfactant monomers in
aqueous phase is maximum.

Formation and Structure of Middle Phase Microemulsion. The l → m →
u transitions of the microemulsion phase as a function of various
parameters are shown in Figure 4. Chan and Shah (31) compared the
phenomenon of the formation of middle phase microemulsion with that
of the coacervation of micelles from the aqueous phase. They concluded that the repulsive forces between the micelles decreases due
to the neutralization of surface charge of micelles by counterions.
The reduction in repulsive forces enhances the aggregation of micelles as the attractive forces between the micelles become predominant. This theory was verified by measuring the surface charge
density of the equilibrated oil droplets in the middle phase (9).
It was observed that the surface charge density increases to a maximum at the salinity at which the middle phase begins to form.
Beyond this salinity, the surface charge density decreases in the

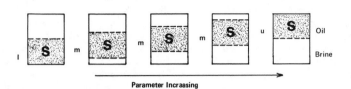

Parameter Increasing

The transition l → m → u occurs by:

1. Increasing Salinity
2. Decreasing oil chain length
3. Increasing alcohol concentration (C_4, C_5, C_6)
4. Decreasing temperature
5. Increasing total surfactant concentration
6. Increasing brine/oil ratio
7. Increasing surfactant solution/oil ratio
8. Increasing molecular weight of surfactant

Figure 4.   Schematic illustration of the l→m→u transition of the microemulsion phase by several variables.

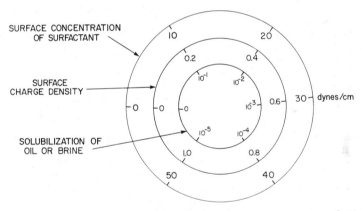

Figure 5.   Schematic presentation of the three components of the interfacial tension.

three-phase region. Based on several observations of different sur-
factant/brine/oil systems, Chan and Shah (31) proposed the mechanism
of middle phase microemulsion formation as shown in Figure 7. In
general, the higher the solubilization of brine or oil in the middle
phase microemulsion, the lower the interfacial tension with the excess phases. The salinity at which equal volumes of brine and oil
are solubilized in the middle phase microemulsion is referred to as
optimal salinity for the surfactant/oil/brine systems under given
physicochemical conditions (21,23). Previous investigators (22,29,
30) have shown that the oil recovery is maximum near the optimal
salinity of the system. Therefore, one can conclude that the middle
phase microemulsion plays a major role in enhanced oil recovery
processes.

Using various physicochemical techniques such as high resolution NMR, viscosity and electrical resistivity measurements, Chan
and Shah (31) have proposed that the middle phase microemulsion in
three-phase systems at and near optimal salinity is a water-external
microemulsion of spherical droplets of oil. Extended studies to
characterize the middle phase microemulsions by several techniques
including freeze-fracture electron micrographs revealed the structure as a water-external microemulsion (31). The droplet size in
the middle phase microemulsion decreases with increasing salinity.
The freeze-fracture electron microscopy of a middle phase microemulsion is shown in Figure 8. This system was extensively studied by
Reed and Healy (21-23). It clearly indicates that the discrete
spherical structure of the droplets in a continuous aqueous phase is
consistent with the mechanism proposed in Figure 7. It should be
pointed out that several investigators (32-39) have proposed the
possibility of bicontinuous structure or coexistence of water-external and oil-external microemulsions in the middle phase. For
very high surfactant concentration (30-40 %) systems, the existence
of anomalous structures which are neither conventional oil-external
nor water-external microemulsions, have been proposed to explain
some unusual behavior of these systems (32-36).

Solubilization. The effectiveness of surfactant formulations for
enhanced oil recovery depends on the magnitude of solubilization.
By injecting a chemical slug of complete miscibility with both oil
and brine present in the reservoir, 100% oil recovery can be
achieved.

The effect of hydrated radii, valency and concentration of
counterions on oil-external and middle phase microemulsions was investigated by Chou and Shah (40). It was observed that 1 mole of
$CaCl_2$ was equivalent to 16-19 moles of NaCl for solubilization in
middle phase microemulsion, whereas for solubilization in oil-external microemulsions, 1 mole of $CaCl_2$ was equivalent to only 4
moles of NaCl. For monovalent electrolytes, the values for optimal
salinity for solubilization in oil-external and middle phase microemulsions are in the order: LiCl > NaCl > KCl > NH_4Cl, which correlates with the Stokes radii of hydrated counterions. The optimal
salinity for middle phase microemulsions and critical electrolyte
concentration varied in a similar fashion with Stokes radii of
counterions, which was distinctly different for the solubilization
in oil-external microemulsions. Based on these findings, it was

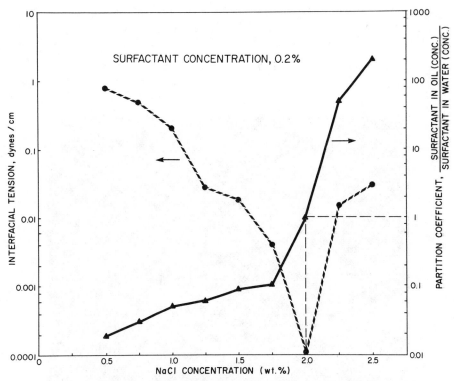

Figure 6.   Effect of salinity on interfacial tension and partition coefficient for TRS 10-80/n-octane system.

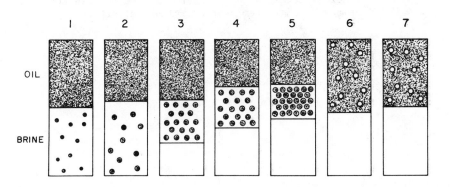

Figure 7.   Schematic illustration for the mechanism of transition from lower---→ middle---→ upper phase microemulsion upon increasing salinity.   Solid circles, oil-swollen micelles (or microdroplets of oil); open circles, reverse micelles (or microdroplets of water).

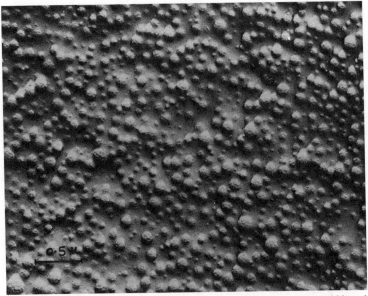

Figure 8. Freeze-fracture electron micrograph of the middle phase
of the Exxon system at the optimal salinity.

concluded that the middle phase microemulsion behaves like a water continuous system with respect to the effect of counterions (40).

The effect of alcohol concentration on the solubilization of brine has been studied in this laboratory (41). It was observed that there is an optimal alcohol concentration which can solubilize the maximum amount of brine and can also produce ultralow interfacial tension. The optimal alcohol concentration depends on the brine concentration of the system. The effect of different alcohols on the equilibrium properties and dynamics of micellar solutions has been studied by Zana (42).

Phase Behavior. The surfactant formulations for enhanced oil recovery consist of surfactant, alcohol and brine with or without added oil. As the alcohol and surfactant are added to equal volumes of oil and brine, the surfactant partitioning between oil and brine phases depends on the relative solubilities of the surfactant in each phase. If most of the surfactant remains in the brine phase, the system becomes two phases, and the aqueous phase consists of micelles or oil-in-water microemulsions depending upon the amount of oil solubilized. If most of the surfactant remains in the oil phase, a two-phase system is formed with reversed micelles or the water-in-oil microemulsion in equilibrium with an aqueous phase.

The phase behavior of surfactant formulations for enhanced oil recovery is also affected by the oil solubilization capacity of the mixed micelles of surfactant and alcohol. For low-surfactant systems, the surfactant concentration in oil phase changes considerably near the phase inversion point. The experimental value of partition coefficient is near unity at the phase inversion point (28). The phase inversion also occurs at the partition coefficient near unity in the high-surfactant concentration systems (31). Similar results were also reported by previous investigators (43) for pure alkyl benzene sulfonate systems.

Figure 9 shows the effect of surfactant concentration on the volume of the middle phase microemulsion. It is interesting that the plot goes through the origin indicating that even at very low surfactant concentrations, a microscopic amount of the midddle phase microemulsion must exist at the interface between oil and brine.

Salinity Tolerance. As the petroleum reservoir salinity can be very high, the surfactant formulations should be designed for high salt tolerance. The widely used petroleum sulfonates for enhanced oil recovery exhibit relatively low salt tolerance in the range 2-2.5% NaCl concentration, and even smaller for the optimal salinity. The presence of divalent cations in the brine decreases the optimal salinity of surfactant formulations (44).

Since optimal salinity leads to a favorable condition for maximum oil recovery, one would like to design methods to adjust the optimal salinity of a given surfactant formulation (45,46). Figure 10 shows the optimal salinity of a mixed surfactant formulation consisting of a petroleum sulfonate and an ethoxylated sulfonate. It is evident that the optimal salinity of the formulation increases with increasing concentration of ethoxylated sulfonate, the optimal salinity increases from 1% to 24% NaCl brine. Moreover, it is interesting to note that these formulations, when equilibrated with

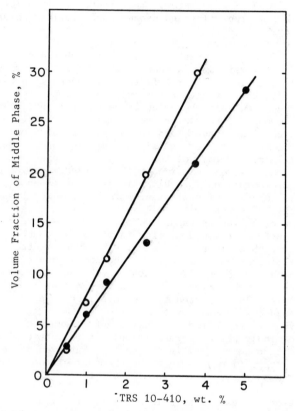

Figure 9.   Effect of surfactant concentration on the volume of
middle phase microemulsion. Key: o, 1.5% isobutanol;
•, 3% isobutanol.

oil, produce middle phase microemulsion with very low interfacial
tension (< 10^{-3} dynes/cm). Similar results were also obtained using
ethoxylated alcohols in surfactant formulations by previous investigators (47). It is interesting that the ethoxylated sulfonate alone
(curve F in Figure 10) is unable to produce ultralow interfacial
tension. However, when mixed with petroleum sulfonate, it produces
very low interfacial tension. Therefore, the surfactant formulations consisting of mixed petroleum sulfonates and ethoxylated sulfonates or alcohols are promising for enhanced oil recovery from
high salinity reservoirs.

Macroemulsions in Enhanced Oil Recovery Processes

As the surfactant slug is injected into the reservoir, the mixing of
injected slug with reservoir components takes place. The mixing of
surfactant with reservoir oil and brine often produces emulsions.
Moreover, the reservoir parameters such as porosity, pressure,
temperature, composition of connate water and crude oil as well as
gas-oil ratio affect the formation of oil field emulsions.

Usually, the formation of stable macroemulsions in the oil
fields is considered undesirable and can cause severe problems.
Previous investigators (48-50) have reported poor oil recovery due
to problems associated with stable emulsions. In addition, water-
in-oil emulsions should be resolved before the oil refining process
because of the presence of considerable amounts of emulsified water
in crude oil increases the cost of oil transportation and leads to
other maintenance problems. Because of corrosion problems, most
pipelines do not accept curde oil with significant amount of emulsified water. On the other hand, McAuliffe (51) has shown a bene-
ficial effect of macroemulsions in oil recovery as their injection
into sandstone cores increased sweep efficiency. According to this
concept, the emulsion droplet enters a pore constriction smaller
than itself. For an emulsion to be most effective, the oil droplets
in the emulsion must be larger than the pore-throat constrictions in
the porous media. The injected emulsion enters the highly permeable
zones, which in turn reduces the channeling of water. Therefore,
water starts to flow into low permeable zones, resulting in a greater sweep efficiency. Based on this concept, a field test was conducted which showed an improvement in oil recovery by macroemulsion
flooding process. In order to understand the effect of macroemulsions on the oil recovery process, the following parameters should
be discussed.

Electrophoretic Mobility. The electrophoretic mobility of the crude
oil droplets as a function of caustic concentration has been determined in relation to enhanced oil recovery (52). It was observed
that a maximum in electrophoretic mobility corresponds to a minimum
in interfacial tension at the crude oil/caustic interface (Figure
11). The maximum electrophoretic mobility at minimum interfacial
tension can be attributed to the ionization of carboxyl groups present in the crude oil, which in turn determine the charge density
at the crude oil/caustic interface, depending on NaOH concentration.

The experimental procedure used for the measurement of interfacial tension and the electrophoretic mobility to determine the

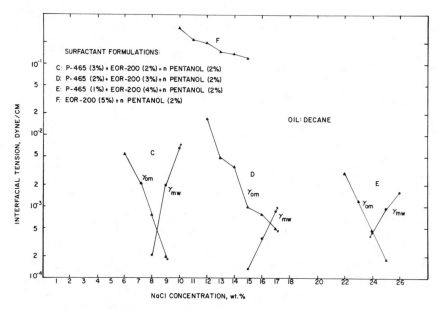

Figure 10. Increase in optimal salinity of formulations consisting of a mixture of a petroleum sulfonate (P - 465) and an ethoxylated sulfonate (EOR-200).

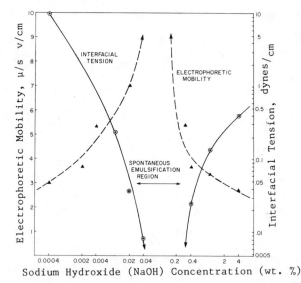

Sodium Hydroxide (NaOH) Concentration (wt. %)

Figure 11. A correlation between electrophoretic mobility and interfacial tension for crude oil-caustic solutions.

optimum caustic concentration is extremely time consuming and laborious. Therefore, optical transmission or absorbance measurements were made on diluted emulsions to overcome this problem. There was a very good correlation between the ultralow interfacial tension and absorbance measurements. It has been observed for several crude oils that the caustic concentration which yields the maximum absorbance also shows a maximum in electrophoretic mobility and a minimum in interfacial tension. These parameters play a prominent role in the oil recovery by caustic flooding process.

Transient Processes. There are several transient processes such as formation and coalescence of oil drops as well as their flow through porous media, that are likely to occur during the flooding process. Figure 12 shows the coalescence or phase separation time for hand-shaken and sonicated macroemulsions as a function of salinity. It is evident that a minimum in phase separation time or the fastest coalescence rate occurs at the optimal salinity (53). The rapid coalescence could contribute significantly to the formation of an oil bank from the mobilized oil ganglia. This also suggests that at the optimal salinity of the system, the interfacial viscosity must be very low to promote the rapid coalescence.

The flow through porous media behavior of various macroemulsions was studied by measuring the pressure drop across a porous medium (Figure 13). It is obvious that a minimum in pressure drop occurs near the optimal salinity of the surfactant formulation. One can conclude that the interfacial tension is an important parameter which influences the pressure drop across porous media (53).

Figure 14 shows a very interesting and an important correlation between the rate of coalescence in macroemulsions and the apparent viscosity in the flow through porous media. It was observed that a minimum in apparent viscosity for the flow of macroemulsions in porous media coincides with a minimum in phase separation time at the optimal salinity. This correlation between the phenomena occurring in the porous medium and outside the porous medium allows us to use coalescence measurements as a screening criterion for many oil recovery formulations for their possible behavior in porous media. It is very likely that a rapidly coalescing macroemulsion may give a lower apparent viscosity for the flow in porous media (53).

The variation in the shape of an oil drop (n-Octane) upon contacting a surfactant formulation consisting of 0.05% TRS 10-80 in 1% NaCl as a function of time is shown in Figure 15. It is evident that as surfactant molecules migrate from the aqueous phase to the interface and subsequently, to the oil phase, the interfacial tension decreases and the spherical drop gradually flattens out. This flattening time reflects the rate at which surfactant molecules accumulate at the oil-brine interface. As shown in Table 1, there is a good correlation between the flattening time, interfacial tension and the oil recovery. The reduction in flattening time leads to favorable oil recovery efficiency (54). Table 2 shows the effect of alcohol concentration on various parameters and oil recovery in porous media. It is evident that the flattening time decreases strikingly in the presence of alcohol suggesting that the alcohol promotes the mass transfer to the interface and a rapid reduction in the magnitude of the interfacial tension.

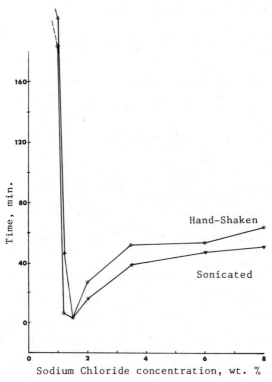

Figure 12. Effect of salinity on the phase separation or coalescence rate of hand-shaken and sonicated macroemulsions.

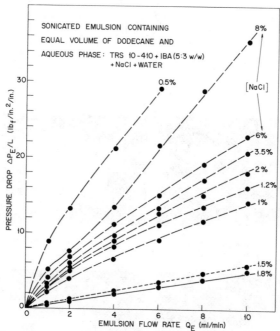

Figure 13. Effect of salinity on the pressure drop-flow rate curves of sonicated macroemulsions.

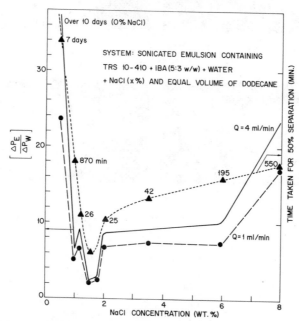

Figure 14. A correlation between the apparent viscosity and coalescence rate of sonicated macroemulsions.

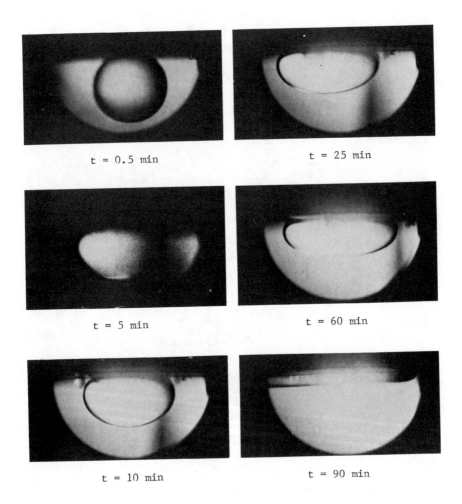

t = 0.5 min   t = 25 min

t = 5 min   t = 60 min

t = 10 min   t = 90 min

Figure 15. An illustration of the drop flattening process for an octane drop upon contacting 0.05% TRS 10-80 in 1% NaCl.

Table I.   IFT, Flattening Time and Oil Recovery Efficiency of 0.05%
           TRS 10-80 in 1% NaCl vs. n-Octane at 25°C

SYSTEM	IFT (mN/m)	FLATTENING TIME* (Seconds)	OIL RECOVERY[+] (%OIP)
I. Fresh Oil/1% NaCl	≈50.8**	∞	61–63
II. Fresh Oil/Equilibrated Surfactant Solution	0.731	6600	44–52
III. Fresh Oil/Fresh Surfactant Solution	0.627	480	75–77
IV. Equilibrated Oil/% NaCl	0.121	900	83
V. Equilibrated Oil/Equilibrated Surfactant Solution	0.0267	240	94
VI. Equilibrated Oil/Fresh Surfactant Solution	0.00209	15	–

*Flattening time is defined as the time required for the n-octane
drop to gradually flatten out.

**Octane/H_2O, 20°C, IFT = 50.8 mN/m, "Interfacial Phenomena",
Davies and Rideal, Chapter 1, p. 17 Table 1, Academic Press,
N.Y. 1963.

[+]Sandpack dimensions: 1.06" dia. x 7" long: Permeability = 3
darcy; flow rate: 2.3 ft./day.

Table II.   The Effect of IBA on Flattening Time, IFT, IFV, Partition Coefficient, and Oil Displacement Efficiency

System	0.1% TRS 10-410 in 1.5% NaCl vs. n-Dodecane	0.1% TRS 10-410 + 0.06% IBA in 1.5% NaCl vs n-Dodecane	0.05% TRS 10-80 in 1% NaCl vs. n-Octane	0.05% TRS 10-80 + 0.04% IBA in 1% NaCl vs. n-Octane
Run	S100-48	S100-43	S100-02	S100-44
Flattening Time	90 sec	<1 sec	420 sec	<1 sec
IFT (dynes/cm)	0.086	0.088	0.025	0.024
Interfacial Viscosity (s.p.)	0.096	0.086	0.023	0.018
Partition Coefficient	0.010	0.009	0.3	1.36
Secondary Recovery By Brine Flooding	–	–	61.2%	60.08%
By Surfactant Soln. Flooding	84.37%	98.32%	60%	91%
Tertiary Recovery	–	–	0	76.84%
Final Oil Saturation	11.73%	1.28%	30%	5.36%

*All displacement experiments are carried out with nonequilibrated systems in sand packs at 25°C; Dimensions and flow rates same as given in Table 1.

Secondary and tertiary oil recovery values are percent of oil-in-place, whereas final oil saturation is percent of total pore volume.

The mixing of surfactant and polymer in the porous medium occurs due to both dispersion and the excluded volume effect for the flow of polymer molecules in porous media, which in turn could lead to the phase separation. Figure 16 illustrates the schematic explanation of the surfactant-polymer incompatibility and concomittant phase separation. We propose that around each micelle there is a region of solvent that is excluded to polymer molecules. However, when these micelles approach each other, there is overlapping of this excluded region. Therefore, if all micelles separate out then the excluded region diminishes due to the overlap of the shell and more solvent becomes available for the polymer molecules. This effect is very similar to the polymer depletion stabilization (55). Therefore, this is similar to osmotic effect where the polymer molecule tends to maximize the solvent for all possible configurations.

In oil recovery processes, the formation of an oil bank is very important for an efficient oil displacement process in porous media. This was established from studies on the injection of an artificial oil bank followed by the surfactant formulation which can produce ultralow interfacial tension with the injected oil. We observed that the oil recovery increased considerably and the residual oil saturation decreased with the injection of an oil bank as compared to the same studies carried out in the absence of an injected oil bank (54). Figure 17 schematically represents the oil bank formation and its propagation in porous media, which is analogous to the snowball effect. If an early oil bank is formed then it moves through the porous medium accumulating additional oil ganglia resulting in an excellent oil recovery, whereas a late oil bank formation will result in a poor oil recovery.

In summary, several phenomena occurring at the optimal salinity in relation to enhanced oil recovery by macro- and microemulsion flooding are schematically shown in Figure 18. It is evident that the maximum in oil recovery efficiency correlates well with various transient and equilibrium properties of macro- and microemulsion systems. We have observed that the surfactant loss in porous media is minimum at the optimal salinity presumably due to the reduction in the entrapment process for the surfactant phase. Therefore, the maximum in oil recovery may be due to a combined effect of all these processes occurring at the optimal salinity.

Demulsification. It is necessary to demulsify the macroemulsions formed due to surfactant flooding. The problem of separating two immiscible liquids when one is dispersed within the other is frequently encountered in petroleum technology. Demulsification by definition is agglomeration and coalescence of dispersed phase, eventually resulting in a breakdown of the macroemulsion into two separate phases. A wide variety of materials such as cotton, wool, glass fibers and Teflon have been used to promote the coalescence rate of macroemulsions. The addition of acids or bases apparently causes neutralization of the particle charge and subsequently, leads to coagulation of droplets.

Several patents have been granted for methods to demulsify the crude oil macroemulsions. In the electrical method, the imposition

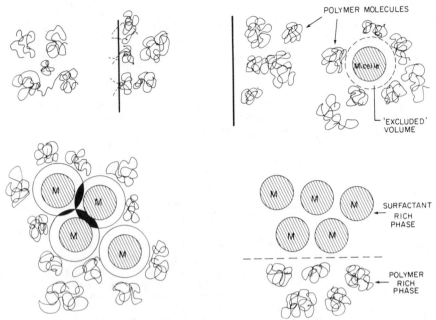

Figure 16. Schematic illustration of surfactant-polymer incompatibility leading to phase separation in mixed surfactant-polymer systems.

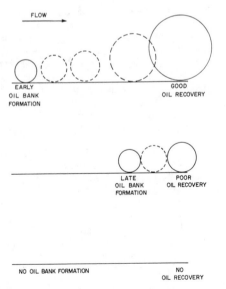

Figure 17. Schematic illustration of the injection of an oil bank
and the subsequent "snowball effect" in enhanced oil
recovery.

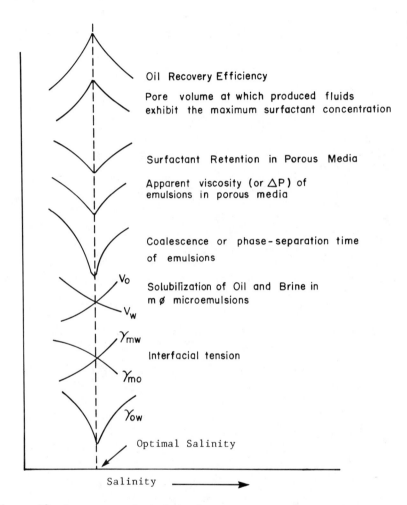

Figure 18. A summary of various phenomena occurring at the optimal salinity in relation to enhanced oil recovery by surfactant-polymer flooding.

170

of high potentials on macroemulsions leads to coagulation of water
droplets. In enhanced oil recovery methods by chemical flooding,
the macroemulsions are stabilized by the surfactants. In order to
break these macroemulsions, it is necessary to rupture the interfacial film. This can be achieved to some extent by heat treatment.
A widely used method for the breaking of oil field macroemulsions is
the use of elevated temperatures (27,28). By the action of heat,
the surfactant becomes more soluble in either the aqueous phase or
oil phase, which destabilizes the interfacial film, and the separation of two phases occurs.

Literature Cited

1. Shah, D.O.; Ed. "Surface Phenomena in Enhanced Oil Recovery";
    Plenum Press, New York, 1981.
2. Shah, D.O.; Schechter, R.S.; Eds. "Improved Oil Recovery by
    Surfactant and Polymer Flooding"; Academic Press, Inc., New
    York, 1977.
3. Lissant, K.J.; Ed. "Emulsions and Emulsion Technology, Part
    III"; Marcel Dekker Press, New York, 1984.
4. Becher, P.; Ed. "Encyclopedia in Emulsion Technology Part I";
    Marcel Dekker Press, New York, 1983.
5. Wagner, O.R.; Leach, R.O.; Soc. Pet. Eng. J., 1966, 6, 335.
6. Taber, J.J.; Soc. Pet. Eng. J., 1969, 9, 3.
7. Melrose, J.C.; Brader, C.F.; J. Cand. Pet. Tech., 1974, 13, 54.
8. Foster, W.R.; J. Pet. Tech., 1973, 25, 205.
9. Chan, K.S.; Ph.D. Dissertation, University of Florida, Gaines-
    ville, (1978).
10. Chiang, M.Y.; Chan, K.S.; Shah, D.O.; J. Cand. Pet. Tech.,
    1978, 17(4), 1.
11. Wagner, O.R.; Leach, R.O.; Trans AIME, 1959, 216, 65.
12. Morrow, N.R.; 27th Annual Technical Meeting of Pet. Soc. of
    CIM, Calgary, Alberta, June 7-11, 1976.
13. Emery, L.W.; Mungan, N.; Nicholson, R.W.; J. Pet. Tech., 1970,
    22, 1569.
14. Leach, R.O.; Wagner, O.R.; Wood, H.W.; Harpke, C.F.; J. Pet.
    Tech., 1962, 14, 206.
15. Slattery, J.C.; Oh, S.C.; ERDA Symposium on Enhanced Oil and
    Gas Recovery, Tulsa, September 9-10, 1976.
16. Bansal, V.K.; Shah, D.O. in "Micellization, Solubilization
    and Microemulsions"; Mittal, K.L., Ed., Plenum Press, New
    York, 1977, p. 87.
17. Taber, J.J. in "Surface Phenomena in Enhanced Oil Recovery";
    Shah, D.O., Ed., Plenum Press, New York, 1981, p. 13.
18. Chiang, M.Y.; Shah, D.O.; SPE 0988 presented at the SPE 5th
    International Symposium on Oilfield and Geothermal Chemistry,
    Stanford, CA, May 28-30, 1980.
19. Cayias, J.L.; Schechter, R.S.; Wade, W.H.; J. Colloid Interface
    Sci., 1977, 59, 31.
20. Wilson, P.M.; Murphy, C.L.; Foster, W.R.; SPE 5812 presented at
    SPE Improved Oil Recovery Symposium, Tulsa, OK, 1976.
21. Healy, R.N.; Reed, R.L.; Soc. Pet. Eng. J., 1974, 14, 451.
22. Reed, R.L.; Healy, R.N. in Improved Oil Recovery by Surfactant
    and Polymer Flooding (Shah, D.O.; Schechter, R.S., Eds.),
    Academic Press, New York, 1977, p. 383.

23. Healy, R.N.; Reed, R.L.; Stenmark, D.G., Soc. Pet. Eng. J., 1976, 16, 147.
24. Shah, D.O.; Chan, K.S.; Bansal, V.K., 83rd National Meeting of AIChE, Houston, TX, March 20-24, 1977.
25. Baviere, M.; SPE 6000, 51st Annual Fall Technical Conference and Exhibition of the Society of Petroleum Engineers of AIME, New Orleans, October 3-6, 1976.
26. Chan, K.S.; Shah, D.O., J. Disp. Sci. Tech., 1980, 1, 55.
27. Blair, C.M.; Chem. Ind. (London), 1960, 538.
28. Lissant, K.J.: "Emulsions and Emulsions Technology"; Part 2, Dekker, New York, 1974, p. 71.
29. Shah, D.O., Proc. European Symp. on Enhanced Oil Recovery, Elsevier, Lausanne, Switzerland, 1981, p. 1-41.
30. Boneau, D.F.; Clampitt, R.L.; J. Pet. Tech., 1977, 29, 501.
31. Chan, K.S.; Shah, D.O., SPE 7869, SPE-AIME International Symposium on Oilfield and Geothermal Chemistry, Houston, January 22-24, 1979.
32. Shinoda, K., J. Colloid Interface Sci., 1967, 26, 70.
33. Shinoda, K.; Saito, H., J. Colloid Interface Sci., 1968, 26, 70.
34. Miller, C.A.; Hwan, R.; Benton, W.J.; Fort, T., Jr., J. Colloid Interface Sci., 1977, 61(3), 554.
35. Friberg, S.; Lapezynska, I.; Gilberg, G., J. Colloid Interface Sci., 1976, 56, 19.
36. Hawn, R.; Miller, C.A.; Fort, T., Jr., J. Colloid Interface Sci., 1979, 68, 221.
37. Scriven, L.E., Nature, 1976, 263, 123.
38. Ramachandran, C.; Vijayan, S.; Shah, D.O., J. Phy. Chem., 1980, 84, 1561.
39. Scriven, L.E., in "Micellization, Solubilization and Micro-emulsions", Mittal, K.L., Ed., Plenum Press, New York, 1977, Vol. II, p. 877.
40. Chou, S.I.; Shah, D.O., J. Colloid Interface Sci., 1981, 80(2), 311.
41. Hsieh, W.C.; Shah, D.O., SPE 6594, SPE-AIME International Symposium on Oilfield and Geothermal Chemistry, La Jolla, June 27-28, 1977.
42. Zana, R.; in "Surface Phenomena in Enhanced Oil Recovery", Shah, D.O., Ed., Plenum Press, New York, 1981, p. 521.
43. Wade, W.H.; Morgan, J.C.; Jacobson, J.K.; Salager, J.L.; Schechter, R.S., SPE 6844, SPE-AIME 52nd Annual Fall Technical Conference and Exhibition, Denver, October 9-12, 1977.
44. Healy, R.N.; Reed, R.L., SPE 5817, SPE Improved Oil Recovery Symposium, Tulsa, March 22-24, 1976.
45. Bansal, V.K.; Shah, D.O., Soc. Pet. Eng. J., June, 1978, 167.
46. Bansal, V.K.; Shah, D.O., J. Colloid Interface Sci., 1978, 65(3), 451.
47. Dauben, D.L.; Froning, H.R., J. Pet. Tech., 1971, 23, 614.
48. Strange, L.K.; Talash, A.W., J. Pet. Tech., 1977, 29, 1380.
49. Whitley, R.C.; Ware, J.W., J. Pet. Tech., 1977, 29, 925.
50. Widmeyer, R.H.; Satter, A.; Graves, R.H.; Frazier, G.D., J. Pet. Tech., 1977, 29, 933.
51. McAuliffe, C.D.; J. Pet. Tech., 1973, 25, 729.
52. Bansal, V.K.; Chan, K.S.; McCallough, R.; Shah, D.O., J. Canadian Pet. Tech., 1978, 17, 1.

53. Vijayan, S.; Ramachandra, C.; Doshi, H.; Shah, D.O., in "Sur-
     face Phenomena in Enhanced Oil Recovery", Shah, D.O., Ed.,
     Plenum Press, New York, 1981, p. 327.
54. Chiang, M.Y., Ph.D. Dissertation, University of Florida, Gaines-
     ville, FL, 1978.
55. Cash, R.L.; Cayias, J.L.; Hayes, M.; McAlister, D.J.; Schares,
     T.; Wade, W.H., J. Pet. Tech., September, 1976, 985.

RECEIVED December 11, 1984

Role of the Middle Phase in Emulsions

F. M. FOWKES, J. O. CARNALI, and J. A. SOHARA[1]

Department of Chemistry, Lehigh University, Bethlehem, PA 18015

We have studied some soap-stabilized oil-in-water emulsions which flocculate at higher salinities without
coalescense, developing flat planes of contact. The
contact angles between the oil/water interfaces and the
plane of contact increase with salinity up to 60°,
indicating that the tensions in the plane of contact
are less than the net oil/water interfacial tension.
It is shown that these contact angles develop when
middle phase films (M) coat the oil droplets and fill
the planar spaces between flocculated droplets. The
net interfacial tension is $\gamma_{MO} + \gamma_{MW}$ and the contact
angle θ_W allows calculation of each of these two
tensions;

$$\cos\theta_W = \gamma_{MO} / (\gamma_{MO} + \gamma_{MW})$$

A small angle X-ray scattering study showed that the
middle phase films between flocculated oil droplets
were about 90 nm in thickness, and spinning drop
interfacial tension measurements were found to be able
to centrifuge middle phase films off the oil drops
along the axis of rotation.

Such contact angle measurements are recommended as
general tools for middle phase studies because they
are sensitive to very low middle phase concentration
and measure the relative hydrophilic or oleophilic
character of the middle phase.

In a number of studies of middle phase phenomena in oil-water-
emulsifier systems it has been shown that middle phases form when
the emulsifier is about equally partitioned into the oil and water
phases Healy et al.,(1), Salager et al., (2), Shah et al., (3),
Kuneida and Shinoda, (4). In these studies it has been shown that

[1] Current address: Atlas Powder Company, Tamaqua, PA 18252

the interfacial tension of the middle phase versus oil (γ_{MO}) decreases as the emulsifier content of the oil increases, while γ_{MW}
decreases as the emulsifier content of the water increases
(shown in Figure 1). In these studies the partition coefficient
for the emulsifier has been systematically varied by changes in
emulsifier carbon number or ethylene oxide number, by changes in the
carbon number of the oil phase, by changes in the salinity of the
aqueous phase, by changes in temperature (with nonionic emulsifiers) or by changes in pH with soap or amine salt emulsifiers.
In many of these studies the structure of the middle phase is not
established, but it is clearly immiscible in water or oil and its
electrical conductivity is closer to water than oil. Phase diagram
studies of oil-water-emulsifier systems Ekwall, (5), indicate
that surfactant-rich phases immiscible in oil or water have rod-
shaped or lamellar micelles with some degree of optical anisotropy
or flow birefringence, and these phases have much greater elec-
rical conductivity than oil. Figure 1 illustrates that the middle
phase composition varies smoothly from a water-rich composition to
an oil-rich composition as the emulsifier partition changes from
mostly water-soluble to mostly oil-soluble. If lamellar structures
are present the relative thickness of oleophilic and hydrophilic
layers must vary smoothly from the water-rich compositions to the
oil-rich compositions.
     Middle phase studies are generally conducted with emulsifier
concentrations high enough that the volume of the middle phase is
easily observed, typically 5-10%. Middle phases may be just as
important at lower concentrations (less than 1%) but have been
difficult to observe. However, the interfacial tension has been
found to go through a minimum Chan and Shah, (6) just as in
Figure 1.
     In research unrelated to these middle phase studies, emulsions
of mineral oil in water stabilized with 0.5% of sodium soaps were
observed to flocculate in an unexpected fashion upon increasing
salinity Princen et al., (7), Aronson and Princen, (8).   In
0.1 M sodium chloride the emulsion droplets flocculated into
clusters of spheres, but in 0.3 to 0.5 M salt solutions the flocculated droplets were separated by flat planes as shown in Figure 2.
Where these planes met the oil/water interface a distinct "contact
angle" was observed (56° in 0.5 M salt, as depicted in Figure 2).
Although other explanations were offered initially, we now know
that these contact angles result from thin films of middle phase
which have spread spontaneously over all oil droplets and in the
plane separating flocculated droplets. Such contact angles are
easy to observe and become a sensitive measure for formation of
middle phase films too thin to observe directly.
     In this paper we present three techniques for the study of
thin middle phase films adsorbed on emulsion droplets. Film thicknesses have been measured by small angle X-ray scattering, contact
angles of adjacent droplets have been measured in flocculated
emulsions, and much direct evidence for such films has been observed
visually in the spinning drop interfacial tensiometer.

Experimental Details

Emulsions. A white mineral oil of 125/135 Saybolt viscosity

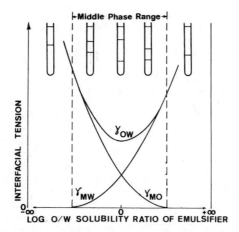

Figure 1. Interfacial tension versus oil/water emulsifier
partition coefficients for systems with middle phases (M).
(Healy et al, 1976). Reproduced with permission from
Healy, R. N., R. L. Reed, and D. G. Stenmark, Soc. Pet. Eng. J.,
June 1976, p. 147; copyright owner: Society of Petroleum
Engineers of AIME.

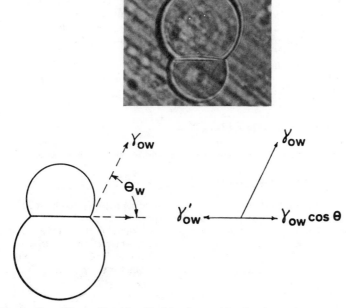

Figure 2. Flocculated oil droplets displaying plane of contact
and contact angles θ_W (0.4 M NaCl).

(Fisher Scientific) was emulsified into aqueous soap solutions of
0.25 w% each of sodium laurate and sodium oleate prepared from
sodium hydroxide, lauric acid (Aldrich Gold Label) and oleic acid
(Fisher Purified). Coarse emulsions were used for microscopy (as
in Figure 2), but fine emulsions (with droplet sizes of about
0.2 microns), used for determination of middle phase film thicknesses, were made by ultrasonication with a cell disruptor.
Sodium chloride contents of these emulsions varied from 0.1 M to
0.5 M.

Interfacial Tensions. Initially interfacial tensions were measured
by the pendant drop technique. Later studies with an EOR spinning
drop interfacial tensiometer gave much information on middle phase
films, as shown in the photographs made with a camera mounted on
our tensiometer.

Small Angle X-Ray Scattering (SAXS). A Kratky camera using the
1.54 $\mathring{A}$ CuKα line in a 7.5 x 0.25 mm beam was made available by the
Western Electric Research Center in Princeton, New Jersey. Scattering at angles of 0.1 milliradian increments were measured for
1000 seconds each. The statistical evaluation of the data,
desmearing and data analysis were performed with the help of the
program FFSAXS, Version 4 by C. G. Vonk of the DSM, Central
Laboratory, Geleen, The Netherlands Fowkes and Carnali, (9).

Contact Angles. Figure 2 shows a photograph of a pair of emulsion
droplets which have flocculated in 0.5 M sodium chloride solution,
and a drawing of such a system to show how the contact angle θ was
determined. The radii of the two drops (R_1 and R_2) and the overall
length of the doublet (L) were determined and from these measurements X, the radius of their circle of contact, was calculated from:

$$(R_1{}^2 - X^2)^{\frac{1}{2}} + (R_2{}^2 - X^2)^{\frac{1}{2}} = L - R_1 - R_2 \tag{1}$$

and the contact angle θ was calculated from:

$$2\theta = \sin^{-1}(X/R_1) + \sin^{-1}(X/R_2) \tag{2}$$

For each contact angle reported several doublets were measured.

Experimental Results and Discussion

Contact Angle and SAXS Studies. The contact angles which developed
at the junction of flocculated oil droplets, illustrated in
Figure 2, can be used together with the measured interfacial tensions (Table 1) in a vector diagram to demonstrate that the interfacial tensions on the outer surface of the oil drops (γ_{OW}) are
appreciably greater than the interfacial tensions (γ'_{OW}) in the plane
of contact between drops:

$$\gamma'_{OW} = \gamma_{OW} \cos\theta_W \tag{3}$$

The contact angles in Figure 2 are measured through the aqueous

phase, and are therefore designated θ_W. Similar contact angles for water droplets in oil would be measured through the oil phase (θ_0).
Table I shows that $\gamma_{OW} - \gamma'_{OW}$ increases from 0 to 0.49 mJ/m^2 as the salinity increases to 0.5 M.

Table I. Interfacial Tensions and Contact Angles for Oil Droplets
in 0.5% Soap Solution vs. Salinity

NaCl Conc.	γ_{OW}(mJ/m^2	θ_W(degrees)	$\gamma_{OW} - \gamma'_{OW}$(mJ/m^2)
0		0	0
0.3 M	1.6	32	0.25
0.4 M	1.2	45	0.35
0.5 M	1.1	56	0.49

The above results are in good agreement with the original findings of Princen and Aronson. Their explantion for the lower interfacial tensions in the plane of contact was that the aqueous layer between the two adsorbed monolayers might be so extremely thin that special attractive forces between oil droplets diminished the effective interfacial tensions. We therefore set out to measure the thickness (H) of this aqueous layer by small angle X-ray scattering (SAXS), using fine emulsions of various salinities. The SAXS findings were analyzed by Guinier plots and by correlation functions in some detail; all results indicated that the layers between flocculated droplets were not at all thin, but relatively thick (90 nm). This measurement clearly demonstrated that no special short-range attractions between oil droplets can exist in this system.

Another explanation for the contact angles and the decreased interfacial tentions between oil droplets is that increasing salinity has induced formation of middle phase (M) which has coated the oil droplets and filled the space between droplets in the plane of contact, as shown in Figure 3. The interfacial tensions between droplets are those between oil and middle phase (γ_{MO}) whereas the tensions between oil and water are the sum of the oil and water interfacial tensions with the thin interfacial film of middle phase ($\gamma_{MO} + \gamma_{MW}$). Thus the apparent γ_{OW} tensions are $\gamma_{MO} + \gamma_{MW}$ tensions and these always exceed the tensions operating in the plane of contact between flocculated oil droplets (γ_{MO}). The contact angle θ_W is therefore a measure of the presence of thin films of middle phase around the oil droplets:

$$\cos \theta_W = \gamma_{MO} / (\gamma_{MO} + \gamma_{MW}) \qquad (4)$$

Table II retabulates the data of Table I according to equation (4) and shows that the measured interfacial tension of the oil/water interface and the contact angle allow determination of γ_{MO} and γ_{MW}.

Table II. Interfacial Tensions of Middle Phase vs. Oil or Water

NaCl Conc.	θ_W	$\gamma_{MO} + \gamma_{MW}$	γ_{MO}	γ_{MW}
0.3 M	32°	1.6 mJ/m^2	1.35 mJ/m^2	0.25 mJ/m^2
0.4 M	45°	1.2	0.85	0.35
0.5 M	56°	1.1	0.61	0.49

The findings of Table II look remarkably like those illustrated
in Figure 1, for increased salinity has decreased γ_{MO} and increased
γ_{MW} as the emulsifier becomes partitioned more strongly into the oil
phase. The contact angle information could become part of the same
diagram, as illustrated in Figure 4. The left edge of the middle
phase region in Figure 4 is where cos θ_W first exceeds zero,(where
the contact angle first appears). The mid-point of the diagram
(where γ_{MO} equals γ_{MW}) is where θ_W is 60° and cosθ = 0.5, as can be
seen by equation (4). On the right side of the diagram water-in-oil
emulsions are formed and contact angles θ_c would be measured through
the oil phase:

$$\cos\theta_0 = \gamma_{MW}/(\gamma_{MO} + \gamma_{MW}) \qquad (5)$$

On the right side of the diagram aqueous droplets will be coated
with middle phase and when the emulsifier is partitioned sufficiently into the oil phase θ_0 approaches zero at the right boundary
of the middle phase region.

In systems with higher emulsifier concentrations the observed
middle phase may well tend to hold trapped droplets. In the water-
rich middle phase compositions on the left side of Figure 4 (where
cosθ_W exceeds 0 and is less than +0.5) oil droplets can easily
become entrapped in the middle phase, for their interfacial tensions
(γ_{MO}) are less than in the aqueous phase ($\gamma_{MO} + \gamma_{MW}$). Similarly
the oil-rich middle phase compositions on the right side of Figure 4
will tend to entrap aqueous droplets. Electron micrographs of
middle phases do indeed show such entrapped droplets (Chan and Shah,
1982).

Spinning Drop Interfacial Tensiometer Studies. In the foregoing
studies with oil-in-water emulsions stabilized with 0.5% of sodium
soaps the volume of middle phase which develops with increased
salinity is ordinarily too small to observe. However, if a single
oil drop is spun in the center of a large excess of the aqueous
soap solution for some hours, middle phase films are seen to be
centrifuged to the ends of the oil drop and to be ejected along
the axis of rotation as shown in Figure 5. In Figure 5a the aqueous
phase is salt-free and contacting droplets have a θ_W of 0°. In
Figure 5b (0.3 M sodium chloride) the end of the oil drop after 24
hours is seen to be narrower and less reflective as the middle phase
film is centrifuged towards that end. In Figure 5c (0.4 M sodium
chloride), after 17 hours of spinning, middle phase material is seen
to be spun off along the axis of rotation, and in Figure 5d (also
0.4 M salt) the droplets in contact show θ_W values of 40-50°.

As middle phase films form in the spinning drop interfacial
tensiometer, the interfacial tension drops. However, as these films

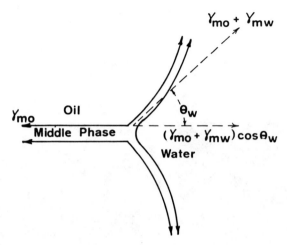

Figure 3.   Diagrammatic sketch of middle phase films surrounding
oil droplets and filling the plane of contact between them.

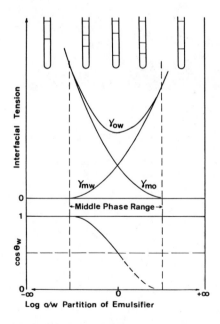

Figure 4. Contact angles and interfacial tensions of oil–water–
emulsifier systems with middle phases as a function of the
emulsifier partition between oil and water.

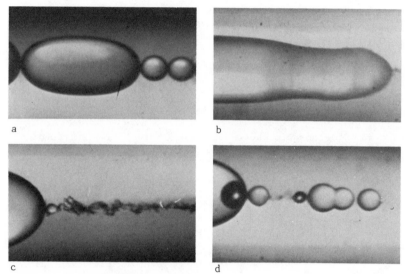

<u>Figure 5</u>. Photographs of oil drops in spinning drop interfacial
tensiometer with 0.5% soap solution:

     (a) no salt present, $\theta_W = 0°$
     (b) 0.3 M salt, spun 24 hours
     (c) 0.4 M salt, spun 17 hours
     (d) 0.4 M salt, $\theta_W = 40-50°$

are spun off, the interfacial tension rises again, as shown in
Figure 6. Without salt present (triangles) no middle phase forms
and there is no rise in interfacial tension. However, with salt
present the resulting middle phase is spun off and the net interfacial tension rises appreciably.

We have sought to see whether these contact angles are observed
in other systems with middle phases and find them to be a general
phenomenon. For instance, in an emulsion system involving long
chain amine salts as emulsifiers, where middle phases occur, θ_W
values of 30–50° were observed.

Conclusions

1. The contact angles previously observed with flocculated oil
droplets in saline emulsions are found to result from thin films of
a middle phase which coats emulsion droplets and fills the planes of
contact between droplets.
2. In these systems the interfacial tensions between oil droplets
and the aqueous phase are the sum of the two tensions of the middle
phase ($\gamma_{MO} + \gamma_{MW}$). Such measured tensions and measured contact
angles θ_W allow calculation of γ_{MO} and γ_{MW}:

$$\cos\theta_W = \gamma_{MO}/(\gamma_{MO} + \gamma_{MW}).$$

3. Small angle X-ray scattering measurements showed that in the
system under investigation the middle phase films between flocculated oil droplets were about 90 nm in thickness.
4. Contact angles between contacting drops are a very sensitive
measure of middle phase behavior and can be used to detect middle
phases in systems with low emulsifier content.
5. The magnitude of the contact angle is a measure of the relative
oil and water content of the middle phase.
6. In the spinning drop interfacial tensiometer the thin films of
middle phase can in time be centrifuged off of oil drops.
7. Contact angles between emulsion drops are fairly easy to observe
in the spinning drop interfacial tensiometer.

Acknowledgments

Support of this project was initiated by a starter grant from the
Department of Energy (Fowkes and Carnali, 1983). The SAXS studies
were made with a Kratky camera at AT&T Western Electric Company's
Engineering Research Center in Princeton, NJ; our thanks to Dr. John
Emerson, who made these arrangements and provided technical advice
to the project. The spinning drop measurements were made with the
support of Atlas Powder Company. Our thanks also to IBM for a
distinguished graduate fellowship for J. O. Carnali.

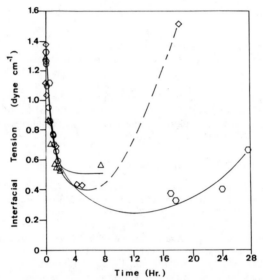

Figure 6. Time-dependence of interfacial tensions of oil drops
spinning in 0.5% soap solutions (25°).   Triangles - no salt.
Hexagons and circles - 0.3 M salt.   Squares - 0.4 M salt.   Interfacial tensions rise as middle phase films are spun off.

Literature Cited

1. Healy, R. N., R. L. Reed, and D. G. Stenmark, Soc. Pet. Eng. J., June, 1976, 147.
2. Salager, J. L., J. C. Morgan, R. S. Schechter, W. Y. Wade, and E. Vasquez, Soc. Pet. Eng. J., April 1979, 107.
3. Shah, K. D., D. W. Green, M. J. Michnick, G. P. Willhite, and R. E. Jerry, Soc. Pet. Eng. J., Dec. 1981, 763.
4. Kundieda, H., and K. Shinoda, Bull. Chem. Soc. Japan, 1982, 55, 1777.
5. Ekwall, P., in "Advance in Liquid Crystals, Vol. 1 (Academic Press, 1975) 1-142.
6. Chan, K. S., and D. O. Shah. In "Surface Phenomena in Enhanced Oil Recovery" Plenum, 1982; pp. 53-72.
7. Princen, H. M., M. P. Aronson, and J. C. Moser, J. Colloid Interface Sci., 1980, 75, 246.
8. Aronson, M. P., and H. M. Princen, Nature, 1980, 286, 370.
9. Fowkes, F. M. and J. O. Carnali, DOE Report #DE-FG19-80ET12267, 1983.

RECEIVED June 8, 1984

Structural Considerations of Lamellar Liquid Crystals Containing Large Quantities of Solubilized Hydrocarbon Alkanes

ANTHONY J. I. WARD[1,3], STIG E. FRIBERG[1], and DAVID W. LARSEN[2]

[1]Chemistry Department, University of Missouri at Rolla, Rolla, MO 65401
[2]Chemistry Department, University of Missouri at St. Louis, St. Louis, MO 63121

The order parameters of aliphatic hydrocarbons solubilized in a lamellar liquid crystal were determined from ^{2}H NMR data. The variation of order parameters along the hydrocarbon chain with varying amounts of hydrocarbon solubilized supports a model with the main part of the hydrocarbon forming a layer between the amphiphilic layers with only a small amount of it penetrating between the amphiphilic molecules.

Large quantities of hydrocarbon oils may be solubilized by aqueous lamellar dispersion of nonionic surfactants of the type n-dodecyl polyethylene glycol ether (1); in the case of the lamellar phase of n-dodecyl tetraethylene glycol ether ($C_{12}E_4$) up to 55% (W/W) of n-hexadecane. This implies a volume ratio of 1.6:1 of hydrocarbon to non-hydrocarbon constituents giving a formally calculated thickness for the hydrocarbon layer (2) of 60 Å.

The stability of a lamellar liquid crystal with such a large amount of oil is an intriguing problem. The main structural entity to be clarified before a serious attempt at an examination of the problem may be made is the degree of order of the hydrocarbon chains. This factor in turn depends on the location of the solubilized hydrocarbon chains; are they penetrating the amphiphilic layer or are they forming a liquid layer between the layers of amphiphilic molecules.

[3]Permanent address: Department of Chemistry, University College, Dublin, Belfield, Dublin 4, Ireland

0097–6156/85/0272–0185$06.00/0

The small angle X-ray data (1) gave little indication of penetration of the hydrocarbon chains into the amphiphilic layer. In fact, the observed increase in interlayer spacing, d, was too large to be accounted for even by allowing the surfactant molecules to adopt a fully extended conformation at the additon of hydrocarbon. Hence, the X-ray results were interpreted as the increased spacing being due to the formation of an oil layer in the hydrocarbon part of the lamellar structure. It is difficult to conceive a reason for stability of such a layer, and, if this interpretation is correct, it raises questions about the stability of the lamellar structure.

Low angle X-ray data alone cannot provide an unambiguous answer to the question of location and order of the hydrocarbon chains. We considered direct determinations of the order parameters of the solubilized hydrocarbon chains to be useful information. A preliminary ^{2}H NMR investigation (3) of alkane solubilized in the lamellar phases formed by $C_{12}E_4$ and water provided tentative support for the model with most of the solubilized hydrocarbon as a liquid layer between the amphiphilic layers and with only insignificant interpenetration between the solubilized hydrocarbon chains and those of the surfactant molecules.

In the present contribution, a detailed description of the composition dependence of the solubilizate order is given and the effect of solubilizate chain-length variation studied in order to characterize the structure and dynamics within the bilayer interior containing these surprisingly thick oil layers.

Experimental

n-dodecyl tetraethylene glycol ether ($C_{12}E_4$) was obtained from Nikko (Japan) Ltd. and was >98% pure by GLC criteria. Perdeuterated n-hexadecane and n-decane were obtained from MSD Isotopes.

Samples were prepared by weighing the components, mixing thoroughly and centrifuging to remove air bubbles. This procedure was repeated in a minimum of two times. Portions of the samples were centrifuged into glass NMR tubes for N.M.R. investigations. ^{2}H NMR spectra were obtained on a spectrometer operating in the Fourier transform mode built at the Chemistry Department, UMSL by one of the authors (DWL).

Results & Discussion

Figure 1 shows the spectrum obtained for n-perdeutero decane solubilized in the lamellar phase of water and $C_{12}E_4$. A comparison with the spectrum of n-perdeutero hexadecane (Fig. 1a) from the preliminary investigations by Ward et al (3) is instructive. In both cases, the spectrum may be simulated by a number of overlapping spectra corresponding to half the number of carbon atoms in the solubilized chain.

A complex variation in the spectra of hexadecane is observed as a function of the oil content of the liquid crystal with a water/$C_{12}E_4$ ratio of 40:60 (Fig. 2), the main features being as follows:
(a) the spectra comprise overlapping "powder" system; (b) at low oil contents, the central components were not always resolvable; (c) with increasing the oil weight fraction, the observed quadrupolar splittings passed through a maximum at relatively low concentrations thereafter being reduced.

These are illustrated in Figure 3 as a plot of quadrupolar splittings versus oil content for hexadecane.

The splittings pass through a maximum with the oil content varying from approximately 20% solubilized hydrocarbon for the three outer carbons in the chain to less than 10% for the second carbon atom counted from the center of the chain.

These results mean a variation of the hydrocarbon chain order parameter profile with the concentration. However, the monotonously decreasing values for higher oil contents give a regularity to the pattern and the values for 10% of oil were chosen to illustrate the variation of order parameter with the position of the carbon atom (Fig. 4).

The assignment of the splittings to segments of the alkane chain was based on preliminary experiments with selectively deuterated analogues and is tentative. It indicates an order profile, which is different to that normally expected for amphiphilic chains (4-7).

The systems reported by Seeling (4), Charvolin (5) and Smith (6) were concerned with amphiphilic solubilizates anchored at the interface by a polar group. Such a structure gives large values of the order parameter; approximately one to two orders of magnitude larger than the presently reported. Furthermore the order parameter values are approximately constant for all the carbon atoms except the three to four ones at the end of the chain. The values earlier reported for hydrocarbons (7) were for short hydrocarbon in liquid crystals with extremely high content of water. The order parameters were approximately constant along the chain justifying an interpretation in the form of the hydrocarbon being a tumbling rigid rotator.

Our present results with low values of Δv_i and the almost linear variation from the center of the chain reflect the absence of an ordering influence at one end of the chain.

The distinction from the order in solubilized amphiphiles is obvious. The order parameters of the hydrocarbon are much smaller and the distribution of order parameter values is symmetrical from the center of the chain.

On the other hand, the distinction is also evident from the values of order parameters in the systems investigated by Reeves and collaborators (7). The hydrocarbons solubilized in the present lamellar phase with moderate amount of water present (<50%) display a large variation in the order parameter with the position of the carbon atom. The ratio between carbon atom #1 and #6 (Fig. 4) is seven, the same order of magnitude as found for the anchored amphiphiles (5-7).

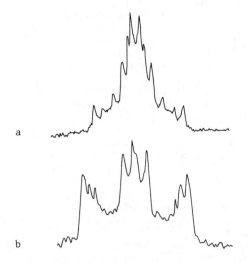

Figure 1.   (a) The deuterium NMR spectrum of a $C_{12}E_4$/n-$C_{16}D_{34}$/ H_2O lamellar dispersion (weight fraction of oil = 10.2%). (b) The deuterium NMR spectrum of a $C_{12}E_4$/n-$C_{10}D_{22}$/H_2O lamellar dispersion (weight fraction of oil = 9.6%).

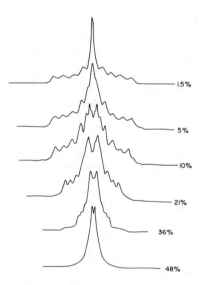

Figure 2.   Deuterium NMR spectra of n-hexadecane-d_{34} solubilized in a lamellar dispersion of $C_{12}E4$/H_2O as a function of oil content (the numbers are weight fraction of oil) at a fixed soap.water ratio (60:40 W/W).

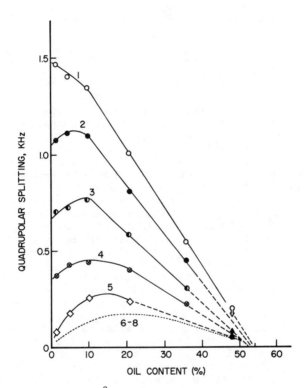

Figure 3. Variation of ^{2}H quadrupolar splittings of the different chain segments of solubilized n-$C_{16}D_{34}$ as a function of oil content at fixed soap/water ration (60:40 W/W).

The fact that the maximum splitting in the solubilized hydrocarbons
are of the same magnitude as the splitting of the terminal groups of
an amphiphilic (5-7) may be interpreted as an indications of a geometrical connection between the two. Also the presence of very small
($\lesssim$ 100 Hz) and unresolved splittings for the 2-3 terminal segments
implies that they are located at the bilayer center and are highly
motional disordered. Some preliminary observations of the $C_{12}E_4$/
n-$C_{10}D_{22}$/H_2O system indicate a similar type of profile (Fig. 5) and
hence, similar location.

The second factor to be evaluated is the variation of order
parameter with oil content.

Figure 3 shows that all the splittings, except that assigned to
the central chain segment show an increase with the initial additions
of oil. This indicates an ordering of the chain segments presumably
arising from restriction from mixing with the amphiphilic chains.
The ordering occurs up to an oil volume fraction, ϕ_H, of ca. 0.18
(as measured for segments 5-8) for a soap.water weight fraction of
0.6. It is interesting to notice that this value corresponds closely
with a value of 0.175 for ϕ_H where the curve of interlayer spacing
against ϕ_H calculated for nonpenetration crosses the experimental
curve (Moucharafieh et al 1979) for the same sample composition.
The results from the two methods support an interpretation of larger
mixing for the initial additions of hydrocarbon molecules.

At oil contents greater than that where the maximum value of $\Delta\nu_i$
occurs, the values show a linear decrease with increasing oil content.
We interpret this as the result of a fast exchange, on the ^2H NMR
time-scale, between n-hexadecane molecules located at the end of the
amphiphile molecules and hence, partially mixed with the surfactant
and those located in the center of the oil layer and undergoing
essentially isotropic motion ($\Delta\nu_i$=0). The extrapolation of the
curves to $\Delta\nu_i$=0 supports this interpretation since all pass through
the oil content axis at a weight fraction of 53-55% coinciding with
the maximum value of oil that may be solubilized observed from the
phase diagram. An interpretation of this oil content dependence of
$\Delta\nu_i$, therefore, for solubilized hexadecane implies that up to an
oil/soap mole ratio of approx. 1:2 the oil chains mix to some extent
with the surfactant chains; additional oil goes into the center of
the bilayer essentially in the form of an isotropic liquid layer
containing up to a maximum 1 mole oil per mole of $C_{12}E_4$. The preliminary results obtained from the $C_{12}E_4$/$C_{10}D_{22}$/H_2O lamellar phase
supports this conclusion. The solubilized decane shows a complex
^2H NMR spectrum (Fig. 1b) which may be simulated using 5 overlapping
powder spectra, with the central component remaining unresolved
($\Delta\nu \lesssim$ 100Hz) at all oil contents studied. The order profile (Fig.5)
is similar to that observed for hexadecane and shows that the
terminal methyl segments are not mixed with the soap chains having
almost isotropic behavior. A comparison of the $\Delta\nu_i$ for the central
values segment for different alkanes is given in Table I.

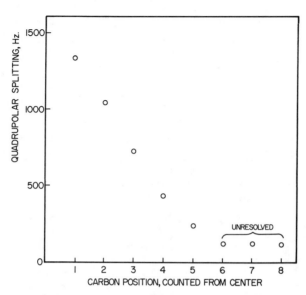

Figure 4. Order profile for solubilized $n-C_{16}D_{34}$ in terms of quadrupole splittings determined from the central chain segment (#1) to the terminal methyl segments (#8) at a concentration of 10% oil.

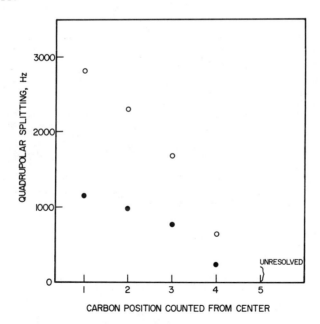

Figure 5. Order profile for solubilized $n-C_{10}D_{22}$ in terms of quadrupole splittings determined from the central chain segment (#1) to the terminal methyl segments (#5). Key: weight fraction of oil/soap 9.6% (o), 35% (●).

Table I. Comparison of $\Delta\nu_i$ for the central values segment for different alkanes

Oil	$\Delta\nu_i$ (Hz)
$C_{10}D_{22}$	2730
$7,7'-C_{13}H_{26}D_2$	2250
$C_{10}D_{34}$	1350
(for 10:90 oil:soap)	

An increase in splitting with decreased chain length is found consistent with the greater ability of the smaller chains to mix with the soap chains which consequent more ordering.

Conclusion
An interpretation of ^2H NMR data from solubilized alkanes in aqueous lamellar phases of a nonionic surfactant is found consistent with the presence of a layer of oil located at the center of the bilayer. This supports a similar novel conclusion derived from small angle X-ray data (1). The presence of such an oil layer will have consequences, as yet not understood, on the osmotic pressures within the bilayer and the interlayer interactions responsible for the stability of the lamellar phase.
It should be observed that the temperature stability of this liquid crystalline phase with liquid oil layers of large dimensions does not show a "critical" behavior. Increased temperature leads to a reduced thickness of the layers in an orderly fashion (8).

Literature Cited
1. Moucharafieh, N., Friberg, S. E. and Larsen, D. W., Mol. Cryst. Liq. Cryst., 1979, 53, 189.
2. Fontell, K., "Liquid Crystals and Plastic Crystals", Vol. 2, Ellis Harwood: London, 1974, p. 80.
3. Ward, A. J. I., Friberg, S. E., Larsen, D. W., and Rananavare, S. B., J. Phys. Chem. (In press).
4. Seeling, J., Progr. Coll. & Polym. Sci.., 1978, 65, 172.
5. Charvolin, J., Manneville, P. and Deloche, B., Chem. Phys. Letters, 1973, 23, 345.
6. Smith, I. C. P., Stockton, G. W., Tulloch, A. P., Polnaszek, C. F., and Johnson, K. A., J. Colloid Interface Sci., 1977, 58, 439.
7. Forrest, B. J. and Reeves, L. W., Chem. Rev., 1981, 1.
8. Friberg, S. E., unpublished data.

RECEIVED December 31, 1984

Use of Videomicroscopy in Diffusion Studies of Oil-Water-Surfactant Systems

KIRK H. RANEY, WILLIAM J. BENTON, and CLARENCE A. MILLER

Department of Chemical Engineering, Rice University, Houston, TX 77251

Aqueous solutions containing anionic surfactants
and alcohol cosurfactants were contacted with various
oils. A microscope which utilized a vertical sample
orientation and a video camera was used to observe and
record the resulting diffusional processes. As a
result, an improved and detailed viewing of
intermediate phase growth, interface motion, and
spontaneous emulsification was achieved.
The theory of diffusion paths, extended to allow
for diffusion in a two-phase dispersion, was used to
solve the diffusion equations for a model, pseudoternary system. Predicted diffusion paths were
generally consistent with the experiments in regard to
the number and type of intermediate phases formed and
the occurrence of spontaneous emulsification. Also,
except in cases of convection or highly nonuniform
growth of liquid crystal myelins, interface positions
varied with the square root of time, as predicted by
theory.

When a surfactant-water or surfactant-brine mixture is carefully
contacted with oil in the absence of flow, bulk diffusion and, in
some cases, adsorption-desorption or phase transformation kinetics
dictate the way in which the equilibrium state is approached and
the time required to reach it. Nonequilibrium behavior in such
systems is of interest in connection with certain enhanced oil
recovery processes where surfactant-brine mixtures are injected
into underground formations to diplace globules of oil trapped in
the porous rock structure. Indications exist that recovery
efficiency can be affected by the extent of equilibration between
phases and by the type of nonequilibrium phenomena which occur
(1). In detergency also, the rate and manner of oily soil removal
by solubilization and "complexing" or "emulsification" mechanisms
are controlled by diffusion and phase transformation kinetics (2-
3).
In both applications, it is important to know whether

0097-6156/85/0272-0193$08.50/0
© 1985 American Chemical Society

intermediate phases form at the boundary between the phases initially contacted and, if so, their manner and rate of growth. Whether or not spontaneous emulsification occurs is also of interest. We describe here an experimental technique which provides such information by allowing microscopic observation of the region of contact between phases in thin, rectangular, optical capillary cells.

A novel feature of the technique is that the cells are maintained in a vertical orientation with the various interfaces horizontal during the entire experiment. This scheme represents a significant improvement over the previous method where the cells were vertical during the initial contacting process but where rotation to the horizontal position was required for observation with a conventional microscope (4). Although much useful information was obtained from the earlier technique, the inevitable flow associated with oil overriding the aqueous phase after rotation obscured some features of intermediate phase development and precluded measurement of the rates of growth of these phases.

As in the previous study (4), the experiments involve brine-alcohol-petroleum sulfonate mixtures brought into contact with oil. In these systems, which are the type used for enhanced oil recovery, the initial mixtures are, depending on the salinity, either stable dispersions of lamellar liquid crystal and brine or a single liquid crystalline phase. The latter is formed at higher salinities than the former, in accordance with the general pattern of phase behavior in such systems described elsewhere (5). Indeed, one aspect of our work which differs from studies made by others of spontaneous emulsification in enhanced oil recovery processes (6-7) is emphasis on the need to understand the role of liquid crystals in the overall nonequilibrium process. Related studies of spontaneous emulsification in other systems are reviewed elsewhere (4).

After describing the experimental technique in the next section, we report our observations of intermediate phase formation and spontaneous emulsification in three parts corresponding to three types of equilibrium phase behavior found when equal volumes of oil and the surfactant-alcohol-brine mixtures are equilibrated. The three types are well known (8-9) and, in order of increasing salinity, are a "lower" phase, oil-in-water microemulsion in equilibrium with excess oil, a "surfactant" or "middle" phase, probably of varying structure, in equilibrium with both excess oil and excess brine, and an "upper" phase, water-in-oil microemulsion in equilibrium with excess brine.

Finally, we present an interpretation of our observations in terms of diffusion paths. Basically, the diffusion equations are solved for the case of two semi-infinite phases brought into contact under conditions where there is no convection and no interfacial resistance to mass transfer. Other simplifying assumptions such as uniform density and diffusion coefficients in each phase are usually made to simplify the mathematics. The analysis shows that the set of compositions in the system is independent of time although the location of a particular composition is time-dependent. The composition set can be plotted on the equilibrium phase diagram, thus showing the existence of intermediate phases and, as explained below, providing a method for predicting the occurrence of spontaneous emulsification.

The diffusion path method has been used to interpret nonequilibrium phenomena in metallurgical and ceramic systems (10–11) and to explain diffusion-related spontaneous emulsification in simple ternary fluid systems having no surfactants (12). It has recently been applied to surfactant systems such as those studied here including the necessary extension to incorporate initial mixtures which are stable dispersions instead of single thermodynamic phases (13). The details of these calculations will be reported elsewhere. Here we simply present a series of phase diagrams to show that the observed number and type of intermediate phases formed and the occurrence of spontaneous emulsification in these systems can be predicted by the use of diffusion paths.

Materials and Methods

Apparatus. As mentioned above, a microscope was designed and built that allows samples to remain in a vertical orientation, as shown in Figure 1. A schematic diagram of the microscope assembly is shown in Figure 2. Basically, the assembly can be divided into three functional sections. These include the devices for producing, filtering, and condensing the transmitted light, the microscope body which contains the magnifying optics, and the peripheral devices used to observe and record the contacting experiments.

Light is produced by a 100-watt quartz-halogen bulb contained in a lamp housing (Nikon). The light intensity is controlled with a 12-volt regulating transformer. Attached to the front end of the lamp housing are two bars which serve as a sliding mount for filters and a condenser.

Light first passes through two filters. A hot mirror reflects most of the heat contained in the light, and a diffusion filter gives the light field a uniform intensity. Next, the light passes through an iris diaphragm and condenser. These are used to concentrate light on the sample for maximum image resolution and contrast. Located between the diaphragm and condenser is a rotating polarizer. A two-way mechanical stage holds the sample in the transmitted light.

The microscope body (Bausch and Lomb) has a four-objective revolving turret. Flat-field objectives, ranging from 4x to 40x, are used which allow all sections of the visual field to be simultaneously focused. The trinocular body permits the image to be either viewed with 10x wide-field eyepieces or sent to a video camera system. Also, the microscope body contains an analyzing polarizer that can be activated for observation of birefringence in the sample.

Attached to the microscope by a C-mount adapter is a video camera (RCA TC-1000L). A 9-inch video monitor (Panasonic) is used for real-time viewing of contacting experiments. Recording of certain phenomena for later observation is made with a reel-to-reel video recorder (SONY AV-3600).

All equipment except the monitor and recorder are positioned on an instrument bench. Coarse-scale focusing is accomplished by adjusting the position of the microscope body. The instrument bench is located on a vibration-isolation table.

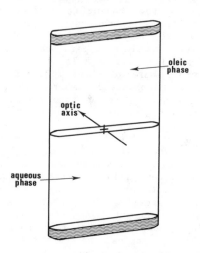

Figure 1.   Vertical sample configuration.

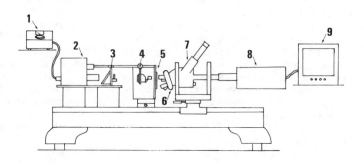

1 - Regulating Transformer   6 - 4-Objective Nosepiece
2 - Lamp Housing   7 - Trinocular Head
3 - Hot Mirror, Diffusion Filter 8 - Video Camera
4 - Aperture Diaphragm, Condenser 9 - Monitor
5 - Mechanical Sample Stage

Figure 2.   Schematic diagram of microscope assembly.

Procedure. Rectangular glass capillaries (Vitro-Dynamics) were used as sample cells in the contacting experiments. Their shape prevented optical distortion due to curved surfaces. The capillaries were 50 mm in length and 2 mm in width. Various thicknesses are available, but 200-micron pathwidth capillaries were found to be best suited for overall observation of interfaces. These were also thick enough to allow discrimination of weak birefringence in the samples.

Initially, a sample cell was half-filled with surfactant solution by capillary action. Next, the cell was sealed at the bottom and attached to a microscope slide with fast-setting epoxy. If movement of interfaces was to be measured, a reference mark was placed next to the capillary near the meniscus.

With the capillary maintained in a vertical orientation, the slide was taped to the mechanical stage of the microscope assembly. The sample was positioned in the viewing field, the light intensity adjusted, and the image focused. To minimize mixing of the phases, a syringe was used to carefully fill the rest of the capillary with oil. This procedure allowed microscopic observation even during the initial contacting period. Finally, the top of the capillary was sealed to prevent evaporation of the liquid within.

Measurements of interface positions were made directly from the video monitor. To allow for interface irregularities which were sometimes observed, readings were made at equally spaced intervals along the interface and averaged.

Properties of Systems Studied. Contacting experiments at ambient conditions (23±4°C) were performed with two commercial petroleum sulfonate systems. The preparation procedure for the aqueous solutions is described elsewhere (5).

PDM-337/TAA. In this system, aqueous solutions, prepared from d ionized and triple-distilled water, contained 9.0% by volume of a 63/37 weight-ratio mixture of PDM-337 surfactant (Exxon) and reagent grade tert-amyl alcohol (TAA). The surfactant is 84% active and is primarily the monoethanolamine salt of dodecyl orthoxylene sulfonic acid. Samples were prepared in the salinity range of 0.4 to 2.6 gm NaCl/dl surfactant solution. The NaCl was reagent grade.

Diffusion studies were made using an Isopar M/Heavy Aromatic Naptha (IM/HAN) 9:1 oil mixture (Exxon). Isopar M and HAN are refined paraffinic and aromatic oils, respectively. Figure 3 shows equilibrium salinity scans measured in the laboratory for equal-volume mixtures of the surfactant solution and oil. Since room temperature varied somewhat, the effect of temperature on phase behavior was determined. As Figure 3 shows, there is a small temperature effect, especially at the lower salinities. However, it is not large enough to have influenced the basic results of the contacting experiments. Optimum salinity, where equal volumes of oil and brine are contained in the middle phase, is approximately 1.4 gm/dl.

The structure of the oil-free aqueous surfactant solutions is shown at the bottom of the diagram. S is a dispersion of lamellar liquid crystalline spherulites in brine, L is a homogeneous lamellar liquid crystalline phase, and S+L is a transition region between them where liquid crystal gradually becomes the continuous phase. Optimum salinity is located near the center of the S+L range, where no gross phase separation occurs.

TRS 10-410/IBA. Solutions containing 5.0% by weight TRS 10-410 surfactant (Witco) and 3.0% reagent grade isobutyl alcohol (IBA) were used in the salinity range of 0.7 to 2.3 gm/dl. The surfactant is a 61.5% - active mixture of petroleum sulfonates with an equivalent weight of 424.

Contacting experiments were performed with reagent grade n-dodecane (C12) and n-undecane (C11) (Humphrey). An experimentally determined phase volume scan with n-dodecane at 23°C is shown in Figure 4. The middle phase is smaller in volume, but three-phase equilibria occur over a wider salinity range than in the PDM system. Optimum salinity is approximately 1.7 gm/dl. At this salinity, the aqueous solution structure is almost entirely liquid crystal. Note that at somewhat higher salinities the aqueous solution separates into two phases, the liquid crystal L and an isotropic phase C which scatters light and exhibits streaming birefringence. This phase is discussed further elsewhere (5).

Based on literature correlations (14), an estimate of the phase volume scan with n-undecane is also plotted. The decrease in hydrocarbon chain length moves the optimum to approximately 1.5 gm/dl salinity, which, like the PDM system, is near the center of the S+L regime. Also, the volume of the middle phase is predicted to increase, while the three-phase salinity range is predicted to shrink slightly.

Results

Low-Salinity Diffusion Phenomena. A study was made of the diffusion phenomena in the salinity range below the three-phase regime (see Figures 3 and 4). Both systems exhibited the same basic behavior, shown in Figure 5, during the contacting experiments. Initially, the aqueous surfactant "solution" was a dispersion of spherulitic liquid crystalline particles in an isotropic aqueous solution. Immediately after the oil was contacted with this dispersion, an isotropic phase began to develop near the surface of contact. The phase grew in both directions but most rapidly toward the aqueous end. Figure 6 is a photograph showing the intermediate phase for the PDM system with 0.6 gm NaCl/dl solution. The drops in the oil phase at the top of the figure resulted from initial mixing rather than from diffusion.

The boundary between the intermediate phase and liquid crystal dispersion was not actually an interface but rather a moving front where liquid crystal dissolved. This structure was evident because the boundary was quite often irregular, especially in the period immediately after contacting of the phases. However, it tended to flatten out as time proceeded. An interesting phenomenon which occurred in all experiments was increased birefringence near the dissolving front. This phenomenon resulted from a buildup of liquid crystalline material, which is shown schematically in Figure 5. The buildup is also evident in Figure 6.

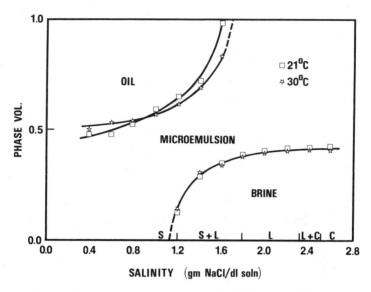

Figure 3. Salinity scans for the PDM-337/Isopar M/HAN system.

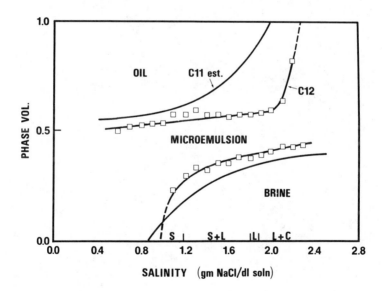

Figure 4. Salinity scans for the TRS 10-410/n-undecane (C11) and the TRS 10-410/n-dodecane (C12) systems at 23°C.

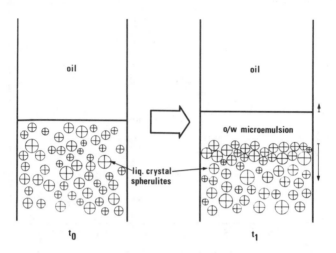

Figure 5.   Diagram of low-salinity diffusion phenomena.

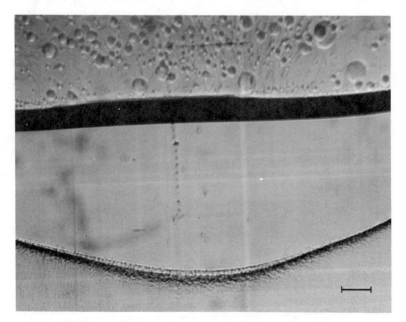

Figure 6. 0.6 gm/dl–salinity PDM system 1 hour after contact
(Bar equals 0.1 mm).

Positions of the phase boundaries were measured at increasing time intervals for up to ten days. The positions of dispersion boundaries for the PDM system are shown in Figure 7 as functions of $t^{1/2}$ where t is the elapsed time of the experiment. Similar results were obtained for other PDM samples and for the TRS samples. The data were fit with straight lines using least-squares analysis, the extrapolated position at t = 0 being used as a reference. (Experimental determination of an initial position was complicated by curvature at the pre-injection air-aqueous solution meniscus and by small amounts of unavoidable mixing during the injection of oil.) The best-fit slopes and correlation coefficients r for both boundaries are given in Table I.

Diffusion path theory predicts a linear relationship between interface positions and $t^{1/2}$ when there is no convection. Based on the correlation coefficients in Table I, this relationship appears to hold for the systems at these conditions. The low coefficients for the upper interfaces resulted from the measurement uncertainty (± 0.05 mm) being the same order of magnitude as the total movement of the interfaces, which is also why these plots are not included in Figure 7. Division of the best-fit slopes by $2t^{1/2}$ gives an estimate of the interface velocities at any elapsed time t.

The identity of the intermediate phase formed at these conditions can be deduced from the relative movement of the interfaces. Because the phase grew quickly in the direction of the aqueous surfactant solution, it contained predominantly brine. Although small in quantity, some oil did diffuse into it. From this information and from its isotropic appearance, one can conclude that the intermediate phase was an oil-in-water microemulsion. Additional support for this conclusion is that this type of microemulsion is an equilibrium phase at low salinities.

Intermediate-Salinity Diffusion Phenomena. From the low-salinity end of the three-phase region to approximately halfway to optimum salinity, the systems exhibited behavior like that seen at low salinities (Figure 5). As is evident from Table II, good linear relationships exist between interfacial position and $t^{1/2}$, indicating negligible convection in both of the systems. Most runs showed some deviation from linearity at small elapsed times (t < 9 hrs.). This deviation resulted from the formation of a thin layer of a second intermediate phase, presumably brine, below the microemulsion phase. The brine disappeared after a few hours, indicating that it formed as a result of some slight mixing during oil injection rather than by diffusion.

The increased relative velocities of the microemulsion-oil interfaces at these conditions (as evidenced in Table II) indicate that the microemulsion contained a significant amount of oil. As a result, the microemulsion was probably a middle phase. This conclusion is supported by the formation of the brine phase, which exists in equilibrium with this type of microemulsion at these conditions.

As salinity was increased toward optimum, the formation of a permanent brine layer between the liquid crystal dispersion and microemulsion occurred. For the PDM system, the layer was stable except for some penetration by small projections of surface liquid crystal. No noticeable buildup of liquid crystal occurred at the dispersion boundary.

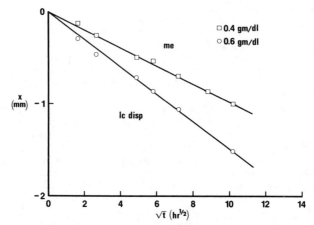

Figure 7. Dispersion boundary positions for the PDM system at low salinities.

Table I. Least-Squares Slopes and Correlation Coefficients r for Low-Salinity Study

System	Salinity (gm NaCl/dl soln)	Liquid Crystal Boundary		Microemulsion-oil Interface	
		Slope ($mm\ hr^{-1/2}$)	r*	Slope ($mm\ hr^{-1/2}$)	r
PDM/IM/HAN	0.4	-0.098	-0.996	0.0074	0.695
"	0.6	-0.138	-0.996	0.0070	0.704
"	0.7	-0.264	-0.996	0.0037	0.784
"	0.8	-0.119	-0.999	0.0069	0.865
"	0.9	-0.174	-0.991	0.0151	0.906
TRS/C11	0.7	-0.150	-0.996	0.0019	0.704
"	0.9	-0.063	-0.983	0.0040	0.661
TRS/C12	0.9	-0.114	-0.999	0.0005	0.135
"	1.0	-0.066	-0.993	0.0450	0.466

* r = +1.0 (or -1.0 if slope is negative) for perfect linearity

Table II. Least-Squares Slopes and Correlation
Coefficients r for Intermediate-Salinity
Study with no Brine Formation

System	Salinity (gm NaCl/dl soln)	Liquid Crystal Boundary		Microemulsion-oil Interface	
		Slope $-\frac{1}{2}$ (mm hr $\frac{1}{2}$)	r	Slope $-\frac{1}{2}$ (mm hr $\frac{1}{2}$)	r
TRS/C11	1.0	-0.185	-0.996	0.052	0.989
"	1.1	-0.134	-0.993	0.087	0.991
TRS/C12	1.1	-0.073	-0.996	0.006	0.801
"	1.3	-0.157	-0.990	0.071	0.996
PDM/IM/HAN	1.0	-0.122	-0.997	0.030	0.987
"	1.1	-0.126	-0.999	0.015	0.965

Figure 8 shows the microemulsion interface positions for the PDM system in this regime. Convection is not indicated. The dispersion front boundaries were very irregular in shape, and, therefore, those positions are not plotted. However, estimates of the relative dispersion front velocity in each experiment are given in the first two entries of Table III. As is evident from the small difference between these values and those for the brine interface, the brine layer grew very slowly at these salinities.

In contrast, the brine phase tended to form large pockets in the TRS system, causing extensive movement of the liquid crystal to the brine-microemulsion interface. This behavior is schematically illustrated in Figure 9. Not surprisingly, interface movements were inconsistent with diffusion path analysis. Figure 10 shows such a plot for the TRS/C12 system at 1.5 gm/dl salinity. The nonlinearity results from convection, which speeds up equilibration. Experimentally, the inconsistency with diffusion path theory was evident from the time-dependent appearance of the upper microemulsion interface, an indication of variable interface compositions.

Another type of interfacial instability occurred in both systems whenever liquid crystal penetrated the brine to contact the brine-microemulsion interface. At high magnification (40x), rapid convection of liquid crystal particles to the interface was observed at volcano-like instabilities (Figure 11). Reported earlier for the same systems (4), this type of instability forms convection currents in the surrounding brine phase. After times ranging from a few seconds to a few hours, the instabilities "choke-off." The mechanism by which this small-scale convection is initiated, maintained, and terminated is as yet unknown.

An interesting phenomenon occurred at optimum salinity in the TRS/C12 system. The intermediate microemulsion phase formed initially as discrete drops along the interface between brine and oil. Figures 12(a)-(c) show the coalescence which occurred among the drops as time elapsed. Although initially ascribed to interface motion resulting from the positioning of the sample cell in a horizontal configuration (4), this phenomenon apparently results from the middle-phase type of microemulsion being a nonwetting phase. As such, it cannot form a uniform layer until a sufficient amount of it is present. In this experiment, a complete layer did not form until approximately ten minutes after initial contact. Also apparent in Figures 12(a)-(c) are examples of the previously described volcano-like instabilities which appear as liquid crystal cones under each microemulsion drop.

As salinity was increased past optimum, large-scale convection continued to occur in the TRS system. Because brine formed rapidly at these salinities, the brine pockets grew quickly. The end result was rapid equilibration (a few days) of the samples.

The dynamic behavior of the PDM system in the salinity range between optimum salinity and the upper-phase microemulsion region was similar to that below optimum in that intermediate brine and microemulsion phases formed. Figure 13 shows these phases for the 1.5 gm/dl-salinity PDM system. The middle-phase type of microemulsion, being high in oil content at these salinities, grew more rapidly in the direction of the oil phase than it did at low

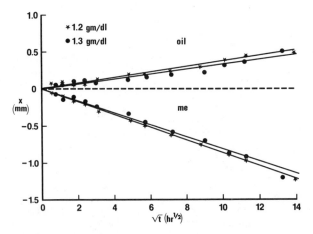

Figure 8. Brine-microemulsion interface positions (bottom) and microemulsion-oil interface positions (top) for the PDM system at salinities below optimum.

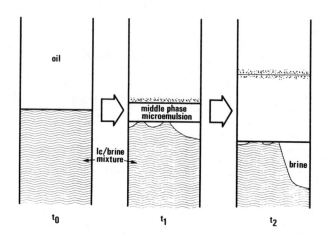

Figure 9. Diagram of intermediate-salinity diffusion phenomena in the TRS system.

14. RANEY ET AL. *Use of Videomicroscopy in Diffusion Studies* 207

Table III. Least-Squares Slopes and Correlation Coefficients
r for Intermediate-Salinity Study

System	Salinity (gm NaCl/dl soln)	Liquid Crystal Boundary Slope (mm hr$^{-1/2}$)		Brine–microemulsion Interface Slope (mm hr$^{-1/2}$)	r	Microemulsion–oil Interface Slope (mm hr$^{-1/2}$)	r
PDM/IM/HAN	1.2	-0.092	*	-0.088	-0.999	0.037	0.989
"	1.3	-0.092	*	-0.084	-0.995	0.032	0.992
"	1.4**	-0.060	*	-0.043	-0.994	0.116	0.995
"	1.5	-0.070	*	-0.049	-0.997	0.045	0.997
"	1.6	-0.094	*	-0.063	-0.986	diffuse	
"	1.7	-0.087	*	-0.035	-0.995	diffuse	

* not approximated by least-squares analysis because of irregularity

** optimum salinity

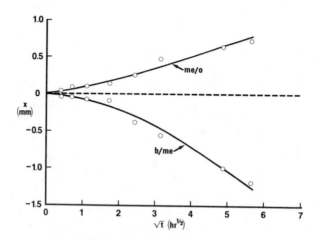

Figure 10. Microemulsion interface positions for the 1.5 gm/dl-salinity TRS/C12 system.

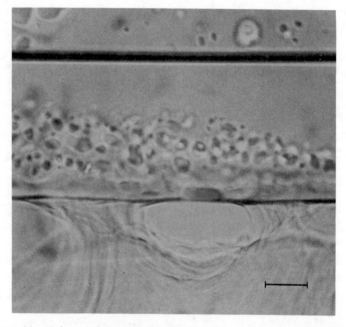

Figure 11. Volcano-like instabilities in the 1.7 gm/dl-salinity TRS/C12 system 30 minutes after contact (Bar equals 0.05mm).

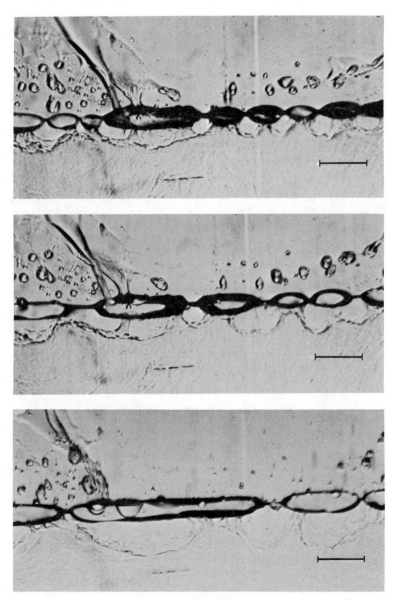

Figure 12. Formation of the microemulsion phase as discrete drops in the 1.7 gm/dl-salinity TRS/C12 system; views (a), (b), and (c) illustrate coalescence of drops with increasing time during the first five minutes after contact (Bars equal 0.2 mm).

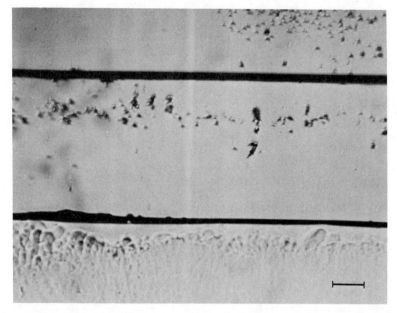

Figure 13. 1.5 gm/dl—salinity PDM system 18.5 hours after contact
(Bar equals 0.1 mm).

salinities. The slopes of the microemulsion interface positions as functions of $t^{1/2}$ are given in the last entries of Table III.

The initial aqueous structure of the PDM solutions at these salinities was predominantly lamellar liquid crystal. Swelling of the liquid crystal with brine during the contacting experiments produced nonequilibrium tubular structures called myelinic figures. This process is shown schematically in Figure 14. When the brine phase was sufficiently thick, the myelins would reach lengths of up to 0.5 mm. In fact, a single myelinic figure would occasionally penetrate the brine to contact the microemulsion phase. Formation of large myelins was not observed in the TRS system because convection allowed insufficient time for brine penetration into the liquid crystal.

At approximately optimum salinity, spontaneous emulsification of brine drops in the oil phase began in both systems. This phenomenon resulted from local supersaturation of the oil phase, as explained in the discussion section below. The amount of emulsification tended to increase with increasing salinity. As a result, the cloud of emulsion drops began to obscure the interface between the microemulsion and oil, making interface position measurements difficult. These observations of spontaneous emulsification confirm the results of the earlier contacting experiments performed in the horizontal configuration (4).

High-Salinity Diffusion Phenomena. Just as at intermediate salinities, the TRS system exhibited extensive convection at high salinities. The rate of phase equilibration was extremely rapid. Typically, the interfaces in this system moved further in a few hours than those in the PDM system moved in two weeks.

The transition from the three-phase to two-phase region in the PDM system was marked by a sudden increase of spontaneous emulsification in the oil phase. Because formation of an intermediate microemulsion ceased at this point, the emulsion drops remained near the brine interface rather than rapidly moving away to form a single-phase region above the brine. An example of this behavior is shown in Figure 15 for the 2.1 gm/dl-salinity PDM system.

As indicated by Figure 16, which shows the positions of the brine-oil interface for two PDM experiments as functions of $t^{1/2}$, the oil phase grew in volume with time. This solubilization of brine into the oil contrasts with the behavior at lower salinities where the oil phase was consumed by the microemulsion. Based on equilibrium phase behavior, one can conclude that conversion of oil to a water-in-oil microemulsion was occurring above the brine interface. Also, as shown in Table IV, the position of the interface between brine and this oil-continuous phase varied as the square root of time, indicating no extensive convection in these samples.

The evolution of the liquid crystalline phase in the PDM system at high salinities is depicted in Figure 17. Initially at t_0, the liquid crystal was entirely lamellar in structure. Some of the previously described volcano-like instabilities occurred for a short period (t_1) after initial contacting while the brine phase was still thin. Eventually, long and thin myelinic figures formed which slowly began folding over and merging together (t_2-t_4).

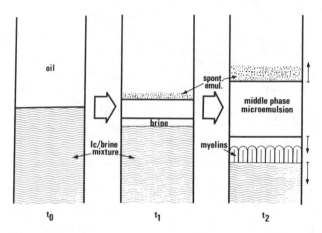

Figure 14. Diagram of intermediate-salinity diffusion phenomena in the PDM system.

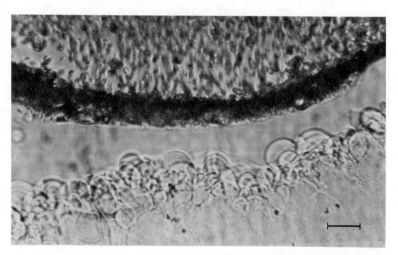

Figure 15. 2.1 gm/dl-salinity PDM system 45 minutes after contact (Bar equals 0.05 mm).

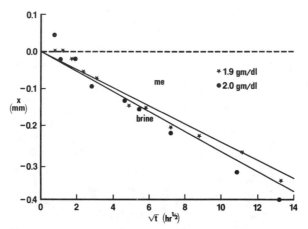

Figure 16. Brine–microemulsion interface positions for the PDM system at high salinities.

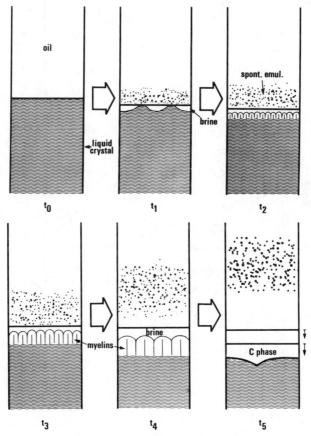

Figure 17. Diagram of high-salinity diffusion phenomena in the PDM system.

Table IV. Slopes, Correlation Coefficients r, and C-Phase
Formation Times for High-Salinity Study

System	Salinity (gm NaCl/dl soln)	Liquid Crystal-Brine Interface		Brine-microemulsion Interface		
		Slope $\frac{1}{2}$ (mm hr^{-})	r	Slope $\frac{1}{2}$ (mm hr^{-})	r	C-Phase Form Time (days)
PDM/IM/HAN	1.9	-0.087	*	-0.024	-0.989	7
"	2.0	-0.071	*	-0.027	-0.987	5
"	2.1	-0.098	*	-0.024	-0.995	3
"	2.2	-0.079	*	-0.037	-0.986	1
"	2.6	-0.040	*	-0.013	-0.962	initially present

* not approximated by least-squares analysis because of irregularity

Round nodules of a nonbirefringent phase then formed at the tips of the projections. Finally, the nodules grew together to form a uniform layer of the new phase at t_5.

This slowly-forming phase is believed to have been a C phase, the isotropic phase with streaming birefringence mentioned earlier as existing at high salinities in oil-free solutions (Figures 3 and 4).   This   conclusion   is   partially   based   on   the   phase's nonbirefringent appearance. Also, formation of a C phase has been observed when small amounts of oil (less than 5%) are equilibrated with high-salinity samples of the same PDM solutions used in these experiments (15). Presumably, enough oil diffused into the liquid crystal for formation of the C phase to occur.

The times required for initial formation of the C phase are given in the last column of Table IV. A trend is evident showing faster formation at higher salinities. This trend correlates with experimental observations which show that less oil is needed to form a C phase as salinity is increased (15).

An extended experiment was performed for the 2.0 gm/dl-salinity PDM system to determine the effect on relative interface velocities of the formation of myelinic figures and the C phase. With reference to Figure 18, the formation of a uniform layer of C phase at $t^{1/2}$ = 12 $hr^{1/2}$ (6 days) corresponds to a decrease in relative velocity of the brine-microemulsion interface. The layer of C phase grew uniformly after its initial formation. In Figure 18, all positions are plotted relative to the same reference position. The offset of the liquid crystal interface at t = 0 indicates brine formation due to initial mixing.

The aqueous solution at the highest salinity studied was entirely C in structure. No myelinic figures formed during the contacting experiment. The interface between C and brine remained smooth throughout. From Table IV, one can see that the interface motion was very much slower than that for lamellar aqueous structures.   No   contacting   experiments   were   performed   with solutions that were initially C+L because of problems with phase separation.

Discussion

Theoretical diffusion path studies were made with a model system for comparison to the experimentally observed phenomena.   A pseudoternary representation was chosen for modeling the phase behavior,   and brine and oil were chosen as the independent diffusing species.   For simplicity and because their exact positions and shapes were not known, phase boundaries in the liquid crystal region were represented as straight lines.   Actually, studies indicate a rather complex transition from liquid crystal to microemulsions as system oil content is increased, especially near optimum salinity (15-16). A modified Hand scheme was used to model the equilibria of binodal lobes (14,17). Other assumptions are described in detail elsewhere (13).

At low salinities, the pseudoternary phase behavior is like that shown in Figure 19. Tie lines indicate a preferential solubility of the surfactant-alcohol mixture in brine. Also, the initial aqueous structure at composition D is a dilute dispersion of liquid crystal in an isotropic aqueous solution. The calculated

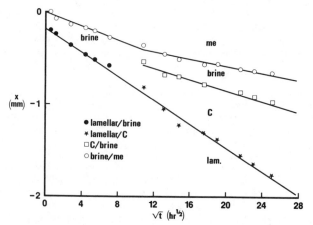

Figure 18. Interface positions for the 2.0 gm/dl-salinity PDM
system showing the effects of C-phase formation.

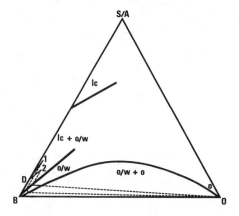

Figure 19. Low-salinity paths showing the effect of varying the
diffusion constant ratio in the microemulsion; path 1 represents
$D(brine)/D(oil) = 5.0$; path 2 represents $D(brine)/D(oil) = 1.0$.

diffusion paths shown in Figure 19 indicate formation of an intermediate microemulsion phase and of a moving dispersion front where liquid crystal dissolves. This predicted sequence of phases as well as the calculated relative velocities of the two boundaries agree with the experimental observations mentioned previously.

The formation of a dispersion front can be described in terms of the diffusion path. As oil diffuses into the liquid crystal dispersion, the continuous phase gradually increases in oil content from an initial composition which would ordinarily be called a micellar solution to compositions which would be called microemulsions. Therefore, no true interface forms at the dispersion front where dissolution of liquid crystal occurs, although there is a rapid change in liquid crystal content over a short distance.

The ratio of the relative diffusion constants of brine and oil in the microemulsion phase was found to be very important. Figure 19 shows the effect of varying this ratio. When oil diffusion is less than that of brine, as would be expected in an oil-in-water microemulsion, the fraction of liquid crystal at the dispersion front increases over that in the bulk dispersion. This situation corresponds to diffusion path 1 in Figure 19. Essentially, brine diffuses out of the dispersion faster than oil can diffuse in, causing a decrease in overall brine concentration at the interface and hence the formation of additional liquid crystal, the phase having the lower brine content. As mentioned previously, this buildup of liquid crystal was observed experimentally.

In the three-phase regime of equilibrium phase behavior, the diffusion path studies were based on a dimensionless parameter S which indicates the position of the system within the regime. Thus, S = 0 corresponds to the salinity where the three-phase region first appears via a critical tie line (14), and S = 1 corresponds to the salinity where it disappears into another critical tie line. The optimum salinity occurs at S = 0.5.

At very small values of S, the three-phase triangle is small, and the same type of path as at low salinities is predicted. However, calculations show that as the triangle grows rapidly with increasing S, such a path no longer satisfies the governing equations. From this point until S reaches a value of about 0.2, it appears that the diffusion path has the form shown in Figure 20. Note that it passes through the three-phase triangle, indicating that an interface forms with a middle-phase microemulsion on one side and a dispersion of liquid crystal in a water-continuous phase on the other. Such a three-phase region has been observed in this range of concentrations for the PDM/IM/HAN system (15). This type of multiphase interface, which is different from the dispersion front found at low salinities, is occasionally seen in metallic and ceramic systems (11) and could presumably form in liquid systems as well.

At approximately S = 0.2 (this value is somewhat dependent on other phase behavior parameters), passage of the calculated diffusion path through the brine phase becomes possible. This change corresponds closely to the point at which brine began forming by diffusion in the contacting experiments and, as a result, indicates that formation of two intermediate phases is preferred over the formation of a single microemulsion phase. A sample of this type of diffusion path is shown in Figure 21 for S = 0.5 (optimum salinity). In this diagram, the size of the brine phase is exaggerated for descriptive purposes. Passage of the diffusion path through the brine corner allows only limited diffusion in the liquid crystal dispersion. As a result, no significant buildup of liquid crystal is predicted to occur, matching experimental observations.

An explanation of why convection occurred when brine formed in the TRS system can be given by interface stability analysis (18). During the experiments, slight tipping of the sample cells indicated that the intermediate brine phases were more dense than the mixtures of liquid crystal and brine below them. This adverse density difference caused a gravitational instability for which the smallest unstable wavelength λ is given by

$$\lambda = 2\pi \left[\frac{\gamma}{(\rho_B - \rho_{LC})g} \right]^{1/2}$$

Here, γ is the interfacial tension, ρ_b and ρ_{LC} are the densities of the intermediate brine phase and liquid crystal, and g is the gravitational constant. This analysis assumes that the liquid crystal forms a continuous phase, which was probably the case for most samples in the salinity range of interest. Assuming a low interfacial tension of 0.01 dyne/cm between the brine and liquid crystal-rich mixture and a density difference of 0.1 gm/cm³, one obtains λ = 0.063 cm. Since this value is the same order of magnitude as the sample width in the experiments (0.2 cm), the possibility of this type of instability occurring is confirmed. Generally, only one brine pocket formed in the TRS samples, as shown in Figure 9, indicating that the smallest unstable wavelength was comparable to the width of the capillary.

In the PDM experiments, no disturbances in the form of brine pockets occurred. Perhaps the smallest unstable wavelength was sufficiently long to prevent their formation. Another possibility is that the high viscosity of the aqueous PDM solutions (over 100 cp (19)) drastically slowed the growth of any disturbances that did form, preventing their detection during the contacting experiments.

The initiation of spontaneous emulsification in the oil phase was observed experimentally at optimum salinity. This phenomenon can be explained in terms of Figure 22, which is an expanded view of the oil corner of the ternary diagram. As salinity is increased past optimum, the surfactant concentration in the excess oil phase increases, causing the oil-phase triangle vertex to lift off the base. A corner of the brine-oil two-phase region under the triangle is then exposed to crossing by the diffusion path. Increasing salinity exposes more of the two-phase region, and an increase in spontaneous emulsification is predicted. The extent of penetration of the diffusion path into the two-phase region is also affected by the diffusion constants in the oil phase, which determine the shape of the diffusion path, and by the slope of the brine-oil phase boundary. This type of spontaneous emulsification differs from that normally seen in liquid systems in that the emulsified brine drops are separated from the bulk brine phase.

The transition that occurs in diffusion paths near the high-salinity end of the three-phase regime is not as complex as that at the low-salinity end. As the triangle shrinks, passage through the

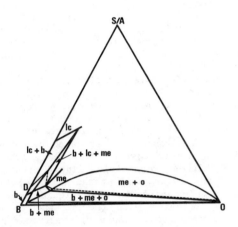

Figure 20. Proposed diffusion path for the low-salinity range of the three-phase regime.

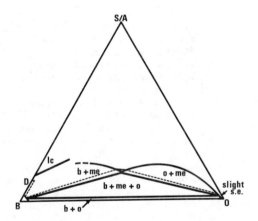

Figure 21. Calculated diffusion path at optimum salinity.

middle-phase microemulsion continues until a path under the triangle is possible. Figure 23 is an example of this type of diffusion path calculated for conditions of high salinity, when the three-phase region has disappeared. For initial compositions D and 0, two interfaces are predicted to form: one between liquid crystal and brine and another between brine and a water-in-oil microemulsion. In contrast to the lower salinity regimes, the oil phase is predicted to grow under these conditions. Also, spontaneous emulsification is indicated where the microemulsion segment of the diffusion path crosses the large two-phase region. This form of spontaneous emulsification is similar to that studied previously for toluene-water-solute systems (12).

The morphology of liquid crystal phase transformations was an important aspect of the diffusion phenomena observed at high salinities. The liquid crystal-brine interface was found not to be flat and horizontal as assumed in diffusion path theory. Instead, myelinic figures of liquid crystal developed, forming a two-phase region of finite thickness containing liquid crystal and brine. A uniform layer of the isotropic C phase eventually formed in this region. Since the changes in the liquid crystal phase were inconsistent with diffusion path theory, the formation of the C phase could not be predicted. However, theory did predict movement of the various interfaces away from the oil phase with velocities in approximately the correct ratio.

Conclusions

A comparison between experimental and theoretical results shows that diffusion path analysis can qualitatively predict what is observed when an anionic surfactant solution contacts oil. Experimentally, one or two intermediate phases formed at all salinities. The growth of these phases was easily observed through the use of a vertical-orientation microscope. Except when convection occurred due to an intermediate phase being denser than the phase below it, interface positions varied as the square root of time. As a result, diffusion path theory could generally be used to correctly predict the direction of movement and relative speeds of the interfaces.

Calculated diffusion paths also successfully predicted the occurrence of spontaneous emulsification in the systems. Near optimum salinity where this phenomenon first appeared, brine drops spontaneously emulsified in the oil but were isolated from the bulk brine phase by a microemulsion. At high salinities, a more common type of spontaneous emulsification was seen with brine emulsifying in the oil directly above a brine layer.

Although the contacting experiments were performed with surfactant systems typical of those used in enhanced oil recovery, application of the results to detergency processes may be possible. For example, the growth of oil-rich intermediate phases is sometimes a means for removing oily soils from fabrics. Diffusion path theory predicts that oil is consumed fastest in the oil-soluble end of the three-phase regime where an oil-rich intermediate microemulsion phase forms.

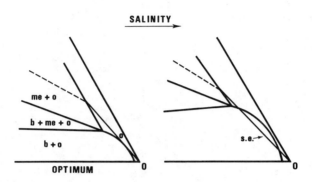

Figure 22. Expanded view of the oil corner showing the onset of spontaneous emulsification at optimum salinity.

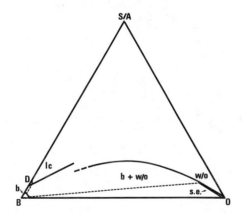

Figure 23. Calculated diffusion path at high salinity.

Acknowledgments

This work was supported by grants from Shell Development Company, Exxon Production Research Company, Amoco Production Company, and Gulf Research and Development Company. K. Raney thanks Cities Service Company for personal financial support during the period this work was performed.

Literature Cited

1.   Lam, C.S.; Schechter, R.S.; Wade, W.H. Soc. Pet Eng. J. 1983, 23, 781.
2.   Stevenson, D.G. In "Surface Activity and Detergency"; Durham, K., Ed.; Macmillan: London, 1961; Chap. 6.
3.   Lawrence, A.S.C. In "Surface Activity and Detergency"; Durham, K., Ed.; Macmillan: London, 1961; Chap. 7.
4.   Benton, W.J.; Miller, C.A.; Fort, T., Jr. J. Disp. Sci. Technol. 1982, 1, 1.
5.   Benton, W.J.; Miller, C.A. J. Phys. Chem. 1983, 87, 4981.
6.   Cash, R.L.; Cayias, J.L.; Hayes, M.; MacAllistar, D.J.; Schares, T.; Schechter, R.S.; Wade, W.H. SPE Paper 5564, 50th Annual Fall Meeting of SPE, Dallas, Texas, September 1975.
7.   Hirasaki, G.J. SPE Paper 8841, First Joint SPE/DOE Symposium on Enhanced Oil Recovery, Tulsa, Oklahoma, April 1978.
8.   Winsor, P.A. "Solvent Properties of Amphiphilic Compounds"; Butterworths: London, 1954.
9.   Healy, R.N.; Reed, R.L.; Stenmark, D.G. Soc. Pet. Eng. J. 1976, 16, 147.
10.   Kirkaldy, J.S.; Brown, L.C. Can. Met. Quart. 1963, 2, 89.
11.   Christensen, N.H. J. Amer. Ceram. Soc. 1973, 62, 293.
12.   Ruschak, K.J.; Miller, C.A. Ind. Eng. Chem. Fundam. 1972, 11, 534.
13.   Raney, K.H. M.S. Thesis, Rice University, Houston, Texas, 1983.
14.   Bennett, K.E.; Phelps, C.H.K.; Davis, H.T.; Scriven, L.E. Soc. Pet. Eng. J. 1981, 21, 747.
15.   Ghosh, O.; Miller, C.A. J. Colloid Interface Sci., in press.
16.   Kunieda, H.; Shinoda K. J. Disp. Sci. Technol. 1982, 3, 233.
17.   Hand, D.B. J. Phys. Chem. 1930, 34, 1961.
18.   Chandrasekhar, S. "Hydrodynamic and Hydromagnetic Stability"; Clarendon Press: Oxford, 1961.
19.   Benton, W.J.; Baijal, S.K.; Ghosh, O.; Qutubuddin, S.; Miller, C.A.   SPE/DOE Paper 12700, SPE/DOE Fourth Symposium on Enhanced Oil Recovery, Tulsa, Oklahoma, April 1984.

RECEIVED June 8, 1984

15

Effects of Polymers, Electrolytes, and pH on Microemulsion Phase Behavior

S. QUTUBUDDIN[1], C. A. MILLER[2], W. J. BENTON[2], and T. FORT, JR.[3]

Chemical Engineering Department, Carnegie-Mellon University, Pittsburgh, PA 15213

Polymer addition induces phase separation in liquid-crystalline solutions of the type injected underground during surfactant flooding. Volume restriction effect is proposed to explain this phenomenon. Added polymer also causes phase separation in an otherwise single-phase oil-in-water microemulsion. Brine is transferred from the microemulsion to the polymer phase as salinity is increased. A thermodynamic treatment is presented which predicts the partitioning of water between the microemulsion and polymer phases. The sensitivity of microemulsion phase behavior to salinity and pH is reported using a mixture of synthetic sulfonate and carboxylic acid. Under appropriate conditions, the effect of salinity on phase behavior can be counterbalanced by pH. Adding electrolyte makes the surfactant hydrophobic while increasing pH can make it hydrophilic due to ionization. Various phase transitions are observed depending on pH and salinity. Middle phase microemulsions can exist over a wide range of salinity. Increasing temperature shifts the middle phase region to higher salinities.

Surfactant-polymer flooding involves successive injections into the reservoir of an aqueous surfactant-cosurfactant solution and a dilute aqueous solution of a high molecular weight polymer. The primary purpose of the surfactant slug is to reduce the interfacial

[1] Current address: Chemical Engineering Department, Case Western Reserve University, Cleveland, OH 44106
[2] Current address: Rice University, Houston, TX 77251
[3] Current address: California Polytechnic State University, San Luis Obispo, CA 93407

0097-6156/85/0272-0223$08.25/0
© 1985 American Chemical Society

tension between oil and brine. Polymer is added to increase the viscosity of the drive water and thereby preserve the stability of various fluid banks in the reservoir. Surfactant-polymer incompatibility and sensitivity of interfacial tensions to salinity are two major problems associated with the surfactant-polymer technique of enhanced oil recovery (EOR).

The efficiency and economics of oil recovery can be adversely affected by interactions between surfactant aggregates and polymer. Such interactions occur because of mixing at the boundary between surfactant and buffer solutions, and because residual surfactant adsorbed on the rock surface may later desorb into polymer solution. Mixing of polymer and surfactant may also occur throughout the surfactant bank because of the "polymer inaccessible pore volume" effect (1). Large polymer molecules are excluded from the smaller pores in the reservoir rock, and travel faster than the surfactant. Thus, polymer molecules enter into the surfactant slug.

The significance of phase separation or incompatibility in polymer-aggregate systems is not limited to EOR. Systems containing more than one colloid are numerous. A range of industrial and agricultural products such as inks, paints and milk fall under that category. Many biological units such as blood cells and ribosomes can be treated as aggregates which interact with other colloidal and non-colloidal components of the living system. Therefore, the study of polymer-aggregate incompatibility is important both industrially and physiologically.

Phase behavior studies of oil-brine-surfactant systems have shown that the ultralow interfacial tension (less than 0.01 dyne/cm) necessary for EOR is very sensitive to salinity changes (2,3). Such low tensions are obtained only within a small range of salinity near the point of "optimum salinity" where equal amounts of oil and brine are solubilized. The tolerance of ultralow tensions to divalent ions is still less.

The term "microemulsion" will be used as defined by Healy and Reed (2): "a stable translucent micellar solution of oil, water that may contain electrolytes, and one or more amphiphilic compounds (surfactants, alcohols, etc.)". Microemulsions have been classified as lower phase (ℓ), upper phase (u), or middle phase (m) in equilibrium with excess oil, excess water, or both excess oil and water respectively. On increasing salinity, phase transitions take place in the direction of lower → middle → upper phase microemulsions.

The effects of pH on microemulsions have been investigated by Qutubuddin et al. (4,5) who have reported a model pH-dependent microemulsion using oleic acid and 2-pentanol. It has been shown that the effect of salinity on phase behavior can be counterbalanced by pH adjustment under appropriate conditions. Added electrolyte makes the surfactant system hydrophobic while an increase in pH can make it hydrophilic by ionizing more surfactant. Based on the phase behavior of pH-dependent systems, a novel concept of counterbalancing salinity effects with pH is being proposed. The proposed scheme for reducing the sensitivity of ultralow interfacial tension (IFT) to salinity is to add some carboxylate or similar surfactant to a sulfonate system, and adjust the pH. The pK and the concentration of the added surfactant are variables that may be selected according to the field conditions for efficient recovery of additional oil.

The goals of this work have been to determine the effect of polymers on the phase behavior of aqueous surfactant solutions, prior to and after equilibration with oil, to understand the mechanism of the so-called "surfactant-polymer interactions" (SPI) in EOR, to develop a simple model which will predict the salient features of the phase behavior in polymer-microemulsion systems, and to test the concept of using sulfonate-carboxylate mixed microemulsions for increased salt tolerance.

Previous Work on Surfactant-Polymer Interactions

The earliest investigations of SPI in EOR were done by Trushenski and coworkers (1,6). They reported that high mobility and phase separation can occur due to SPI. Szabo (7) studied several surfactant-polymer systems, and found that mixtures of sulfonates and polymer solutions separated into two or three phases. The above groups investigated aqueous solutions only, and the mechanism of interaction was not clearly defined.

The effect of polymers on microemulsions phase behavior has been reported by Hesselink and Faber (8). They have described the surfactant-polymer phase separation in terms of the incompatibility of two different polymers in a single solvent, considering the microemulsion as a pseudo-polymer system. The effect of polymers on the phase behavior of micellar fluids has been recently studied by Pope et al. (9) and others (10,11).

The volume restriction effect as discussed in this paper was proposed several years ago by Asakura and Oosawa (12,13). Their theory accounted for the instability observed in mixtures of colloidal particles and free polymer molecules. Such mixed systems have been investigated experimentally for decades (14-16). However, the work of Asakura and Oosawa did not receive much attention until recently (17,18). A few years ago, Vrij (19) treated the volume restriction effect independently, and also observed phase separation in a microemulsion with added polymer. Recently, DeHek and Vrij (20) have reported phase separation in non-aqueous systems containing hydrophilic silica particles and polymer molecules. The results have been treated quite well in terms of a "hard-sphere-cavity" model. Sperry (21) has also used a hard-sphere approximation in a quantitative model for the volume restriction flocculation of latex by water-soluble polymers.

Materials and Methods

Materials. Two different classes of polymers have been used: a biopolymer known as Xanthan gum, and a partially hydrolyzed polyacrylamide. These represent the main categories of water-soluble polymers now being used in EOR. The polymers have very high molecular weights, usually in the range of 10^6 to 10^7. Xanthan gum was obtained from two different sources: Abbott (Xanthan Broth, 2.91% polymer) and Pfizer (Flocon Biopolymer 1035, 2.77% polymer). The partially hydrolyzed polyacrylamide used was Dow Pusher 500 (Dow).

The synthetic sulfonate used as surfactant was TRS 10-410 (60% active) obtained from Witco. It has an average molecular weight of 420. The cosurfactant used with it is a short-chain alcohol, isobutyl alcohol (IBA). The carboxylic acids used with TRS 10-410 for mixed surfactant microemulsions were octanoic or decanoic acid of at least 99% purity obtained from Sigma. Reagent grade sodium hydroxide (J.T Baker) was used to adjust the pH in the mixed surfactant system. The electrolyte used was reagent grade sodium chloride. Reagent grade n-decane and n-hexadecane (Humphrey) were used as the hydrocarbons. Water was triple-distilled, once in a metal container, then in glass, and finally from a potassium permanganate solution in glass.

Solution Preparation and Equilibration. Solutions were mixed in Teflon-capped glass tubes. Stock solutions were prepared in glassware cleaned with chromic acid and thoroughly rinsed in double-distilled water. For studies of the effects of polymer, aqueous solutions were made with and without polymer. Stock solutions were made of (i) surfactant and alcohol in a 5:3 (w/w) ratio (unless otherwise indicated), (ii) aqueous polymer, and (iii) NaCl in distilled water. These were blended to give the desired surfactant, alcohol, polymer, and salt concentrations. The polymers were made at high concentrations by stirring slowly in water with a magnetic bar for a long time. This has been recommended for hydrolyzed polyacrylamides to avoid shear degradation (22). The stock polymer solution was passed through a glass fritted filter and then a Millipore filter. The aqueous sytems were gently heated to 60°C, agitated, first with a vortex mixer and then by ultrasonication, and finally allowed to equilibrate at a constant temperature in an environmental room. The procedure for solution preparation was similar for the mixed surfactant system, except that some sodium hydroxide was also added.

The aqueous solutions were contacted with an equal volume of the hydrocarbon. The mixtures were gently heated to 60°C and thoroughly agitated using a vortex mixer. The mixing was repeated after 24 hours, and the samples allowed to equilibrate at constant temperature in an environmental room. The phase behavior was observed using the Polarized Light Screen described below. Equilibrium was assumed to have been achieved when the volumes and the appearance of the different phases did not show any sign of change with time as observed under polarized light. The equilibration time ranged from a few weeks for polymer-free systems to several months for systems containing polymer.

Polarized Light Screen (PLS). The PLS technique (23,24) was used extensively to obtain immediate information about the macroscopic phase behavior of solutions. Diffuse light is transmitted through a polarizer and an analyzer. The sample is placed between the two polarizers, and system behavior is observed as shown in Figure 1. The PLS technique makes isotropy, birefringence, and scattering easy to identify, and shows interfacial phenomena (such as critically diffuse regions) at a constant temperature.

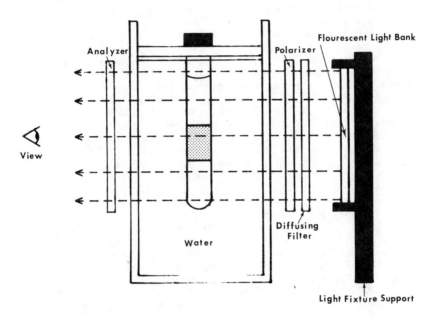

Figure 1. A schematic cross-section of the PLS system. Reproduced with permission from Ref. 29, Figure 1. Copyright 1983, American Chemical Society.

Microscopy. Selected samples were prepared in sealed rectangular optical capillaries (25) and viewed by polarized and Hoffman Modulation Contrast optics. The magnification used was 100-X.

Results and Discussion

Phase Behavior of Aqueous Surfactant Solutions. The aqueous solutions contained 5 gm/dl TRS 10-410 as surfactant, and 3 gm/dl isobutyl alcohol as cosurfactant, unless otherwise indicated. The polymer concentration was varied from zero to 1500 ppm. The aqueous phase behavior in the absence of polymer is shown in Figure 2. The salinity is varied from 0.8 to 2.2 gm/dl NaCl in increments of 0.2 gm/dl. The phase behavior at lower salinities will be discussed later. The general trend is similar to the changes in textures reported for other commercial and model sulfonate solutions (26,27).

A dispersion of spherulitic liquid crystalline particles in brine exists between 0.8 gm/dl NaCl (Figure 2(a), first sample on the left) and 1.2 gm/dl. As the salinity is increased to about 1.4 gm/dl NaCl, the amount of liquid crystals as well as the birefringence increase, and the texture observed using PLS is intermediate between those of the spherulite (S) and lamellar (L) structures. The aqueous solution is a homogeneous lamellar phase between 1.6 and 1.8 gm/dl NaCl. The surfactant molecules form bilayers with their polar heads toward the brine. Figure 3(a) shows the lamellar structure as observed by polarized microscopy at 1.6 gm/dl salt and without any polymer. The bands represent "oily streaks" in a planar background.

At 2.0 gm/dl NaCl the solution separates into two phases: the upper is lamellar, while the lower is an isotropic phase which scatters light and exhibits "streaming" or "flow" birefringence when gently agitated. The latter phase has been designated as the "C" phase (28,29). Two phases exist at 2.2 gm/dl NaCl or higher salinity: a "C" phase (top) in equilibrium with a clear brine phase (bottom). There is a narrow one-phase region of the "C" phase between the two-phase regions. When no salt is present and at very low salinities [less than 0.5 gm/dl NaCl] the aqueous solutions are isotropic. A precipitate is formed at 0.5 gm/dl NaCl (Figure 4).

Figures 2(b) and 2(c) show the effect of Xanthan gum (Abbott) on the phase behavior when the polymer concentrations are 750 ppm and 1500 ppm, respectively, and salinity is varied from 0.8 to 2.2 gm/dl NaCl. The textures are significantly different from those of the polymer-free system shown in Figure 2(a). Different phases and structures observed with the PLS are indicated in Table 1. It may be pointed out that the structures in the different phases were not investigated in detail in this study.

The primary effect of adding polymer to the aqueous solution is that phase separation occurs in the salinity range of 0.8 to 2.0 gm/dl where a liquid crystalline phase exists in the polymer-free system. Phase separation does not occur immediately and may be very slow depending on the composition. Thus, a rather long time is required by the polymer-containing systems to come to equilibrium. The results of this study were obtained after more than twelve months of equilibration. Preliminary observations were reported earlier (30).

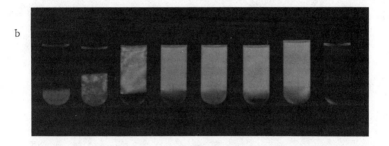

Figure 2. Aqueous phase behavior in polarized light of TRS 10-410/IBA systems with a) no polymer, b) 750 ppm Xanthan, and c) 1500 ppm Xanthan. The composition is 5 gm/dl IBA with salinity varied between 0.8 and 2.2 gm/dl NaCl at 0.2% intervals. T=22°C.

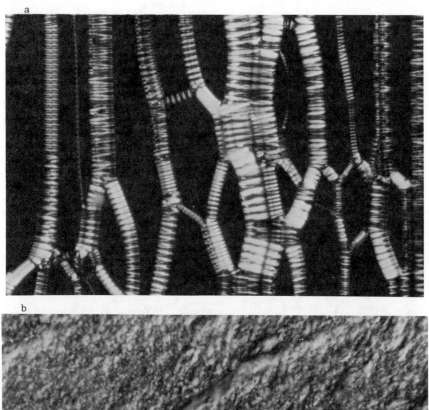

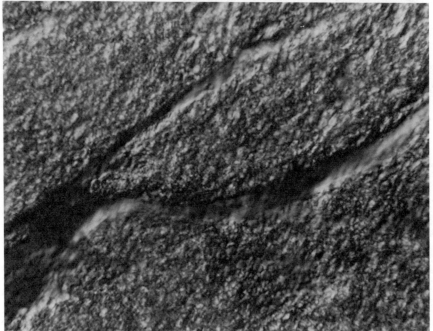

Figure 3. a) Lamellar structure for the TRS 10-410/IBA system
as observed under polarized microscope with no polymer. b) Aqueous
phase structure for the TRS 10-410/IBA system as observed one day
after 750 ppm Xanthan was added.

Figure 3. Continued on next page

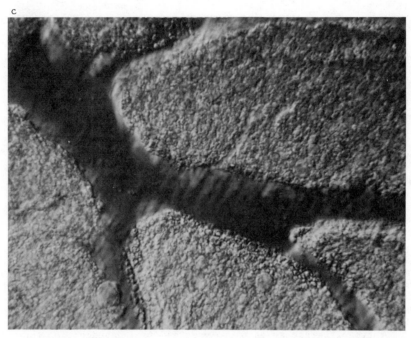

Figure 3. c) Aqueous phase structure for the TRS 10-410/IBA
system as observed one week after 750 ppm Xanthan was added.

The dynamics of polymer-induced phase separation have been observed with polarized microscopy. Figures 3(b) and 3(c) show the textures when 750 ppm Xanthan gum was added to the system of 3(a). Figure 3(b) was observed one day after addition of polymer, and Figure 3(c) a week later. As expected, the polymer phase represented by the darker region has separated out to a greater extent after a week.

Table I shows that at least one of the phases is birefringent between 0.8 and 2.0 gm/dl NaCl. Three phases exist for certain compositions with a lamellar phase at the top, a streaming birefringent phase in the middle and an isotropic phase at the bottom. Such is the case when the salinity is 1.4 or 1.8 gm/dl and the polymer concentration is 750 ppm Xanthan (Figure 2(b)). With or without polymer, two phases are present at 2.2 gm/dl NaCl. The added polymer appears to remain mostly in the lower phase and consequently increases the viscosity. When no salt is present, the aqueous solutions containing polymer are isotropic phases up to a polymer concentration of 1000 ppm. The aqueous solutions at 1500 ppm and higher concentrations of polymer show streaming birefringence.

The effect of two different polymers on the phase behavior at low salinities is shown in Figure 4. No polymer is present in Set B (blank) while Sets D and X contain Dow Pusher 500 and Xanthan (Pfizer) polymers, respectively. The solutions are simple isotropic phases at 0.1 and 0.3 gm/dl NaCl whether or not any polymer is present.

The above results of phase behavior as summarized in Table I suggest that the observed macroscopic phase separation is directly related to the structure of the surfactant solution. There is no phase separation when the solutions are isotropic. But the added polymer induces phase separation only when a single liquid crystalline phase or a stable dispersion of liquid crystal particles in brine is present in the absence of polymer. The effect is the same for two different types of polymer and is independent of the polymer concentration over a fair range of composition.

Further evidence to support the above hypothesis on the role of structure in phase separation of aqueous solutions is provided by the effect of additional alcohol. The amount of alcohol was increased from 3.0 to 5.0 gm/dl, the surfactant concentration kept constant, and the salinity varied. The addition of alcohol extended the range of salinity where the aqueous solutions are isotropic to 0.8 gm/dl NaCl. According to the above hypothesis, no phase separation should take place on addition of polymer to the isotropic solutions existing up to 0.8 gm/dl NaCl. Indeed, no phase separation was observed when as much as 1500 ppm Xanthan was added at such compositions. Thus, the addition of alcohol increases the critical electrolyte concentration for phase separation, an effect seen also by others (9).

It is speculated that the effect of temperature on the critical electrolyte concentration is similarly related to the effect of temperature on the structure of aqueous solutions. An increase in temperature has been shown to extend the range of micellar solutions to a higher salinity in anionic surfactant systems (31). Hence, polymer-aggregate incompatibility would be less when the temperature is increased. However, addition of alcohol or change in temperature

Table I.   Phase behavior of polymer/surfactant-cosurfactant aqueous
solutions as observed in polarized light

NaCl conc. (gm/dl)	Polymer conc., ppm					
	0	250	500	750	1000	1500
0	I	I	I	I	I	SB
0.1	I*	–	–	I*	–	–
0.3	I*	–	–	I*	–	–
0.5	I/ppt*	–	–	I/ppt*	–	–
0.8	S	I/B	I/B	I/SB	I/B	I/B
1.0	S	I/B	B/I	I/B	B/I	B/I
1.2	S	B/I	B/I	B/I	B/I	B/I
1.4	S+L	S+L/B	S+L/B	L/SB/I	B/I	L/B
1.6	L	L/B	L/B	L/SB	L/I	L/B
1.8	L	L/B	L/B	L/SB/I	L/SB/I	L/B
2.0	L/C	L/B	L/B/I	L/SB	L/SB/I	C/P
2.2	C/BR	C/P	C/P	C/P	C/P	C/P

Legend: B, non-lamellar birefringent; BR, brine; C, "C" phase
described in text; I, isotropic; L, lamellar; P, polymer-rich
brine; S, spherulite; S+L, spherulite plus lamellar; SB, streaming
birefringent.
*Experiments done with two polymers, Xanthan (Pfizer and Dow
Pusher 500)

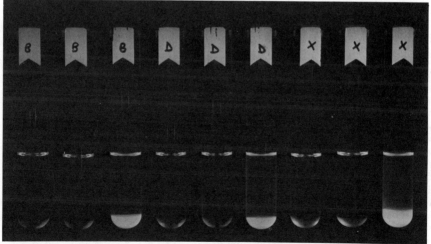

Figure 4.   Aqueous phase behavior in polarized light of three
different systems at low salinity. B: blank (no polymer), D:
Dow Pusher (750 pppm), and X: Xanthan gum (750 ppm). The salinities in each set are 0.1, 0.3 and 0.5 gm/dl NaCl from left to
right. T=22°C.

at constant salinity may shift the system away from optimum phase
behavior in terms of ultralow interfacial tensions on equilibration
with oil.

Phase Behavior on Equilibration with Oil. Microemulsions are formed
when the aqueous surfactant–cosurfactant solutions are mixed with
oil, and allowed to equilibrate. Figure 5(a) shows the phase behavior when 5 ml aqueous surfactant solutions (without any polymer)
were equilibrated with equal volumes of n–dodecane. The salinity
was varied from 0.8 to 2.2 gm/dl NaCl in 0.2 gm/dl increments. At
low salinities a lower phase microemulsion exists in equilibrium
with excess oil. The middle phase microemulsion appears at about
1.05 gm/dl NaCl and is in equilibrium with both excess oil and
excess brine. At 2.2 gm/dl and higher salt concentrations the
system becomes an upper phase microemulsion. Ultracentrifuge
studies (27,31) have shown that the lower phase microemulsion is
water–continuous while the upper phase is oil–continuous.
     The effect of 750 ppm Xanthan gum on the microemulsion phase
behavior is shown in Figure 5(b). The observed phase behavior is
similar except that the extent of the three–phase region is widened.
Thus at both 0.8 and 1.0 gm/dl salt concentrations there exists a
polymer–containing brine phase in equilibrium with the
microemulsion phase. When no polymer is present, the microemulsion
phase is in equilibrum with only excess oil. The volumes of the
polymer phases are small and the interface between the polymer phase
and microemulsion is difficult to detect in Figure 5(b). However,
phase separation is clearly visible in Figure 5(c), which illustrates the oil–equilibrated phase behavior at a higher polymer concentration of 1500 ppm.
     Figure 6 shows the volume fraction of the different phases in
equilibrium as a function of salinity for systems containing no
polymer and 750 ppm Xanthan. Similarly, Figure 7 shows the volume
fraction for systems containing no polymer and 1500 ppm Xanthan. It
is clear that the salient effect of the polymer is in the low salinity range where a highly viscous polymer phase separates out at the
bottom. The volume of the polymer phase at a given salinity has been
found to decrease with decreasing polymer concentration. But phase
separation has been observed at low salinities even with as low as
100 ppm Xanthan (Abbott).
     The effect of changing the chain length of the hydrocarbon on
polymer–aggregate incompatibility is shown in Figure 8.   On
increasing the carbon chain length from twelve to sixteen, the optimum salinity in the polymer–free system changes slightly to maintain
the "hydrophilic–lipophilic" balance. The round–bottom samples in
Figure 8 contain no polymer, but the hydrocarbon is now n–hexadecane. Lower phase microemulsions exist at 0.4 and 1.0 gm/dl NaCl.
The phase transition points are shifted to higher salinities because
of greater hydrophobicity of the oil. Thus, middle phases occur at
1.6 as well as 2.2 gm/dl salt. It may be recalled that with n-
dodecane the microemulsion is upper phase at 2.2 gm/dl salt. The
effect of polymer addition is shown on the right–hand side of Figure
8 (flat–bottom tubes). The basic phase behavior is similar. But,
as before, there is phase separation at low salt concentrations
which corresponds to the lower phase region in the polymer–free

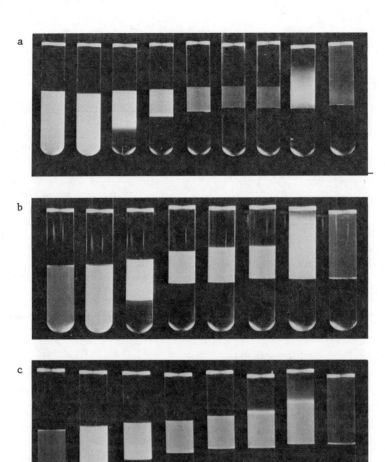

Figure 5. Oil-equilibrated phase behavior in polarized light
of 5 gm/dl TRS 10-410, 3 gm/dl IBA/n-dodecane systems with a) no
polymer, b) 750 ppm Xanthan, and c) 1500 ppm Xanthan. The salinity is varied from 0.8 to 2.2 gm/dl NaCl with 0.2 gm/dl increments. T=22°C.

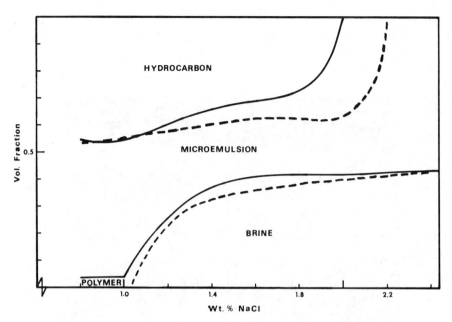

Figure 6. Volume fraction versus salinity plot for 5 gm/dl
TRS 10-410, 3 gm/dl IBA/n-dodecane systems with 750 ppm Xanthan
(solid line) and no polymer (dotted line). T=22°C.

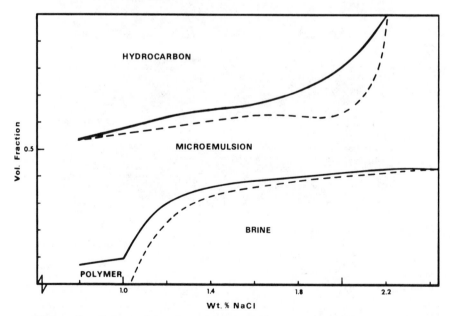

Figure 7. Volume fraction versus salinity plot for 5 gm/dl
TRS 10-410, 3 gm/dl IBA/n-dodecane systems with 1500 ppm Xanthan
(solid line) and no polymer (dotted line). T=22°C.

system. Therefore, phase separation due to polymer-aggregate incompatibility is a general phenomena, and the mechanism is independent of the carbon-chain length of the oil.

Qualitative Explanation for Polymer-Aggregate Incompatibility. The major reason for phase separation in aqueous systems, as well as in lower phase microemulsions, is postulated to be the volume restriction effect. Polymer molecules lose configurational entropy when they are forced into narrow spaces between particle surfaces. The polymers used in EOR have very high molecular weight, so the size of the macromolecule is very large. Estimates from literature values of the viscosity of partially hydrolyzed polyacrylamides indicate that the polymer size exceeds a micron for molecular weights on the order of 10^6. For the Xanthan gum polymer, laser light-scattering results are consistent with the molecular model of a rigid rod having a length of about a micron (32,33).

The results from aqueous solution phase behavior studies show that phase separation takes place only when aqueous solutions are liquid crystalline. A schematic of a lamellar phase is shown in Figure 9. For the lamellar structure of the systems investigated, the bilayer thickness is about 10 to 30 °A while the thickness of the brine layer does not exceed 100 to 200 °A. Hence, the size of the polymer molecule is too large to be accommodated into the brine spacing in lamellar structure.

Using the mathematical formalism for Brownian motion of particles within a bounded region, Casassa (34-36) has evaluated the partitioning of random-flight and rigid-rod polymer chains between a dilute macroscopic solution phase and liquid within cavities of different shapes. The conformational freedom of a polymer is decreased in a cavity due to the volume restriction effect. Figure 10 (36) illustrates the slab model with unbroken rod for simplicity. For the case of a slab the value of the distribution coefficient (K) decreases exponentially with the ratio ℓ/a where ℓ is the length of the rigid rod and a is the slab width. The value of K is 0.1 for ℓ/a of about 10 as shown in Figure 11 (36).

An analogy may be drawn between the above slab model and the lamellar structure shown in Figure 9. For the polymer molecules studied, the ratio ℓ/a is very large, on the order of 100, and, hence, the distribution coefficient as given by the slab model is very small, approaching zero. Similar results are found for flexible chains (36). This explains in a simple manner why the polymer molecules partition out of the lamellar liquid crystalline phases and form a separate phase. When there is a dispersion of liquid crystals, e.g. in the spherulitic region, the polymer molecules are again larger than the spacing between the particles. Hence, coexistence would be entropically unfavorable.

A similar argument based on volume restriction also explains why the macromolecules do not enter the microemulsion phase and separate out into an excess phase. The size of the macromolecule is about an order of magnitude larger than the oil droplets in o/w microemulsion, and the spacing between the droplets is small. There would be a considerable loss of entropy if the macromolecules were forced between the droplets. The drop size and spacing are, however, dependent on parameters like salinity, alcohol concentration,

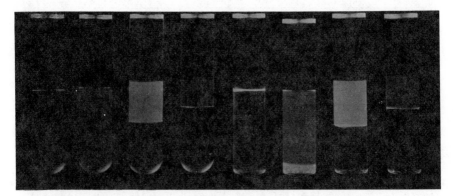

Figure 8. Microemulsion phase behavior in polarized light of
5 gm/dl TRS 10-410, 3 gm/dl IBA/n-hexadecane systems with and
without 750 ppm Xanthan (Pfizer). Flat bottom tubes with
polymer and round bottom tubes without polymer. Salinities are
0.4, 1.0, 1.6 and 2.2 gm/dl NaCl from left to right. T=22°C.

Lyotropic Liquid Crystal

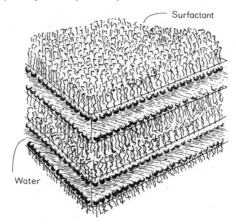

Figure 9. A schematic of a lamellar phase.

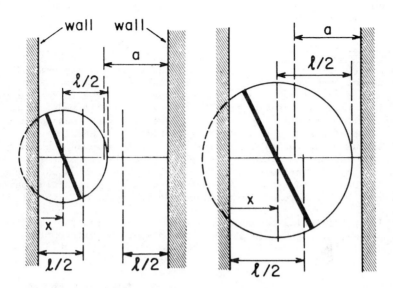

Figure 10. Unbroken rod inside a slab-shaped cavity. Left:
l<2a; Right: l>2a [36]. Reproduced with permission from Ref.
36, Figure 1. Copyright 1972, John Wiley & Sons, Inc.

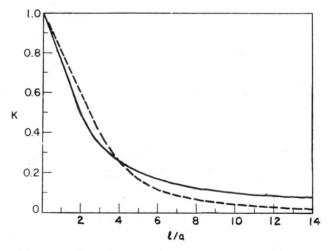

Figure 11. Distribution coefficient K for the slab model.
Rigid rod (solid line) and once-broken rod (dotted line).
Reproduced with permission from Ref. 36, Figure 2, Copyright
1972, John Wiley & Sons, Inc.

and temperature. Hence, the observed phase separation depends on
such parameters.

Thermodynamic Model for Phase Equilibrium between Polymer Solution
and O/W Microemulsions. Figures 6 and 7 show that when phase
separation first occurs, most of the water is in the microemulsion.
With an increase in salinity, however, much of the water shifts to
the polymer solution. Thus, a concentrated polymer solution becomes
dilute on increasing salinity. The objective of this model is to
determine the partitioning of water between the microemulsion and
the polymer-containing excess brine solution which are in
equilibrium. For the sake of simplicity, it is assumed that there
is no polymer in the microemulsion phase, and also no microemulsion
drops in the polymer solution. The model is illustrated in Figure
12. The model also assumes that the value of the interaction parameter (χ) or the volume of the polymer does not change with salinity.

    The partitioning of water may be determined from the equality
of the chemical potentials of water in polymer and microemulsion
phases existing in equilibrium. The two chemical potentials may be
obtained as follows:

    (a) In the polymer solution the simple Flory-Huggins theory
can be applied. The chemical potential is given by:

$$\mu_w^p = \mu_w^o \ (T,P) + RT\{\ln \gamma_w + \ln X_w + \chi\phi_2^2\} \qquad (1)$$

where

$$\ln \gamma_w = \ln \ \{1-\phi_2(1-1/m)\} + \phi_2(1-1/m) \qquad (2)$$

with m the number of segments per polymer molecule; μ_w^o (T,P) the
chemical potential of pure water; μ_w^o the chemical potential of
water in polymer phase; ϕ_2 the volume fraction of polymer ($\ll$1); χ
the Flory-Huggin interaction parameter; ϕ_w the volume fraction of
water; and

$$X_w = m \ \phi_w / \ \{1+(m-1)\phi_w\} \qquad (3)$$

    Since m approaches infinity in the case of very high molecular
weights, the value of X_w approaches 1, and the term RT $\ln X_w$ may be
neglected. Thus, Equation 1 reduces to:

$$\mu_w^p = \mu_w^o \ (T,P) + RT \ \{\ln(1-\phi_2) + \phi_2 + \chi\phi_2^2\} \qquad (4)$$

    (b) In microemulsion systems, the theory presented by Miller
et al. (37,38) can be applied. Let G_m denote the free energy of
microemulsions. Then,

$$G_m = G_m \ (T,P,n_w,n_s^d,n_o^d,N) \qquad (5)$$

where the subscripts o, s and w represent oil, surfactant and water respectively, and the superscript d denotes values for drops. The parameter n is the number of moles, so that n_s^d is, for example, the total amount of surfactant in all the drops. ^sN denotes the total number of drops.

The chemical potential in the microemulsion phase is equal to:

$$\mu_w^m = (\delta G_m / \delta n_w)_{T,P,n_s^d,n_o^d,N} \tag{6}$$

The total free energy has three contributions as shown below:

$$G_m = Ng^d + n_w \mu_w^o + G_m^d \tag{7}$$

where g^d is the free energy of an individual drop, and G_m^d is the free energy associated with dispersion and includes entropy of drops and interaction among drops. From Equations 6 and 7,

$$\mu_w^m = \mu_w^o (T,P) + (\delta G_m^d / \delta n_w)_{T,P,n_s^d,n_o^d,N} \tag{8}$$

Now G_m^d may be evaluated as follows:

$$G_m^d = (U - TS) (N v^d + N_w v_w) \tag{9}$$

In Equation 9, S denotes the configurational entropy, U the potential energy per unit volume, and v^d the volume per drop. The first term, (U-TS) is the free energy per unit volume, F, while the second term is the total volume of microemulsion.

For small volume fractions the configurational entropy can be obtained using the procedure of Ruckenstein and Chi (39). It is based on dividing the solution into close-packed spherical cells of diameter D equal to $2a_o (0.75/\phi)^{1/3}$ where a_o is the drop radius and ϕ is the volume fraction. The drops are separated by a distance of h equal to $(D - 2a_o)$. The entropy S is given by:

$$S = 3k \phi/4 \pi a_o^3 \left\{ \ln \left\{ (4 \pi a_o^3)/(3 v_o) \right\} + 3 \ln [\{0.74/\phi\}^{1/3} - 1] \right\} \tag{10}$$

where v_o is the volume of a solvent molecule.

Following Miller et al. (34), the potential energy is given by:

$$U = 1/2 \text{ X } 12 \text{ X } (3\phi / 4 \pi a_o^3) \text{ X } u(D-2a_o) \tag{11}$$

The potential energy function, u(h), where h is the distance of separation, consists of London-van der Waals attractive forces, u_{vw} and double-layer repulsion forces, u_{el} as given below:

$$u_{vw} = - A/12 \left\{ 1/(\zeta^2 + 2 \zeta) + 1/(\zeta^2 + 2 \zeta + 1) + 2 \ln [(\zeta^2 + 2 \zeta)/(\zeta^2 + 2 \zeta + 1)] \right\} \tag{12}$$

where A is the Hamaker constant, and $\zeta = h/2a_o$; and

$$u_{el} = 8 \; \varepsilon a_o (kT/ze)^2 \tanh^2 (ze \psi_o /4kT) \exp(-\kappa h) \tag{13}$$

where κ^{-1} is the double-layer thickness dependent on the salinity, ψ_o the surface potential of the drop, z the valence of the ions, ε the dielectric constant of the solution, and e the charge per electron.

Therefore, for a given volume fraction of drops, ϕ, one can compute F, the free energy per unit volume of the microemulsion:

$$F(\phi) = U(\phi) - T\ S(\phi) \tag{14}$$

A plot of F as a function of ϕ can then be constructed. The conditions for which phase separation occurs, i.e. when the second derivative of F is negative for some range of ϕ, can be determined along with the composition of the phases in equilibrium.

The above treatment, however, requires knowledge of the Hamaker constant, and a, the area per surfactant ion. To obtain these quantities the experimental data of the polymer-free microemulsion system containing 5 gm/dl TRS 10-410 and 3 gm/dl IBA have been utilized. From the solubilization data, the volume fraction of drops at different salinities was calculated. This is indicated by the dotted line in Figure 13. The radius of the droplets has been reported as a function of salinity by Mukherjee (26). For the salinity range of interest, the radius a_o in °A is empirically related as follows:

$$a_o = 2083\ M - 187.5 \tag{15}$$

where M is the salinity in moles/liter. Using the method of Hwan et al. (37), the area per ion was calculated to be 62.4 °A^2 per ion.

Phase separation for the Witco system takes place at about 0.19M salinity corresponding to a volume fraction of 0.17. Using this information, the Hamaker constant A was estimated to be 2.3 x 10^{-13} ergs. Knowing A, it is possible to predict the phase separation curve which is shown by the solid line in Figure 13. Point M corresponds to a critical point.

The above values of A, a and a_o were used in calculating the partitioning of water between microemulsion and polymer solution. Using Equation 9, G_m^d can be computed. Also, $(\mu_w^m - \mu_w^o)$ can be evaluated according to Equation 8 at a given salinity for different values of Z, the fraction of total water in the microemulsion phase. Similarly, $(\mu_w^p - \mu_w^o)$ can be computed as a function of Z. Equating the two chemical potentials, the value of Z at a given salinity was found.

The results are shown in Figure 14 for χ equal to zero (no enthalpic interaction) and an arbitrary value of 0.3. Comparison with the experimental curve shows that the agreement is better when interaction is considered. The simple model is able to predict the sharp decrease in the amount of water in the microemulsion phase as salinity is increased.

The shift of brine from microemulsion to the polymer solution is the primary effect observed experimentally. Qualitatively, with increasing salinity the electrical repulsion between microemulsion drops decreases and the drop size increases. A concentrated phase is favored, and the brine is driven out of the microemulsion into the polymer solution. Considering the many simplifying assumptions

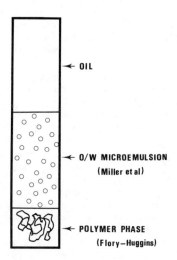

Figure 12.   A schematic of the model O/W microemulsion in equilibrium with an excess oil phase and a polymer phase.

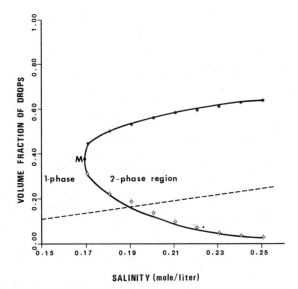

SALINITY (mole/liter)

Figure 13.   Volume fraction versus salinity: Calculated phase boundary (solid line) and experimental data for total volume fraction of drops (dashed line).

in the model, the agreement between theoretical and experimental
curves is satisfactory. It may be possible to improve the agreement
by removing some of the assumptions in the model. Also, one may use
a hard-sphere approximation to compute the free energy of dispersion. But the overall behavior predicted would roughly be the same.

Mixed Carboxylate-Sulfonate Systems. Decanoic and octanoic acids
were selected as carboxylic acids in view of their intermediate
chain lengths. Some initial experiments using a short-chain carboxylic acid (isobutyric acid) were not very promising, apparently
because only a fraction of the acid would partition to the drop surface.

The molar ratio of cosurfactant to active sulfonate is 5.8 for
5 gm/dl TRS 10-410 - 3 gm/dl IBA mixture. The molar ratio of decanoic acid to active sulfonate was chosen to be 1.67. A sufficient
amount of IBA was added to the aqueous solution to keep the molar
ratio of IBA to total surfactant (carboxylate plus sulfonate) the
same as in the carboxylate-free system. Smaller amounts of cosurfactant resulted in unfavorable precipitation. Based on the aqueous
phase, the final composition was 5 gm/dl TRS 10-410, 8 gm/dl IBA and
2 gm/dl decanoic acid. The salinity was varied between 0.8 and 8
gm/dl NaCl for this surfactant composition.

When the NaOH-free aqueous solutions were equilibrated with
equal amounts of dodecane, the resulting systems contained two phases, an upper phase microemulsion and an excess brine phase, at
salinities exceeding 0.8 gm/dl NaCl. Three phases were observed at
0.8 gm/dl NaCl, an excess-oil phase, a middle-phase microemulsion
and an excess-brine phase. Small aliquots of concentrated NaOH were
added to the other systems to observe the upper → middle → lower
phase transitions where possible.

The effect of pH on phase behavior of microemulsions has been
discussed in a different paper (4). In general, an increase in pH
by addition of NaOH at constant salinity makes surfactant more
hydrophilic by ionizing the carboxylic acid. Therefore, under
appropriate conditions, the effect of salinity which is to make the
surfactant hydrophobic, can be counterbalanced by an appropriate
change in pH. The amount of NaOH, or equivalently, the pH needed
for an upper phase microemulsion to shift to a middle phase
increases with increasing salinity. Thus, the concentrations are
0.03M and 0.1M NaOH for 2 and 7 gm/dl NaCl, respectively. The
upper → middle → lower phase transitions were observed with pH
adjustment for salinities less than 5 gm/dl NaCl. For higher salinities, the microemulsion remained as a middle phase even with an
excess of NaOH. All the surfactant molecules are ionized in such a
situation, and the salinity is too high to be counterbalanced by pH
adjustment only.

The amount of cosurfactant necessary for optimum microemulsion
formation can be reduced by changing the chain length of the carboxylic acid to slightly lower values. Thus, middle phases were
obtained with 5 gm/dl TRS 10-410, 5 gm/dl IBA and 1 gm/dl octanoic
acid by adjusting the pH and salinity. Figure 15 illustrates the
phase behavior at 1.5 gm/dl NaCl with varying concentrations of
NaOH. The upper → middle → lower phase transitions were observed as
expected. Middle phases were obtained, but not lower phases, for

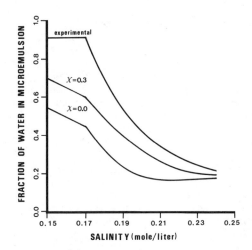

Figure 14. Experimental and theoretical curves of the fraction of water in microemulsion versus salinity. Polymer concentration is 750 ppm. T=22°C.

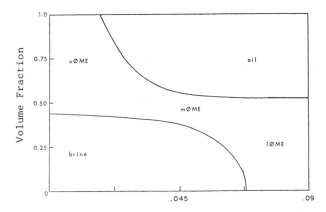

Molar Conc. of NaOH

Figure 15. A plot of volume fraction versus NaOH added for microemulsions containing 5 gm/dl TRS 10-410, 1 gm/dl octanoic acid, 5 gm/dl IBA and 1.5 gm/dl NaCl. T=22°C.

salinities of 3 and 6 gm/dl NaCl. The is illustrated in Figure 16
for 6 gm/dl NaCl. The salinities are too high to be counterbalanced
by changes in pH for the given surfactant composition.

It is possible to extend the middle phase region by increasing
the concentration of carboxylic acid, coupled with an increase in
alcohol concentration. The phase behavior of a mixed microemulsion
system containing 5 gm/dl TRS 10-410, 1.5 gm/dl octanoic acid and 8
gm/dl IBA at 22°C is shown in Figure 17. Between 1.5 and 4 gm/dl
NaCl, if only NaOH concentration is varied then upper → middle →
lower phase transitions occur, apparently due to an increase in the
ionization of the surfactant. Extension of salinity and NaOH concentrations in Figure 17 is expected to show upper → middle → lower
→ middle, and other phase transitions as observed in the oleic acid
system (4).

The effectiveness of the proposed concept of utilizing pH
dependence to counterbalance salinity is best illustrated by citing
a salt scan for the mixed surfactant system [5 gm/dl TRS 10-410, 1.5
gm/dl octanoic acid and 8 gm/dl IBA] at .09M NaOH concentration.
Middle phase microemulsions are formed over a wide range of salinity
(2 to 13 gm/dl NaCl). IFT measurements show that the tensions are
on the order of 10^{-2} dyne/cm near the optimum salinity. By comparison, the TRS 10-410 system without any added carboxylic acid has
a middle phase range of only 1.2 gm/dl NaCl and IFT values on the
order of 10^{-3} dyne/cm near the optimum. Hence, the pH manipulation
has extended the middle phase region by about an order of magnitude,
and yet maintained reasonably low IFT values. It may be noted that
the systems investigated were not optimized. It is possible that pH
manipulation in systems using other surfactants and cosurfactants,
and having optimum composition may lead to better tolerance of phase
behavior and ultralow IFT to salinity changes. The implications of
the results presented in this paper are great in terms of increasing
the salt tolerance, and hence the efficiency of oil recovery by surfactant or caustic flooding.

The effect of temperature on the mixed surfactant system can be
evaluated by comparing Figure 17 with Figure 18 which shows the
phase behavior for the above system at 30°C. Upper → middle → lower
→ middle phase transitions are observed on increasing the molar concentration of NaOH at a fixed salinity such as 5 gm/d NaCl. On
increasing salinity, the lower→ middle → upper phase transitions are
observed as usual. Figure 18 shows that at a constant NaOH concentration, e.g. 0.2M, the lower phase region extends to a higher
salinity at 30°C as compared to 22°C.

An increase in temperature also reduces the middle phase
region. For example, at 0.09M NaOH middle phase microemulsions are
observed between 3.5 and 7.5 gm/dl NaCl. This salinity range of 4
gm/dl is lower than the 11 gm/dl range observed at 22°C, but is
still higher than the middle phase range of about 1 gm/dl observed
at 22°C using only TRS 10-410 surfactant. An increase in temperature would also reduce the middle phase range in conventional
sulfonate systems.

Conclusions

Polymer addition induces phase separation in liquid crystalline

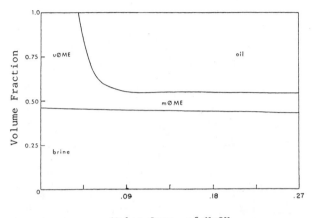

Figure 16. A plot of volume fraction versus NaOH added for microemulsions containing 5 gm/dl IBA and 6.0 gm/dl NaCl. T=22°C.

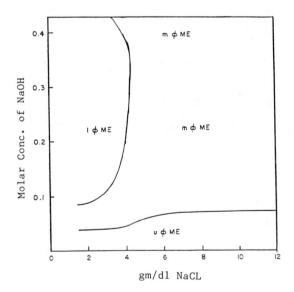

Figure 17. Effect of NaCl and NaOH concentrations on microemulsions containing 5 gm/dl TRS 10-410, 1 gm/dl octanoic acid, and 5 gm/dl IBA. T=22°C.

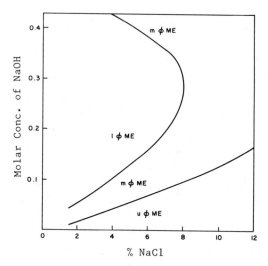

Figure 18.   Effect of NaCl and NaOH concentrations on microemulsions containing 5 gm/dl TRS 10-410, 1 gm/dl octanoic acid, and 5 gm/dl IBA.   T=30°C.

aqueous solutions. No phase separation is observed when the aqueous phase is isotropic containing no liquid crystalline aggregates. The lower phase microemulsion are affected by the polymer which separates into a highly concentrated phase. The effects in both aqueous surfactant systems and microemulsion can be attributed to volume restriction arising from the large size of the macromolecules. Based on simplifying assumptions, a simple thermodynamic treatment can predict the partitioning of water between the microemulsion and the separated polymer solution.

The phase behavior of microemulsions containing a mixture of synthetic sulfonate and a carboxylic acid has been investigated. At a constant sodium chloride concentration, one may observe upper → middle → lower → middle phase transitions by increasing sodium hydroxide concentration. Increasing the temperature shifts the middle phase region to higher salinities. Under appropriate conditions, middle phase microemulsions can be obtained over a wide range of salinity. The salinity range where middle phases exist can be extended by about an order of magnitude using the pH-dependent microemulsions containing mixed surfactants instead of just synthetic sulfonate. This is particularly relevant to enhanced oil recovery.

Acknowledgments

The authors acknowledge Professors E.F. Casassa and G. Berry of Carnegie-Mellon University for useful discussions. This work was supported in part by the U.S. Department of Energy, and in part by grants from Amoco Production Company, Gulf Research and Development Company, Exxon Production Research Company, and Shell Development Company. The authors acknowledge Pfizer, Abbott, Dow and Witco for supplying samples.

Literature Cited

1. Trushenski, S. P.; Dauben, D. L.; Parrish, D. R. Soc. Pet. Eng. J. 1974, 14, 633.
2. Healy, R. N.; Reed, R. L. Soc. Pet. Eng. J. 1977, 17, 129.
3. Qutubuddin, S. "Polymer-Aggregate Incompatibility, Phase Behavior and Electrophoretic Laser Light-Scattering Investigations of Microemulsions with Ultralow Interfacial Tensions", Ph.D. Thesis, Carnegie-Mellon University, Pittsburgh, 1983.
4. Qutubuddin, S.; Miller, C. A.; Fort, Jr. T. "Phase Behavior of pH-Dependent Microemulsions", paper presented at AIChE National Meeting, Houston, March 1983. J. Colloid Interface Sci. (in press).
5. Qutubuddin, S.; Berry, G.; Miller, C. A.; Fort, Jr. T. "An Electrophoretic Laser Light-Scattering Study of Microemulsions", In "Surfactants in Solution", Mittal, K. L., Ed.; Volume 3, Plenum (in press).
6. Trushenski, S. P. In "Improved Oil Recovery by Surfactant and Polymer Flooding"; Shah D. O.; Schechter, R. S.; Eds.; Academic Press, New York, 1977.

7. Szabo, M. T. Soc. Pet. Eng. J. 1979, 19, 1.
8. Hesselink, F. Th.; Faber, M. J. In "Surface Phenomena in
    Enhanced Oil Recovery"; Shah, D. O., Ed.; Plenum, 1981; pp.
    721-740.
9. Pope, G. A.; Schechter, R. S.; Wang, B. "The Effect of Several
    Polymers on the Phase Behavior of Micellar Fluids", SPE 8826,
    paper presented at the 1st Joint SPE/DOE Symp. on Enhanced Oil
    Recovery, April 1980.
10. Siano, D. B.; Bock, J. Polymer Preprints, 1981, 22(2), 61.
11. Desai, N. W.; Shah, D. O. Polymer Preprints, 1981, 22(2), 39.
12. Asakura, S.; Oosawa, F. J. Chem. Phys. 1954, 22, 1255.
13. Asakura, S.; Oosawa, F. J. Polym. Sci. 1958, 31, 183.
14. Monaghan, B. R.; White, H. L. J. Gen. Physiol. 1935, 19, 715.
15. Geoghegan, M. J.; Brian, R. C. Biochem. J. 1948, 43, 5.
16. Edmond, E.; Ogston, A. G. Biochem. J. 1968, 109, 569.
17. DeHek, H.; Vrij, A. J. Colloid Interface Sci. 1979, 70, 592.
18. Feigin, R. I.; Napper, D. H. J. Colloid Interface Sci. 1980,
    75, 525.
19. Vrij, A. Pure Appl. Chem. 1976, 48, 471.
20. DeHek, H.; Vrij, A. J. Colloid Interface Sci. 1981, 84, 409.
21. Sperry, P. R. J. Colloid Interface Sci. 1982, 87, 375.
22. Foshee, W. C.; Jennings, R. R.; West, T. J. "Preparation and
    Testing of Partially Hydrolyzed Polyacrylamide Solutions", SPE
    6202, October 1976.
23. Benton, W. J.; Natoli, J.; Mukherjee, S; Qutubuddin, S; Miller,
    C. A.; Fort, Jr. T. "Some Basic Studies of Petroleum
    Sulfonates and Pure Surfactant Systems", 5th Annual DOE Symp.
    on Enhanced Oil Recovery, Tulsa, OK, August 1979.
24. Miller, C. A.; Fort, Jr. T. "Low Interfacial Tension and
    Miscibility Studies for Surfactant Tertiary Oil Recovery
    Processes", Annual Tech. Progress Report to DOE for period Dec.
    1978 - Nov. 1979.
25. Benton, W. J.; Toor, E. W.; Miller, C. A.; Fort, Jr. T. J. de
    Physique. 1979, 40, 107.
26. Benton, W. J.; Fort, Jr. T.; Miller C. A. "Structures of
    Aqueous Solutions of Petroleum Sulfonates", SPE 7579, 53rd Ann.
    Fall Tech. Conf. and Exhib. of SPE-AIME, Houston, Tx. October
    1978.
27. Mukherjee, S. Ph.D. Thesis, Carnegie-Mellon University, 1981.
28. Benton, W. J.; Miller, C. A. "A New Optically Isotropic Phase
    in the Sodium Octanoate - Decanol - Water System".
29. Benton, W. J.; Miller, C. A. In "Surfactants in Solution",
    Mittal, K. L.; Ed.; Volume 3, Plenum (in press). J. Phys.
    Chem. 1983, 87, 4981.
30. Qutubuddin, S.; Benton, W. J.; Miller, C. A.; Fort, Jr. T.
    Polymer Preprints. 1981, 22(2), 41.
31. Benton, W. J.; Natoli, J.; Qutubuddin, S.; Mukherjee, S.;
    Miller, C. A.; Fort, Jr. T. Soc. Pet. Eng. J. 1982, 22, 53.
32. Wellington, S. L. Polymer Preprints. 1981, 22(2), 63.
33. Jamieson, A. M.; Southwick, J. G.; Blackwell, J. "Dynamical
    Properties of Xanthan Polysaccharide in Solutions", paper pre-
    sented at the 55th Colloid and Surface Science Symposium,
    Cleveland, OH, June 1981.
34. Casassa, E. F. J. Polymer Sci. Part B, 1967, 5, 773.

35.   Casassa, E. F. Tagami, Y.; Macromolecules, 1969, 2, 14.
36.   Casassa, E. F. J. Polym. Sci. Part A-2, 1972, 10, 381.
37.   Hwan, R. N.; Miller, C. A.; Fort, Jr. T. J. Colloid Interface
      Sci. 1977, 68, 221.
38.   Miller, C. A.; Neogi, P. AIChE J. 1980, 26, 212.
39.   Ruckenstein, E.; Chi, J. I. Chem Soc. Faraday Trans. II, 1975,
      71, 1690.

RECEIVED June 8, 1984

Influence of Temperature on the Structures of Inverse Nonionic Micelles and Microemulsions

J. C. RAVEY and M. BUZIER

Laboratoire de Physico-Chimie des Colloïdes, Université de Nancy I, Faculté des Sciences-1er cycle, B.P. No. 239, 54506 Vandoeuvre les Nancy, France

Just above the P.I.T., the oil-rich phase diagram of the nonionic systems (C_{12} EO_4 + decane + water) is constituted by two separated one phase realms, where the structures have been proved to be very different. If the temperature is slightly raised, the intermediate two phase domain tends to disappear : then we get a single one phase area whose delineation (the maximum surfactant/water ratio) is also temperature dependent. The morphological determinations have been performed by the small angle neutron scattering. It has been found that the structures mainly depend on the overall composition of the sample (the geometrical constraints). They may or may not exist, according to the temperature, but we always get lamellas for lowest water contents which turn into water-in oil globules.

One of the most typical characteristics of microemulsions and micelles with nonionic surfactants is their high sensitivity towards the temperature : usually a ternary mixture with a given composition of oil, water and nonionic amphiphile remains monophasic and isotropic only for a narrow temperature range. This well known fact finds expression in a great apparent changeability of the ternary phase diagrams. A pioneering attempt of systematization of the evolution of these phase diagrams was proposed some years ago (1,2). Since then a few more detailed phase behavior investigations have been performed. But they remained purely descriptive (3-5).

The phase diagrams reflect the mutual oil-water solubilization properties of the nonionic surfactants, which can be understood, and then also predicted, only if the structures of the microemulsions (or the micellar aggregates) are known with some degree of certainty. Moreover the thermodynamical explanation of these properties in terms of the hydrophile-hydrophobe forces has to be founded on clear structural evidence and this is far from being the case at the present time (6).

Therefore most interesting would be the knowledge of the correlation between the phase diagrams and the structures according to the temperature but also in relation to the chemical nature of the oil

0097-6156/85/0272-0253$06.00/0

and or surfactant molecules : for a given composition does the structure change with any change of the temperature ? And what comes from the difference in the stability of the molecular aggregates specific to two different compositions ?

The present report makes our contribution to the solution of that problem, although it concerns only one particular nonionic system studied in a small but critical range of temperature. Here we are interested in the correlations (structure-phase behavior) of the tetraethylene glycol dodecyl ether $(C_{12}(EO)_4)$, the oil being the decane and thus for temperatures just above the so-called Phase Inversion Temperature (P.I.T.), i.e. when the surfactant becomes preferentially soluble in the oil. Indeed, as far as its phase diagram is concerned, that system exhibits very attractive features between 18°C and 25°C : there is a progressive coalescence of two distinct one phase domains into only one realm. So we get series of "water in oil" systems for which the structural determinations from small angle neutron scattering measurements appear to be less difficult than for the "surfactant phase".

Now let us emphasize that we want all the conclusions to be drawn exclusively from the experimental results (i.e. the neutron scattering spectra), without making use of any theory on microemulsions. In particular, at a given temperature and for a certain overall composition, the structures must be determined independently of any hypothesis about the (oil) dilution. Quite the contrary, these structural results should be actually used to support the eventual validity of that theory or other, which can be proposed to explain the evolution of the phase diagrams.

Materials and Methods

Chemical and Phase Diagram. The nonionic surfactant was tetraethylene glycol dodecylether $C_{12}(EO)_4$. It was purchased from Nikko Chemicals (Japan). It was used without further purification (99 %), although it has been recognized that the solubilization properties of the pure surfactant may be somewhat influenced by the presence of small quantity of impurities. As a matter of fact, the presence of some traces of a polar contaminant only results in a slight temperature shift of the whole phase diagram, whose exact delineation is all the more sensitive that the oil content is very large (more than 95 %). At any rate, in the present work, we are not dealing with samples whose decane concentration exceeds 93 %. Besides all the measurements have been performed on the same batch of surfactant. The corresponding phase diagram at 20°C may not be quite identical to the one published by Friberg, it will not at all change the conclusions of the present paper, since here we are interested in the evolution of the system, and not in some absolute value of the free energy of the surfactant molecule. Anyway that "equivalence" temperature-contaminant could be explained in terms of an equivalent (but slight) modification of the packing constraints inside the micellar aggregates. Indeed, as it will be shown below, the evolution of the phase diagram corresponds to the progressive introduction of some "disorder" into the amphiphile packing.

Both water and oil were mixtures of isotopes. As a matter of fact, the "aqueous component" always contained 80 % D_2O and 20 % H_2O.

And as far as the oil was concerned, we used partially deuterated decane molecules of various grades of deuteration. The isotopic compositions of such solvents was chosen so that the variation contrast method was the most illustrative and powerful. On the other hand, we have noted that the exact delineation of the one phase domains was almost insensitive to any D/H substitution. Let us recall that the H_2O/D_2O ratio was kept constant throughout the present investigations[2] : we only made isotopic substitutions on the decane component. And since the phase diagrams is not noticeably modified, we can guess that the small change in Van der Waals forces must influence only slightly the structures of these micellar aggregates. This will be confirmed by the results of the present report.

The phase diagrams of the system decane-water-$C_{12}(EO)_4$ are shown in figure 1 for six temperatures in the range $17°-25°C$, and for oil concentration above 40 % w/w. Among the samples studied by neutron scattering only a few ones will be considered here. They have overall compositions which are represented by points A1 to A3. They concern samples with surfactant/oil ratio of 0.176 (in weight by weight). Most of the results presented here will concern the two points A2 and A3 : they correspond to the water/surfactant ratios of 0.50 and 1.17, and occupy very peculiar position in the phase diagram. At low temperatures (17°C) two distinct one phase domains exist. At about 20°C, they coalesce in the region very rich in oil : the two "wings" of this new realm remain separated by a long "finger-like" two phase area, which progressively disappears when temperature increases to 25°C ; correlatively the left most limit moves toward the oil-surfactant basis. As a result, at low temperature (i.e. 20°C) only A2 is in the two phase realm. At intermediate temperature (21°5-23°C) each sample is a one isotropic phase system (although at 21°.5C, A2 and A3 belong to different "wings" of the domain). At higher temperatures samples like A3 undergo a demixing and form emulsions. It must be emphasized that in that temperature range, a lamellar liquid crystal always exists for the lower contents of oil (less than about 35 %), as shown in figure 2 which represents a part of the phase diagram at 20°C.

Neutron Scattering Method. The measurements were carried out at the Institute Laue Langevin in Grenoble (France), using the small angle scattering instruments D11 and D17. The sample-detector distances and the wavelengths (λ) were chosen so that the scattered intensities could be measured inside the (q) scattering vector range $0.005 Å^{-1}$ to $0.2 Å^{-1}$. In fact, the most useful q range was 0.02 to 0.2 $Å^{-1}$ [$q = (4\pi/\lambda) \sin(\theta/2)$, θ means the scattering angle]. A presentation and a discussion of the method we use to analyze the experimental data have been presented elsewhere (7,8). That method can be summarized as follows : we have to find the best theoretical spectra corresponding to "water in oil" micellar aggregates which could fit the experimental spectra in the whole range $q_{min} < q < 0.15-0.2 Å^{-1}$, for which the interparticle effects are scarcely appreciable. These fits necessitate to calculate the intensities on the absolute scale, both theoretically and experimentally. Moreover for a given overall composition, oil molecules of various deuteration grades are used. Then the goal is to obtain the good absolute level of the intensities scattered in this whole q range for each isotopic mixture of the

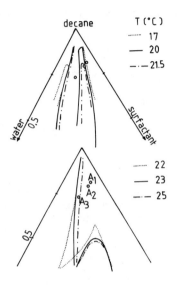

Figure 1. Part of the three phase diagrams of $[C_{12}(EO)_4$-decane-water] for 6 temperatures.

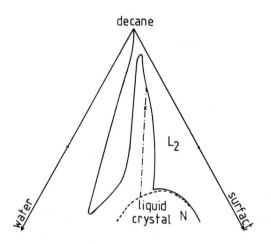

Figure 2. Part of the three-phase diagram of the $C_{12}(EO)_4$-decane-water system. L_2 : isotropic inverse micelle phase ; N : liquid crystal phase (lamellar phase).

decane component. A discussion of the limitations of that method due
to the interparticle effects, the H/D substitution, the polydispersity, the determination of the type of particles (oil in water, water
in oil or bicontinuous) can be found elsewhere (7,8)

Now, the theoretical spectra have to be calculated according to
various models, after physically relevant values have been ascribed
to the various morphological parameters. Several particle shapes have
been considered ; the most pertinent ones in the present study are :
 - The discoid, with or without a water core (figure 3a),
 - The lamella, which is a plane bilayer made of two separate
surfactant layers. A water film may also be intercalated.
 - The "closed" lamella : it is a lamella surrounded by a cylindrical shell made of pure (i.e. without water) surfactant molecules,
which isolates an eventual aqueous central film from the hydrocarbon
solvent (figure 3b).
 - What we call the "hank" is a small bilayer whose each of the
layers interpenetrate the other.

Then we have to decide on the morphological parameters, taking
into account the geometrical constraints in the aggregate whose overall composition is given. The first one is N, the number of surfactant molecules per aggregate. Next are the lengths of both the hydrophile and hydrophobe moieties of the amphiphile, since the configuration of these chains can be more or less extended. Correlatively, we
set the penetration rate of the water α, which is the number of water
molecules per surfactant inside the part of the hydrophile layer
which is effectively hydrated (see the case of the "closed" lamella,
(figure 3b). All the above parameters allow the detailed geometry of
the particle to be known, from which we can infer the rate of penetration of the oil into the hydrophobe layer and the area per polar
head σ (calculated at the hydrophobe-hydrophile interface in the
surfactant layer). Then we compute the intensity scattered by such a
particle and averaged over all the orientations of that scatterer.
The best set of the parameter values are deduced from the best fits
between theoretical and experimental values.

X-Ray Scattering.   A small angle X-ray scattering camera with a
point collimation has been used for the determination of Bragg spa-
cings of many liquid crystal samples. Temperature wa monitored between 20° and 30°C, and the samples were contained into a capillary
tube. The measurements were made by the photographic method on the
first two reflections.

Results

Low Temperature and Low Water Content.   Some results have already
been published which concerned the system at 20°C (8). It was found
that, at this low temperature, there are very small "hank-like"
aggregates in the binary decane-surfactant mixture (N is about 10)
for concentration above 5 %. Progressive addition of water promotes
the formation of larger and larger aggregates. All the water molecules are bound to the oxyethylene sites, and the "swollen micelles"
change from hank-like to lamellar particles. Further addition of
water induces a demixing, one of the phase being a liquid crystal,
as shown in figure 2. This occurs when 10 water molecules are bound

to one polyoxyethylene chain (this corresponds to a weight concentration of 6.8 % along the line of investigation Al–A3). As a matter of
fact, the structure corresponding to Al at 20°C (5 % water) has been
shown to be a lamella where the polyoxyethylene chain is in the extended conformation. The aggregation number is about 600, there are 7.8
water molecules per surfactant, and the area per polar head is
$\sigma = 37 \text{ Å}^2$. On the demixing line (6.8 % water), the same model would
lead to $\sigma = 42 \text{ Å}^2$.

At this point, we have to mention a result of the X-ray investigation of the liquid crystal phase, which exists for oil contents
less than 35 %. Although a detailed presentation of these X-ray measurements will not be shown here, we can point out that this liquid
crystal is a lamellar phase, which may be characterized by an almost
constant value of the area per polar head, whatever its composition
(9). And that value of σ is in fact 40–42 Å^2. This important result
strongly suggests that, at low temperature (20°C), the water solubilization induces the formation of bilayers the size of which increases till the area per polar head is just equal to a maximum value,
which is 40–42 Å^2 at 20°C. Let us emphasize that these swollen micelles do not contain a central film of water, unlike the lamellar phase
if α is greater than 10. Then the demixing phenomena at low temperatures may be understood as follows. All along both the isotropic and
lamellar phase domains, the structures are characterized by a constant value of σ. For a given oil/surfactant ratio, any attempt of
further water solubilization into the lamellar micelles would lead to
the creation of a central water film. But low temperature or/and high
oil content cannot allow the formation of such a film. So we get the
demixing phenomenon. Of course, one of the phase will be constituted
by the finite lamellar aggregates which refuse to accomodate more
water. The other one is the lamellar phase, where the presence of an
infinite water layer is favoured. Then each of the systems in equilibrium corresponds to different water/surfactant ratios, since the
obtention of a lamellar phase with a minimum water layer thickness
(about 5 Å) necessitates the transport of some water from the isotropic phase to the crystalline one. That structural explanation is quite consistent with the delineation of such phase diagrams, and also
with the tie-lines shown in figure 3, or concerning other analogous
systems (10,11).

Yet a water core can exist in swollen micelles or microemulsions
if we make the conditions changed as proved below.

Low Temperature and Higher Water Content.   A few neutron scattering
spectra are shown in figure 4. Experimental (absolute) intensities
are represented by open circles, and concern the sample A3 at 20°C.
Four deuteration rates (0 %, 40 %, 50 % and 80 % in w/w) of the decane component have been used, which give rise to rather different
spectra, both in level of intensity and in shape. The first maximum
is due to interparticle effects which can be roughly taken into account by the hard equivalent sphere model (7,8). Here the scattering
curves become practically independent of the interparticle interac-
tions for $q > q_{min} \simeq 0.03 \text{ Å}^{-1}$. Two theoretical results are shown : while
for some "contrasts" (i.e. oil deuteration rate) the curves only de-
pend on the aggregation number (0 % of deuterated decane), for others,
they may be very dependent of the exact morphology of the micellar

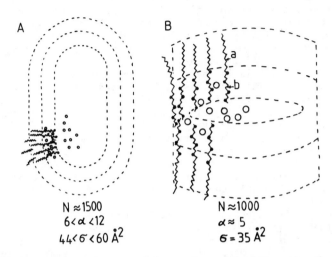

Figure 3. Two particle models used to calculate the theoretical scattered intensity. (A) = the discoid ; (B) = the closed lamella ; (**o**) : water molecule ; (**b**) : oxyethylene chain ; (**a**) : hydrocarbon chain.

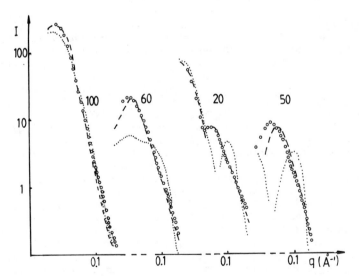

Figure 4. Comparison of the experimental results with the theoretical curves for four contrast values (100, 60, 20 and 50 % of $C_{10}H_{22}$).[% surfactant/ % oil = 0.176, % water = 15]. (**o**) : experimental points. Theoretical curves of the intensity scattered by : (.....) a closed lamella (N = 1500, α = 6, ellipticity = 0.3) ;(— —) a spheroid (N = 1500, α = 6, ellipticity = 0.5).

aggregate (which in the present case can be called a microemulsion, given the amount of water solubilized). As apparent on all the four curves of figure 4, the best fit between theoretical and experimental results is obtained for a globular particle (like the discoid of figure 3a). The surface at the hydrophile-hydrophobe interface is an oblate spheroid whose mean ellipticity is about 0.5. On this surface, σ has a value of 44 $\mathring{A}^2$; 6.5 water molecules are bound to one oxyethylene chain which is in the extended configuration, unlike the hydrophobic chain. This last result is deduced from I(q) for the largest q values : the effective length of the C_{12} chain is probably between L (the zig-zag length) and L/2 ; some incertitude remains due to the uncertainty on these very low I(q) values. Such results are quite in agreement with those previously published (7). As a matter of fact, a further addition of water leads to some increase of N (N $\sim$ 2000), of σ ($\sigma \sim$ 45 to 60 $\mathring{A}^2$) and of α ($\alpha_{max} \sim$ 12-13), till the demixing occurs (in fact, we get emulsions). All other morphological parameters remain nearly constant, in particular the main non spherical shape.

Although the aggregation number N cannot be determined to a high degree of precision because of the interparticle effects, there is no doubt about the type of shape, the rate of hydratation and the conformation of the $(EO)_4$ chain. This is quite apparent in figure 4, where are also drawn the theoretical curves corresponding to the lamellar micelles. We acknowledge that a certain small polydispersity may exist, but it cannot change our main conclusions : indeed the almost perfect fitting of all the four rather different curves by the same model of particle is far from being trivial and prompts us to have confidence in our results.

Higher Temperature and Lower Water Content. For temperatures above 21.°5C, the sample A2 (7 % water) becomes a one isotropic phase system. Most important is the determination of the corresponding structure, since it reflects the direct influence of the hydrophile-hydrophobe forces on the formation of the microemulsions from the swollen lamellar micelles. Here too are shown I(q) spectra for four oil scattering lengths in figure 5. They constitute a very impressive set of fits. Although the use of purely hydrogenated decane allows only the aggregation number to be known whatever the exact morphology of the particles (N $\sim$ 1000), a clear distinction between the possible shapes can be made if the deuteration rate of the oil is 30 % or 40 %. The best fits can be obtained only if we use the model of the "closed" lamella containing a relatively thick central water layer. Indeed the neutron results prove that the EO hydration rate can be only $\alpha = 6$, instead of $\alpha \sim 10.2$ if there were no water film. Correlatively, the area per polar head reduces to $\sigma \sim 35$ $\mathring{A}^2$, which is very like the value of σ corresponding to the sample A1.

On the other hand, the amount of water inside the film is not sufficiently large to induce the formation of globules with a water core : this is best seen from the curve which corresponds to the 40% deuteration rate. Thus when the temperature is raised of about only 2°C, combined hydrophobe-hydrophile forces tend to favour the existence of a water layer, but the amphiphiles keep a dense packing : the temperature change is too small to induce in the surfactant film the disorder necessary for a globular shape. Moreover that water

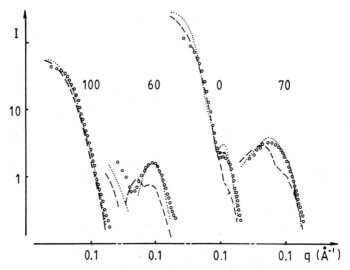

Figure 5.   Comparison of the experimental results with the theoretical curves for four contrast values (100, 60, 0 and 70 % of
$C_{12}H_{22}$). [% surfactant/% oil = 0.176, % water = 7]. (●) : experimental points. Theoretical curves of the intensity scattered
by : (— —) a discoid (N = 1000, α = 6, ellipticity = 0.24) ;
(.....) a lamella (N = 1000, α = 10, ellipticity : 0.17).

layer must be protected from any contact with hydrocarbon. So this lamellar aggregate has to be surrounded by a cylindrical shell made of non hydrated surfactant molecules. A value of 4 Å has been ascribed to the thickness of that shell, this particular choice being not all critical. And here also the $(EO)_4$ chain is in the extended conformation.

Conclusion

The previous direct analysis of neutron scattering data has allowed several structures to be determined according to the temperature and the composition of the system. No major hypothesis was necessary. On the contrary, this set of experimental results would provide some ground for a theoretical and thermodynamical explanation of the evolution swollen micelle-microemulsion. Indeed each type of structure seems to reflect a domination of one or other component of the free energy of these nonionics at room temperature. Although a calculation and a discussion of these energy effects are well beyond the scope of the present paper, we can point out the importance of the forces specific to the hydrocarbon chain and to the oil beside the pure hydration forces. Van der Waals forces would favour the formation of a water layer, while entropic effects seem very important as far as the transitions hank-lamella and lamella-globule are concerned. These effects due to the solvent concentration (but also to the nature of the oil (2,5) are quite evident from the fine evolution of the phase diagrams, especially for water/surfactant ratios in the range 0.5-1.2.

As a summary, the change of the structure due to a small increase of temperature is the formation of a water core resulting from a partial dehydration of the oxyethylene sites. Coorelatively the addition of water promotes the transition between lamellar aggregate and globular microemulsion, till the area per polar head reaches a maximum value which is temperature dependent.

Literature Cited

1. Friberg, S.E.; Lapczynska, I. Progr. Colloid Polym. Sci.,1975, 56, 16.
2. Friberg, S.E.; Buraczewska, I.; Ravey, J.C. "Micellization, Solubilization and Microemulsions", K. Mittal Ed., (1977), vol.11, p.901,Plenum Press, New York.
3. Kunieda, H.; Friberg, S.E. Bull. Chem. Soc.Jap., 1982, 54, 1010.
4. Buzier, M.; RAVEY, J.C. J. Colloïd Interface Sci., 1983, 91, 20.
5. Bostock, T.A.; McDonald, M.P.; Tiddy, G.J.T.; Waring, L. SCI Chemical Society Symposium in Surfactant Agents, Nottingham 1979.
6. Nakamura, M.; Bertrand, G.L.; Friberg, S.E. J. Colloïd Interface Sci., 1983, 91, 516.
7. Ravey, J.C.; Buzier, M. to be published in the Proceedings of the"International Symposium on Surfactants in Solutions", Lund, Sweden, 1982.
8. Ravey, J.C.; Buzier, M.; Picot, C. J. Colloïd Interface Sci. 1984, 97, 9.

9.   Moucharafieh, N.; Friberg, S.E.; Larsen, D.W. Mol. Cryst. Liq.
      Cryst., 1979, 53, 189-206.
10.   Friberg, S.E.; Mandell, L.; Fontell, K. Acta Chem. Scand.,
      1969, 23, 1055.
11.   Friberg, S.E., Flaim, T. ACS Symposium Series n° 177, "Inorga-
      nic Reactions in Organized Media", S.L. Holt Ed. ACS (1982).

RECEIVED June 8, 1984

Sulfones as Unconventional Cosurfactants in Microemulsions

REGINALD P. SEIDERS

Chemical Research and Development Center, Aberdeen Proving Ground, MD 21010

We report here on the use of alkyl sulfones as novel
unconventional cosurfactants in CTAB-stabilized microemulsions. Sulfones, being fully oxidized at sulfur,
have good stability to oxidants such as hypohalite.
The sulfones, sulfolane and 3-methylsulfolane, are
shown to function quite well, as cosurfactants with
CTAB, in the solubilization of both organophosphorus
esters and betahalosulfides. For the organophosphate
used, tributylphosphate, it is shown through pseudo-
three-component phase diagrams that the sulfone
functions as effectively as the alcohol in its role
of cosurfactant. Solubilization of chloroethyl
ethyl sulfide is less effective when the sulfone
cosurfactant is used, but is still a dramatic enhancement over its solubility in water alone. The effect
of added salt on the solubilization is reported, as
well as the effect of changes in the surfactant-
cosurfactant ratio. Preliminary quasielastic light-
scattering measurements are also reported for these
unconventional systems.

Surfactant aggregates (microemulsions, micelles, monolayers, vesicles,
and liquid crystals) are recently the subject of extensive basic
and applied research, because of their inherently interesting
chemistry, as well as their diverse technical applications in
such fields as petroleum, agriculture, pharmaceuticals, and
detergents. Some of the important systems which these aggregates
may model are enzyme catalysis, membrane transport, and drug
delivery. More practical uses for them are enhanced tertiary
oil recovery, emulsion polymerization, and solubilization and
detoxification of pesticides and other toxic organic chemicals.
   Both micellar and microemulsion media have been investigated
as candidates for solubilization and detoxification of nerve
agents. Normal micelles, however, do not appear to bind Sarin
well (1) and oximate functionalized surfactants are severely
limited by their low water solubility (2). These oximate
functionalized surfactants have been shown to be excellent

catalysts for decontamination, but the low solubility would
require the use of logistically burdensome volumes of water
in the field of accomplish dissolution. Microemulsions on the
other hand appear considerably more promising because of their
greatly enhanced solubilizing power. As much as forty to fifty
percent oil can be solubilized by microemulsions in contrast
to the generally less than fifteen percent achievable in micelles.
Much of the pioneering work related to decontamination reactions
in oil-in-water (o/w) microemulsions has been carried out by
Mackay. He has studied, for example, the hydrolysis of phosphate
esters (3, 4), metallation of tetraphenylporphine (5) and
alkylation of substituted pyridines (6).

The research summarized herein has also been directed primarily
at o/w microemulsions because of their potential as improved
aqueous-based decontamination formulations. One of the most
important properties of microemulsions, from a decontamination
viewpoint, is their ability to solubilize large quantities of
hydrophobic materials, i.e. chemical agents. Detoxification
reactions of these solutes with aqueous reagents should thereby
be facilitated. In order to be useful as an oxidative formulation
for certain types of agents a modification of the conventional
types of microemulsion components was deemed necessary. Specifically,
it was decided that the cosurfactant (a low molecular weight
normal alcohol) would have to be eliminated or modified in the
formulation. This decision was based on the susceptibility
of linear alcohols to attack by typical oxidants such as hypohalite.
Dialkyl sulfones were chosen for investigation as possible sub-
stitutes for the alcohol. This report describes our early
research effects which have focused on the enhanced solubilization
effectiveness of microemulsions with the ultimate goal of developing such or system as a possible universal aqueous based decontamination formulation for military use. Pseudo-ternary phase
diagrams for three alcohol-free systems are presented, and the
solubilization of the agent simulants is discussed.

Methods and Materials

All chemicals were obtained reagent grade from Aldrich Chemical
Company. The liquids were passed through a column of Woelm
Act. I alumina for purification, and were subsequently dried
and storaged over 4Å molecular sieves. The cetyltrimethylammonium bromide (CTAB) was twice recrystallized from acetone.

The phase diagrams were prepared at room temperature by
the usual method, where a weighed aliquot of the surfactant/
cosurfactant mixture (E) was diluted with known amounts of water
(W) and then titrated with oil (O) to turbid and clear endpoints.
Alternatively, the dilution of E could be made with oil and
titration with the water. Generally, fifteen to thirty titrations were sufficient to roughly outline the phase maps. Solubility
limits were also determined by titration of the solvent with
the solute (or solute solution) to a cloudy endpoint.

Quasielastic light-scattering (QLS) measurements were carried
out on a fixed angle (90°) laser light-scattering photometer
(Model LSA-1) from Langley-Ford Instruments. The LSA-1 was
modified with a Spectra-Physics 15 mW, 632.8 nm HeNe laser source,

and contained a scattered light collection aperture which defined
approximately one coherence area. The autocorrelation functions
were collected with a Langley-Ford Model 1096 computing correlator.
The analysis program employs the method of cumulants (7) to
fit the correlation function to a single exponential decay directly
in the correlator itself, without the use of a separate external
computer. Calculations are then performed using the results
of the cumulants analysis to yield the average translational
diffusion coefficient and subsequently, from the Stokes-Einstein
equation, the effective diameter of the scattering particles.
The continuous phase was assumed to have the viscosity and refractive
index of pure water at 25°C unless otherwise indicated. Samples
were filtered through a 0.45 um filter directly into the cuvet
to minimize dust incorporation. The samples were then maintained
at 25°+0.1°C unless otherwise stated by means of water circulation
in the cuvet cell block from a constant-temperature bath.

Results and Discussion

Conventional Microemulsions. The pseudo-ternary phase map for
a conventional microemulsion system consisting of CTAB, pentanol,
hexadecane, and water is shown in Figure 1. A conventional
system is defined here as one containing a typical protic cosurfactant. The dramatic enhancement of hexadecane solubilization
is graphically demonstrated by the very large clear region labelled
"I" in this Figure. This single-phase region may be classified
as the bicontinuous type since it stretches from near the W
apex, where aggregates (if they exist) must be in a water-continuous
environment, to the vicinity of the O apex where the system
must be oil continuous. If aggregation exists in the intermediate
regions then at some point inversion between o/w and w/o microemulsions must occur. It is expected that this change in microstructure may have an effect on reactions which may be conducted
in these media (8). However, whatever the structure may be,
it is clear that high solubilization of hexadecane can be achieved
in this system and in other related systems (3). It should
of course be pointed out that the high solubilization of hexadecane
does not necessarily imply similar solubilization of other organic
chemicals, since it is well known that the phase diagram is
often strongly influenced by the type of oil that is solubilized,
as will be shown later in this report.

    Another region of the diagram in Figure 1 that should be
mentioned is labelled "L" near the 60-70% E-W axis. This region
has been observed by others, and is a clear, highly viscous
gel which has liquid crystalline characteristics under crossed
polarized light. In this L region the viscosity decreased drama-
tically on addition to just two to four drops of the oil component.
The highest viscosity is on the E-W line and near the 70% water
area.

Alcohol-Free Microemulsions. Three alcohol-free systems were
prepared in order to determine their solubilization characteristics.
Friberg (9) has suggested that a protic solvent (alcohol, amine,
acid, etc.) is necessary for microemulsion formation, while

Graetzel (10) has suggested that almost any polar solvent could be used. It was decided that sulfones would be studied initially because of their ready availability, stability to oxidants, and ability to form hydrogen bonds to water as evidenced by their notorious hygroscopic nature.

The Sulfolane/CTAB/Water/TBP System. Sulfolane, or tetramethylene sulfone is a high-boiling (130°C/6.5 mm), non-toxic, inexpensive industrial solvent (11) which was used with an equal weight of CTAB to make the emulsifier E. It was necessary to add water to this composition to prevent the CTAB from precipitating. The percent composition due to this water is indicated by the dashed line of Figure 2. Tributylphosphate (TBP) was chosen as the oil phase because it had already (6) been utilized as a nerve agent simulant in a conventional microemulsion system. This prior work indicated that TBP was solubilized very effectively by a CTAB/Butanol/water system to give a spacious clear region very similar to that shown in our system (Figure 2).

As the pseudo-ternary phase map indicates, the sulfolane-containing system used in this work solubilizes the organophosphate TBP very effectively. From the 50% point on the E-W axis, over 60% oil can be added while a clear fluid solution is maintained! Even at 80% water it is possible to add nearly 20% oil before turbidity and subsequent phase separation occurs. It is not known whether a true microemulsion exists in this clear region. However this enhanced solubilization of TBP is not simply a co-solvent effect of the sulfolane and water. This was confirmed by determining the maximum solulbility (6.7%) of TBP in a 2:1 (w/w) mixture of sulfolane and water, respectively.

The Sulfolane/CTAB/Water/CEES System. In contrast to the extremely large clear region with TBP as the oil in Figure 2, the solubilization of the mustard simulant 2-chloroethyl ethyl sulfide (CEES) was less impressive, though still rather dramatic at low water compositions in the clear area of Figure 3. The isotropic clear area here is considerably smaller than it was in the two cases just discussed. This smaller area, however, is not atypical of conventional microemulsions and indeed is larger than, for instance, The Tween 40/pentanol/mineral oil/water system described by Mackay (12). From Figure 3 it can be seen that from the 50% water point on the E-O axis, ca 15% CEES can be solubilized into a homogeneous solution. As the water content increases, solubilization steadily decreases to roughly 2% at 80% water. Although this solubility seems rather low, it is still a twenty-fold increase over the CEES solubility in water only, assuming CEES solubility in water is similar to that of bis(2-chloroethyl) sulfide (0.92 mg/ml) (13). Again, the enhanced solubilization in the sulfolane microemulsion was shown to be far above the co-solvent effect of sulfolane and water. It was found that CEES solubility in sulfolane/water solutions was 2.1% at 40% water and only 1% at 50% water.

Because CEES hydrolyzes at a relatively rapid rate, the microemulsions containing this simulant are very dynamic systems. If the flasks containing turbid solutions, after endpoint determination, were allowed to stand for 20 to 30 min, they would slowly clarify and could then be titrated with more CEES. It is

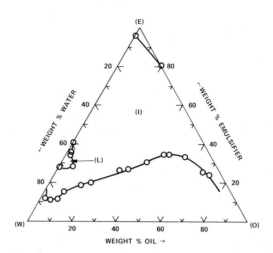

Figure 1. Phase map of the Pentanol/CTAB/Water/Hexadecane System.
The emulsifier (E) [40% CTAB, 60% Pentanol, w/w] plus water (W)
plus Hexadecane oil (O) = 100% by weight. The clear single phase
region is denoted by (I).

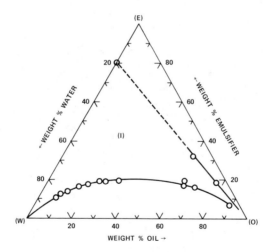

Figure 2. Phase map of the Sulfolane/CTAB/Water/Tributyl-
phosphate (TBP) System. The emulsifier (E) [50% CTAB, 50%
Sulfolane, w/w] plus water (W) plus TBP as the oil (O) = 100%
by weight. The clear single phase region is denoted by (I).

proposed that hydrolysis of CEES is occurring to produce 3-thia-
pentanol which is then acting as a conventional cosurfactant, thereby
enhancing the solubilization effectiveness of the microemulsion.
The time frame is reasonable since the half-life of CEES hydrolysis
is 1 min at 25°C in water (14). This alcohol-formation process
should also be seen with mustard which has a half-life of 4 min at
room temperature (15).

The 3-Methylsulfolane/CTAB/Water/CEES System. A serious problem
with any sulfolane containing microemulsion is the high freezing
point (28°C) of sulfolane which would severely limit use of such
formulations at winter temperatures. A derivative which has
similar aprotic solvent properties as sulfolane, but a much lower
melting point is 3-methylsulfolane. Figure 4 shows the pseudoternary phase diagram that was obtained with 3-methylsulfolane in
place of sulfolane in the emulsifier. It is readily apparent that
the solubilization effectiveness is quite similar for the two
sulfolane systems with CEES. In this present case, from the 50%
water point on the E-W axis it is possible to solubilize ca. 13%
CEES, while at 80% water the amount is down to 2% again.

QLS measurements were attempted in the two sulfolane/CEES
systems at high water compositions of the single phase regions in
Figures 3 and 4. Preliminary experiments give flat autocorrelation
functions that are observed in true molecular solutions. This
implies that there are no aggregates in these systems. QLS data
are sometimes difficult to interpret however, and it is possible
that addition of a salt to screen the charged droplets will clarify
the situation. QLS investigations are continuing in these systems
and will be reported later.

Finally, in two other brief experiments, the effect of seawater
and emulsifier composition on the solubilization of CEES was studied.

Table I summarizes the solubility of CEES in solutions of
emulsifier E (50% CTAB, 50% 3-methylsulfolane) and artifical seawater. Over the compositions investigated (29.4% W, 70.6% E to 52%
W, 48% E) it can be seen, by comparison of the data in Table I with
Figure 4, that the seawater appears to decrease the CEES solubility
1-2%, but this is within the experimental error associated with the
titrations.

Table I. Effect of Seawater on CEES Solubilization

Wt% Seawater	Wt% CEES
29.4	21
33.3	19
40.0	15
52.0	11

Table II tabulates the variation in CEES solubility with
variation in cosurfactant/surfactant ratio from 0.67:1 to 2.33:1.
It is clear from the data that in this system the ratio of cosurfactant to surfactant is not critical for solubilization

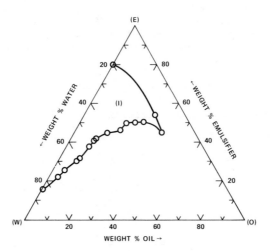

Figure 3. Phase map of the Sulfolane/CTAB/Water/CEES System.
The emulsifier (E) [50% CTAB, 50% Sulfolane, w/w] plus water (W)
plus CEES as the oil (O) = 100% by weight. The clear single
phase region is denoted by (I).

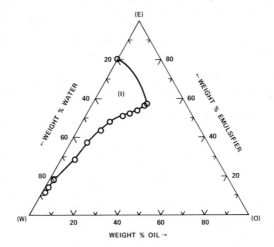

Figure 4. Phase map of the 3-Methylsulfolane/CTAB/Water/CEES
System. The emulsifier (E) [50% CTAB, 50% 3-Methylsulfolane,
w/w] plus water (W) plus chloroethyl ethyl sulfide (CEES) as the
oil (O) = 100% by weight. The clear single phase region is
denoted by (I).

effectiveness. However, it is expected that this ratio will strongly affect viscosity, thermal stability, and other physical properites of the formulation (16, 17). Furthermore, this lack of dependence on the emulsifier composition ratio contrasts with what is often found in conventional systems such as those studied so carefully by Friberg et al. (18).

Table II. Effect of Emulsifier Composition on CEES Solubilization

COS	COS/S	Wt% CEES
Methylsulfolane	2/3	20
	1/1	18
	0.9/1	17
	3/2	18
	2.33/1	16
Pentanol	3/2	36
Butanol	3/2	52

Conclusions.

It has been shown that the organophosphate ester tributylphosphate is very effectively solubilized by a microemulsion system incorporating CTAB and sulfolane as the emulsifier. The single-phase region in this system is similar to that obtained with CTAB and butanol as the emulsifier reported by Mackay (6). Thus it is expected that sulfone-containing microemulsions will solubilize other organophosphates and phosphonates such as pesticides and nerve agents.

Two sulfone containing microemulsion systems were shown to solubilize CEES fairly effectively, thereby implying that these systems might also be useful for solubilization of mustard-type compound. This good solubilization of sulfides, coupled with the expected stability of the sulfone cosurfactant toward oxidants, suggests that these microemulsions should be suitable media for investigation of oxidative decontamination reactions. Further characterization of these microemulsions by QLS is in progress. Small-angle neutron-scattering investigations of these systems is also planned in an attempt to confirm that the sulfolane is located in the microemulsion droplet, and is not simply dispersed throughout the continuous plase.

In the methylsulfolane-containing system, it was found that neither incorporation of salts (as seawater) nor alteration of the cosurfactant/surfactant ratio seriously altered the solubilization of chlorethyl ethyl sulfide. Thus, it seems plausible that low concentrations of inorganic buffers and/or oxidants should not seriously degrade the solubilization effectiveness of microemulsions containing sulfones as cosurfactants.

Acknowledgments.

The author wishes to acknowledge very helpful discussions with Dr. George T. Davis and Mr. Joseph W. Hovanec.

Literature Cited

1. Epstein, J., Kaminski, J. J., Bodor, N., Enever, R., Sowa, R,;
    Higuchi, T. J. Org. Chem. 1978, 34, 2816.
2. Bunton, C. A.; Hamed, F.; Romsted, L. S. Tet. Lett. 1980, 21,
    1217.
3. Hermansky, C.; Mackay, R. A. "Reactions in Microemulsions:
    Phosphate Ester Hydrolysis"; SOLUTION CHEMISTRY OF SURFACTANTS,
    Vol. I., Mittal, K. L., ed. Plenum Press: New York, 1979.
4. Mackay, R. A.; Hermansky, C. J. Phys. Chem. 1981, 85, 739.
5. Mackay, R. A.; Dixit, N. S.; Agarwal, R. In "Inorganic Reactions
    in Organized Media", American Chemical Society Symposium Series
    No. 177, Holt, S., Ed.; Washington, DC, 1982, p. 179-194.
6. "Physicochemical Studies of Solutes in Microemulsions", US Army
    Research Office, June 1979.
7. Chu, B "Laser Light Scattering", Academic Press: New York,
    1974.
8. Shah, D. O.; Manohar, C.; Leung, R. "Microemulsions, Their
    Microstructure and Potential Applications", paper presented at
    the American Oil Chemists' Society Annual Meeting, Chicago,
    Illinois, May 8-12, 1983.
9. Friberg, S. "Emulsions and Microemulsions", AMERICAN CHEMICAL
    SOCIETY SHORT COURSE PRESENTATION, Atlanta, Georgia, April 1981.
10. Graetzel, M., personal communication.
11. Fieser, L. F.; Fieser, M. "Reagents for Organic Synthesis",
    Vol. I., Wiley and Sons Inc.: New York, 1967, p. 1144.
12. Mackay, R. A.; Jacobson, K.; Tourian, J. J. Colloid Interface
    Sci. 1980, 76, 515.
13. Herriott, R. M. J. Gen. Physiol. 1947, 30, 449.
14. Stein, W. H. In "Summary Technical Report of Division 9", NRDC
    Vol. L, Part III, Washington, D.C., 1946, p. 426 and reference
    41d therein.
15. Bartlett, P. D.; Swain, G. C. J. Amer. Chem. Soc. 1949, 71,
    1406.
16. Shah, D. O.; Walker, R. D.; Hsieh, W. C.; Shah, N. J.; Dwivedi,
    S.; Nelander, R.; Pepinski, R.; Deamer, D. W. "Improved Oil
    Recovery Symposium", SOCIETY OF PETROLEUM ENGINEERS OF AIME
    (preprint SPE 5815), Tulsa, Oklahoma, 1976.
17. Shah, D. O.; Bansal, V. K.; Chan, K.; Hsiek, W. C. In "Im-
    proved Oil Recovery by Surfactant and Polymer Flooding", Shah,
    D. O.; Schecter, R. S., Eds., Academic Press: New York, 1981,
    p. 293-337.
18. Friberg, S.; Flaim, T. In "Inorganic Reactions in Organized
    Media", ACS SYMPOSIUM SERIES 177, Holt, S., Ed., American
    Chemical Society: Washington, D.C., 1982, p. 1-17.

RECEIVED June 8, 1984

Transport Properties of Oil-in-Water Microemulsions

KENNETH R. FOSTER[1], ERIK CHEEVER[1], JONATHAN B. LEONARD[1],
FRANK D. BLUM[2], and RAYMOND A. MACKAY[2,3]

[1]Department of Bioengineering, University of Pennsylvania, Philadelphia, PA 19104
[2]Department of Chemistry, Drexel University, Philadelphia, PA 19104

We have studied a variety of transport properties of
several series of O/W microemulsions containing the
nonionic surfactant Tween 60 (ATLAS tradename) and n-
pentanol as cosurfactant. Measurements include dielectric relaxation (from 1 MHz to 15.4 GHz), electrical
conductivity in the presence of added electrolyte,
thermal conductivity, and water self-diffusion coefficient (using pulsed NMR techniques). In addition,
similar transport measurements have been performed on
concentrated aqueous solutions of poly(ethylene oxide)
(PEO), which has the same hydrophilic group as that on
the surfactant and is responsible for stabilizing the
microemulsions. Some, but not all, of these transport
properties are significantly lower than expected from
the Maxwell or Hanai mixture theories. The dielectric
relaxation at microwave frequencies (which reflects
dipolar relaxation of the water) shows a relaxation
time that is significantly longer than that of pure
liquid water, with evidently a broad distribution of
relaxation times present. It appears that these
changes in large part arise from hydration phenomena,
and therefore can be used to study hydration effects
in these systems.

The transport properties of microemulsions are of great interest
both for the information they provide about the physical properties
of the systems, and in industrial applications of these materials.
The transport of matter or energy through oil in water (O/W) microemulsions is determined both by the volume fraction and geometry of
the oil and emulsifier microdroplets (the "structure effect") and by
possible modifications in the transport properties of the continuous
water phase by its interaction with the hydrophilic groups in the
surfactant and cosurfactant that stabilize the microemulsion (the
"hydration effect"). Through the use of appropriate mixture theories, these two effects can in part be separated.

[3]Current address: Chemical Research and Development Center, Aberdeen Proving Grounds,
MD 21005

In previous studies, we examined the dielectric properties
of several ionic and nonionic O/W microemulsions at radio through
microwave frequencies. The dielectric relaxation at microwave
frequencies is due to dipolar relaxation of the water which exhibits, in the pure liquid, a center relaxation frequency of 20 GHz
corresponding to a dipolar relaxation time of 8 ps. In the microemulsions the average dielectric relaxation time of the water is
increased by a factor of 2-10, depending on the total water content
of the system. Moreover, these systems display a distribution of
relaxation times, that suggests the presence of a fraction of water
with dielectric relaxation time 5-10 times that of bulk water, with
the remaining water having the same dielectric relaxation properties
as the pure liquid. These changes parallel those observed in the
conductivity at audio frequencies in a variety of microemulsions
with nonionic surfactant and cosurfactants containing dilute electrolyte as the continuous phase (3). We suggested that other transport properties should show similar changes (2).

In the present study, we have examined other transport properties of O/W microemulsions containing the nonionic surfactant Tween
60 whose dielectric and conductivity properties have been previously characterized. We have chosen properties (water self-diffusion,
ionic conductivity at low frequencies, and thermal conductivity)
that can be analyzed using the same mixture theory, and which therefore can be compared in a consistent way. Limited transport data
are presented from other microemulsions as well.

Materials and Methods

The transport properties were measured at 25°C using a variety of
techniques. An extensive analysis of the dielectric relaxation
measurements is provided by Epstein (4). The other methods are
conventional and will only be summarized briefly:

Ionic Conductivity. The electrical conductivity measurements were
performed using a Hewlett Packard model 4192 impedance analyzer
under computer control, using a conductance cell similar to that
described by Pauly and Schwan (5). The conductivity measurements
were essentially constant between 1-100 kHz, ruling out electrode
polarization or other artifacts. In O/W microemulsions, no appreciable dielectric relaxation effects are expected or observed below
1 GHz (1).

Water Self-Diffusion. The self-diffusion coefficient of the water
was measured using a JEOL FX900 Fourier transform NMR spectrometer
operating at 90 MHz for protons. The pulsed field gradient technique was employed using the homospoil coils to establish the field
gradient, similar to that described by Stilbs (6). The self-diffusion coefficients of the water protons were corrected for exchange
with the hydroxyl protons of the alcohols, but this correction was
insignificant except when the fraction of water was very low.

Thermal Conductivity. The thermal conductivity of the samples was
measured using a thermistor probe technique similar to that described by Balasubramaniam and Bowman (7). A small thermistor head

was immersed in the sample and subjected to a step temperature increase of 2°C within a few milliseconds; the voltage across the
thermistor was subsequently sampled at 25 Hz with 12 bit resolution
by computer. A linear regression of the power dissipated in the
thermistor vs. the inverse square root of time yields the steady
state power dissipation whose inverse is a linear function of the
inverse of the thermal conductivity of the sample.

The microemulsions were identical to those examined in our
previous dielectric studies (1,2). The emulsifier (Tween and n-
pentanol) was mixed with the oil (hexadecane) in the proportion (by
weight) 0.594 Tween:0.306 1-pentanol:0.100 hexadecane. Water or
electrolyte (0.1 N NaCl for the conductivity measurements) was then
added to give the final composition. The physical characterization
of this system is given in References 5,11,13, and 14 of our previous paper (1). In the discussion to follow, the compositional
phase volume p is the volume fraction of oil and emulsifier, and
is equal to $(1 - wg)$, where w and g are the weight fractions of
water and the specific gravity of the microemulsion, respectively.
We will also consider the apparent phase volume p' which is calculated from the mixture theories as the total volume fraction of the
microemulsion that is excluded from the transport. Assuming that
the transport property of the hydration water is negligible compared
to that of the bulk liquid, p' would include the hydration water as
well as the oil and emulsifier.

Theoretical

The basis for interpreting the transport data is mixture theory,
which relates the transport properties of the bulk suspension to
those of the continuous and dispersed phases. Of the many mixture
relations that have been proposed, we employ those of Maxwell and
Hanai (Equations 1 and 2, respectively):

$$\frac{\Lambda_m}{\Lambda_w} = \frac{2\Lambda_w + \Lambda_o - 2p(\Lambda_w - \Lambda_o)}{2\Lambda_w + \Lambda_o + p(\Lambda_w - \Lambda_o)} \qquad \text{(Maxwell)} \qquad (1)$$

$$\frac{\Lambda_m - \Lambda_o}{\Lambda_w - \Lambda_o} \left(\frac{\Lambda_w}{\Lambda_m}\right)^{1/3} = 1 - p \qquad \text{(Hanai)} \qquad (2)$$

In these relations, Λ is the conductivity of the suspension, and
the subscripts m, o, and w refer to the microemulsion, oil and
emulsifier combined, and water. The Hanai expression can be considered to be an extension of the Maxwell theory that more con-
sistently accounts for the presence of neighboring particles (8);
for the O/W microemulsions considered here, the predictions of the
Maxwell and Hanai formulas (as well as various other mixture theories) are not greatly different. Moreover, while these theories
were developed for suspensions of spherical particles, the predictions of the mixture theories are not expected to vary greatly with
the geometry of the dispersed particles, provided that the droplets
are prolate or oblate ellipsoids whose axial ratios are not greatly

different from 1 (9). Consequently, our choice of these particular
mixture theories, while somewhat arbitrary, is not crucial in the
interpretation of the present results.

The mixture theories were originally developed for dielectric properties, but can be applied to other properties that are
governed on a macroscopic level by Laplace's equation (10). Consequently, the generalized conductivity in the above equations can
be the ionic conductivity σ, thermal conductivity K, or complex
electrical conductivity σ^* defined by:

$$\sigma^* = \sigma + j\omega\varepsilon\varepsilon_r$$

where ε is the dielectric permittivity, ω is the frequency in
radians/sec, and ε_r is the permittivity of free space (a constant,
8.85×10^{-14} F/cm). For the self-diffusion measurements, the
generalized conductivity would be the quantity (1-p)D, where D is
the self-diffusion coefficient that is measured by the pulsed NMR
technique, and p is the volume fraction of oil and emulsifier. The
need for the additional factor (1-p) arises from the fact that the
NMR technique measures the mean-square displacement of the water
molecules in the aqueous phase, while the true self-diffusion coefficient is defined (by Fick's law) as the total flux through the
entire volume of the solution (11).

Results

Figure 1 shows the dielectric relaxation properties of the Tween
microemulsions plotted on the complex permittivity plane (from
Foster et al (1). The mean relaxation frequency (corresponding to
the peak of each semicircle) decreases gradually from 20 GHz for
pure water at 25°C to ca. 2 GHz for a concentrated microemulsion
containing 20% water. Since the permittivity of the suspended oil/
emulsifier is 6 or less at frequencies above 1 GHz, this relaxation
principally arises from the dipolar relaxation of the water in the
system. Therefore, the data shown in Figure 1 clearly show that the
dielectric relaxation times of the water in the microemulsions are
slower on the average than those of the pure liquid. The depressed
semicircles indicate a distribution of relaxation times (9), and
were analyzed assuming the presence of two water components (free
and hydration) in our previous studies.

Figures 2-4 show the thermal and ionic conductivity, and water
self-diffusion coefficient measured in these same systems. Also
shown are the transport properties of PEO solutions of molecular
weights ranging from 200 to 14,000 (12). The predictions of the
Hanai and Maxwell relations are indicated, which were calculated on
the assumption that the ionic conductivity or self-diffusion coefficient of the water or suspending electrolyte is equal to that of
the pure liquid and that of the oil and emulsifier combined is zero.
Also shown are similar results from the PEO solutions of various
molecular weights. The thermal conductivity of the microemulsions
and PEO solutions are shown in separate figures because the limiting
thermal conductivity at zero water content is slightly different
(0.27 times that of water for the microemulsion, vs. 0.31 for the
PEO).

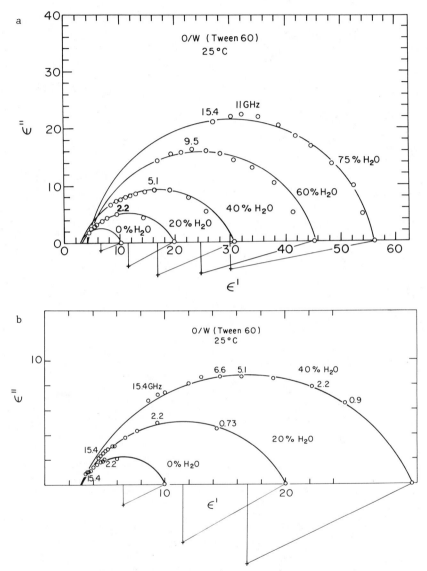

Figure 1. Plots of the complex permittivity of the O/W microemulsions prepared with Tween 60 on the complex dielectric plane ("Cole-Cole" plots), showing the depressed semicircles that indicate a distribution of relaxation times. Figure 1b is an expanded portion of Figure 1a. A few frequencies are indicated for reference. Reproduced with permission from Reference 1. Copyright 1982 Academic Press.

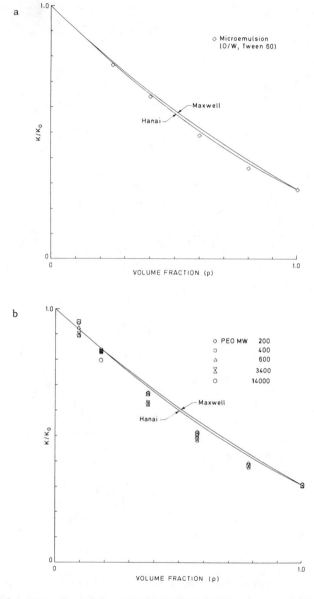

Figure 2. Thermal conductivity of the Tween microemulsions
(Figure 2a) and of a series of PEO solutions of different
molecular weights (2b). The predictions of Equations 1 and 2
are shown for reference. The thermal conductivity has been
normalized to that of water.

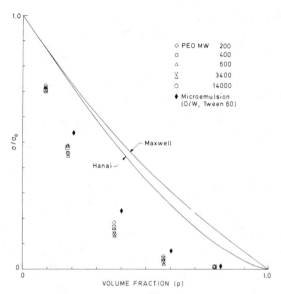

Figure 3. Ionic conductivity of the Tween microemulsions and PEO solutions, compared with Equations 1 and 2. For these experiments, the aqueous phase was 0.1 N NaCl or 0.1 N KCl, and the measured conductivity values were normalized to that of the suspending electrolyte.

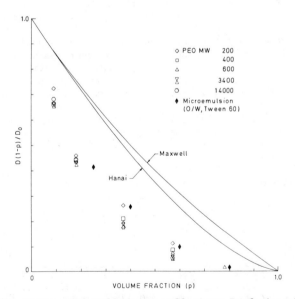

Figure 4. Water self-diffusion coefficient D of the microemulsions and PEO solutions, normalized to that of the pure liquid water. The need for the additional factor (1-p) is described in the text. Also shown are predictions of the Maxwell and Hanai equations.

The striking observation is that the ionic conductivity and water self-diffusion coefficient, but not the thermal conductivity, deviate significantly from the predictions of the mixture theories. This could arise from structural effects, such as a gradual transition from O/W to W/O structure with decreasing water content. We argue instead that these deviations principally result from hydration effects, and not from structural properties of the microemulsions. This would be expected because of the similarity of the data from the microemulsions and PEO, in which structure effects would be quite different.

Discussion

A simple analysis, based on the mixture theory, supports this interpretation. In the following discussion we will assume that the hydration water can be included with a volume fraction p' which is excluded from transport. The difference between the apparent excluded volume fraction hydration p' and p corresponds to values that are in the range expected for the hydrophilic moieties in the surfactant and cosurfactant. Finally, we will suggest a physical interpretation for the observations.

Apparent Hydration of the Suspended Microdroplets. The ionic conductivity and water self-diffusion data, divided by the respective values for the bulk liquid, are summarized in Table I, together with the apparent volume fractions p' that are calculated from the Maxwell and Hanai mixture theories. The similarity in the ionic conductivity and water self-diffusion data is surprising, in view of the greatly different underlying mechanisms for these phenomena.

By hypothesis, the difference between the compositional phase volume p and the total excluded volume p' represents the volume fraction of hydration water. The calculated hydration, expressed as a ratio of (moles hydration water) : (moles EO plus moles OH) is presented in Table II. From thermodynamic studies, the expected hydration of the EO and OH groups are 2 and 3 water molecules, respectively (13,14). While the "hydration" as obtained from the present transport measurements does not reflect stoichiometric binding but rather kinetic effects, the hydration values obtained are quite reasonable.

Table I. Summary of Transport Properties of the Tween 60 O/W Microemulsions

Compositional phase volume p	σ/σ_w	p' (Eq. 1)	p' (Eq. 2)	$(1-p)D/D_w$	p' (Eq. 1)	p (Eq. 2)
0.80	0.002	0.99	0.96	0.016	0.98	0.94
0.60	0.070	0.90	0.84	0.097	0.86	0.79
0.40	0.230	0.69	0.63	0.255	0.66	0.60
0.25	–	–	–	0.413	0.49	0.44
0.20	0.540	0.36	0.34	–	–	–

Table II.   Apparent Hydration of the Microemulsions

compositional phase volume p	(conductivity)		(diffusion)	
	(Eq. 1)	(Eq. 2)	(Eq. 1)	(Eq. 2)
0.80	1.06	0.89	1.00	0.70
0.60	2.23	1.78	1.93	1.41
0.40	3.24	2.57	2.91	2.24
0.25	–	–	4.31	3.41
0.20	3.58	3.13	–	–

(Note: Hydration numbers expressed as moles water per moles EO
plus moles OH.)

It was earlier shown (12) that the hydration values of the PEO
solutions that are calculated in the same manner also agree with
expected values. Since the EO group is the moiety in the surfactant
that is responsible for stabilizing the microemulsion, a comparison
of the transport data of the microemulsions with those of the PEO
solutions is of interest. The slightly higher ionic conductivity
and water self-diffusion coefficients of the microemulsions can be
attributed to the fraction of oil that is not hydrated and consequently can only contribute to the obstruction effect. While
structural changes might be expected in the microemulsions at high
phase volumes, they evidently produce no large changes in the transport properties presently reported.

Physical Mechanisms. The simplest interpretation of these results
is that the transport coefficients, other than the thermal conductivity, of the water are decreased by the hydration interaction. The
changes in these transport properties are correlated: the microemulsion with compositional phase volume 0.4 (i.e. 60% water)
exhibits a mean dielectric relaxation frequency one-half that of
the pure liquid water, and ionic conductivity and water self-
diffusion coefficient one half that of the bulk liquid. In bulk
solutions, the dielectric relaxation frequency, ionic conductivity,
and self-diffusion coefficient are all inversely proportional to
the viscosity; there is no such relation for the thermal conductivity. The transport properties of the microemulsions thus vary
as expected from simple changes in "viscosity" of the aqueous
phase. (This is quite different from the bulk viscosity of the
microemulsion.)

This is, however, a macroscopic explanation of changes that
occur on a molecular level, and is rather superficial. There is
clearly a distribution of dielectric relaxation times in the microemulsion. The timescale of the dielectric relaxation measurement
(tens of picoseconds) is too short for the phenomenon of fast exchange. It would appear, therefore, that the motional restriction
of the water must vary throughout the microemulsion.

Perhaps a more defensible hypothesis is that the rotational and
translational correlation times are both increased, by similar
factors of ten or less, when a water molecule is sufficiently
close or hydrogen bonded to an EO or OH group in the surfactant.

Nevertheless, the concept of "average viscosity", has predictive
value. A more extensive discussion of this problem is presented
elsewhere (12).

Effect of Microemulsion Structure on the Transport Properties. It
appears from the discussion above that the reduction in the ionic
conductivity and water self-diffusion coefficient is primarily
attributable to hydration effects, not principally to changes in
the structure of the microemulsion with higher phase volume.
Either no pronounced changes in structure occur with increased
phase volume (which seems unlikely) or they are of such a nature as
not to greatly affect the transport properties. Since the mixture
theories are not extremely sensitive to the exact shape of the
suspended particles the second possibility seems more likely.
     Water self-diffusion data from other microemulsions suggest an
effect of structural changes on the transport properties. Figure 5
shows the water self-diffusion coefficient in several ionic and nonionic systems (15). The data, while remarkably similar, do show
more variation than would be expected from hydration effects alone.
In particular, the water self diffusion coefficient in the microemulsion prepared with the ionic surfactant SCS appears to be
anomalously low at one composition (p = 0.42). However, that composition corresponded to a point in the phase diagram close to a
region of phase separation (16). The sample exhibited unusually
high bulk viscosity which presumably arose from long range structure. Further NMR studies of the self-diffusion properties of each
species in these systems will be reported (15). Lindman and co-
workers (17) in similar studies have shown how self-diffusion
properties of each species in a microemulsion can yield information
about changes in structure with composition that is difficult to
obtain from measurements of the sort reported here.

Conclusions

To our knowledge, this is the first report of such a wide variety
of transport measurements in a single series of microemulsions.
Some of the properties (water self-diffusion, ionic conductivity,
dielectric relaxtion) are substantially different from predictions
of mixture theories; another property (thermal conductivity) is in
much better agreement. The discrepancies between the ionic conductivity and water self-diffusion coefficient and predictions of
the mixture theories yield hydration values for the microemulsion
that agree with anticipated values. It appears that all of these
changes can be correlated with variation in one property - the viscosity of the suspending water. While it appears that structural
changes with varying composition are also reflected in the transport properties in some cases, hydration effects appear to play a
significant and perhaps dominant role in determining the overall
transport properties. Our results suggest that the usefulness of
dielectric measurements at microwave frequencies, together with the
other transport measurements described here, in studying hydration
phenomena in these complex systems.

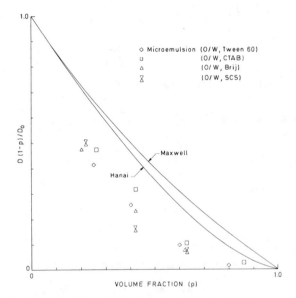

<u>Figure 5.</u> Water self-diffusion coefficients in a variety of ionic and nonionic microemulsions. The compositions of these microemulsions are given in Reference 2.

Acknowledgments

This work was supported in part by National Science Foundation Grant CPE 82-04911, Drexel University Graduate School, Drexel University Research Corporation and the Donors of the Petroleum Research Fund.

Literature Cited

1. Foster, K. R., Epstein, B. R., Jenin, P. C. and Mackay, R. A. J. Colloid Interface Sci. 1982, 88, 233.
2. Epstein, B. R., Foster, K. R. and Mackay, R. A. J. Colloid Interface Sci. 1983, 95, 218.
3. Mackay, R. A. and Agarwal, R. J. Colloid Interface Sci. 1978, 65, 225.
4. Epstein, B. R. Ph.D. Thesis, University of Pennsylvania, Philadelphia, Pennsylvania, 1982.
5. Pauly, H. and Schwan, H. P. Biophys. J. 1966, 6, 621.
6. Stilbs, P. J. Colloid Interface Sci. 1982, 87, 385.
7. Balasubramaniam, T. A. and Bowman, H. F. J. Biomech. Eng. Trans. ASME (Series K), 1979, 99, 148.
8. Hanai, T. "Electrical Properties of Emulsions"; In Emulsion Science, Sherman, P. Ed., Academic Press, NY, 1968, Chap. 5 p. 353.
9. Hasted, J. B. "Aqueous Dielectrics", Chapman and Hall, London, 1973.
10. Chiew, Y. C. and Glandt, E. D. J. Colloid Interface Sci. 1983, 94, 90.
11. Clark, M.E., Burnell, E. E. Chapman, N. R. and Hinke, J. A. M. Biophys. J. 1982, 39, 289.
12. Foster, K. R., Cheever, E., Leonard, J. B. and Blum, F. D. Biophys. J. (in press).
13. Franks, F. In " Water, a Comprehensive Treatise"; Franks, F. Ed.; Plenum Press: New York, 1973, Vol. II, pp. 1-54.
14. Molyneux, P. In "Water, a Comprehensive Treatise"; Franks, F. Ed.; Plenum Press: New York, 1975, Vol. IV, pp. 617-633.
15. Cheever, E., manuscript in preparation
16. Mackay, R. A., Letts, K. and Jones, C. In "Micellization, Solubilization, and Microemulsions"; Mittal, K. L. Ed.; Plenum Press: New York, 1977, Vol. II, pp. 801-815.
17. Lindman, B., Stilbs, P. and Mosely, M. E. J. Colloid Interface Sci. 1981, 83, 569.

RECEIVED June 8, 1984

Micellar Structure and Equilibria in Aqueous Microemulsions of Methyl Methacrylate

J. O. CARNALI and F. M. FOWKES

Department of Chemistry, Lehigh University, Bethlehem, PA 18015

Oil-in-water microemulsions have been formulated using sodium lauryl sulfate (SLS) as the surfactant, n-pentanol or n-hexanol as the cosurfactant and methyl methacrylate (MMA) as the oil component. Phase behavior studies were performed in order to locate the oil-in-water microemulsion region. An extensive L_1 phase was also discovered for the three component system consisting of water, SLS and MMA. The vapor pressure of MMA over these systems, measured by a headspace technique, indicated preferential partitioning of MMA into the micellar phase with a free energy of transfer of -14 kJ/mole. The molecular environments of hexanol and MMA within both the microemulsions and the L_1 phase were followed by a ^{13}C NMR shielding study. The experimental evidence indicated a substantial water interaction on the part of both hexanol and MMA. Using a two-site model for the location of these molecules in the microemulsion, we found that our data were consistent with the swollen micelle model for an oil-in-water microemulsion. An appreciable amount of the hexanol was found to be located within the micelle core while some of the MMA was found at the micelle surface. The partitioning of the alcohol between interface and core is probably a general result for microemulsions and is linked to their thermodynamic stability. The presence of MMA at the micelle surface is probably responsible for the large L_1 phase region observed but is a phenomenon restricted to oils possessing hydrophilic sites.

The current structural model for microemulsions was advanced by Hoar and Schulman (1). These authors pictured the transparent dispersions of oil in water or of water in oil as consisting of small spherical droplets of the dispersed phase within the continuous phase. Later, this model was refined to include an interfacial film of surfactant and cosurfactant coating the droplets (2). It has also been pointed out that the compositions leading to microemulsions could be related

0097-6156/85/0272-0287$06.00/0

to the isotropic aqueous (L_1) and non-aqueous (L_2) phases found in ternary mixtures of surfactant, cosurfactant and water (3). In this context, oil-in-water microemulsions are pictured as oil-swollen micelles while water-in-oil microemulsions are water-swollen inverse micelles. Both views are consistent with a microemulsion model possessing very distinct continuous and dispersed phases. Further, the dispersed phase exists with a very well-defined structure consisting of spherical particles. This structural viewpoint has been supported by a variety of experimental results. For example, the particle nature of microemulsions has been studied with light scattering (4,5,6). Evidence for distinct droplets of dispersed phase has also been observed through neutron scattering (7). Further evidence comes from ultracentrifugation work (2) and from electron microscopy (8,9).

Recently, however, the discrete, particle-like nature of microemulsions has been re-examined in the light of new experimental findings. The apparent molal volume of toluene within a microemulsion phase has been studied as a function of composition (10). This quantity was found to vary considerably, indicating that structural changes were occurring within the homogeneous microemulsion phase. Also, self-diffusion coefficients of the components of several microemulsion systems have been measured with NMR techniques (11). For many systems, no evidence was found for a well-defined structure and oil/water interfaces were argued to be very flexible and dynamic. Similar findings were the result of fluorescence work (12), and interfacial flexibility combined with a random structure has been discussed from a theoretical standpoint (13).

The goal of the present work is to obtain a consistent structural model for a microemulsion system. In particular, we are interested in carrying this model down to the molecular level so that the intermolecular effects which are responsible for the stability of these systems can be elucidated. We have studied the system consisting of water, SLS and MMA with and without n-hexanol or n-pentanol. We have determined the phase boundaries of the isotropic microemulsion and L_1 phases and determined how these are affected by surfactant concentration and alcohol chain length. Measurements were also made of the vapor pressure of MMA over these systems to determine the concentration of MMA in the water surrounding the microemulsion droplets. From these data, the energetics of transfer of the MMA from aqueous to micellar solution were determined. Finally, a ^{13}C NMR chemical shielding study was performed to find how the MMA and the alcohol were distributed within the microemulsion. The combination of the latter two studies allowed a quantitative structural model for the microemulsions to be presented.

Materials and Methods

Materials. Sodium lauryl sulfate was purchased from Aldrich Chemical Company, Inc. and its constitution has been reported in detail (14). The material is a blend of C-12, C-14 and C-16 sodium sulfates. It was purified by extraction with diethyl ether and then by recrystallization from absolute ethanol. The surface tension of aqueous solutions of this material was measured as a function of concentration. A curve with a single break resulted indicating a CMC

of $2.2 \times 10^{-3}\underline{M}$. The methyl methacrylate and n-pentanol were reagent
grade materials from Fisher Scientific Company. The n-hexanol was
reagent grade from Aldrich and the water was once distilled.

Phase Diagram Determination. Phase diagrams were determined from
equilibrated mixtures of MMA, alcohol and aqueous SLS solutions.
Stock solutions of SLS were delivered by a micrometer syringe into
glass ampules. This was followed by various amounts of MMA and alcohol delivered in the same way. The ampules were then sealed,
thoroughly agitated and stored at 23°C. The turbidity of these
samples at a wavelength of 520 nm was measured on a daily basis until
no further change was noted. The boundaries of the isotropic transparent phases were determined from the compositions at which the
turbidity began to increase. The location of phase boundaries was
checked by long term storage of systems on either side of the line
or by centrifugation at 30,000 × g of systems just within the transparent region.

Headspace Analysis. A vapor pressure method was used for determining
the fraction of the MMA which is bound in the micelles (15). In
this method it is assumed that the micelle is a distinct phase with
respect to the continuous solution so that the measured vapor pressure of MMA over micellar solutions can be directly related to its
concentration in the continuous solution. By mass balance, the
fraction of MMA bound to the micelles can then be determined. The
procedure was to load serum bottles with water or SLS solution containing various amounts of MMA. The quantities of MMA were such
that its concentration never exceeded its solubility in the aqueous
solution or its saturation limit in the microemulsion, respectively.
The bottles were then closed off with teflon-coated septa and gently
agitated in a thermostat for at least three days at 23°C. A few
hours before sampling, the thermostat was cooled to two degrees below room temperature to ensure that no condensate would collect on
the inner surface of the septum. After this, 0.5 ml vapor samples
were removed from the headspace of the bottles with a 1.0 ml gas-
tight syringe. The vapor sample was then injected into a Perkin
Elmer Sigma 3 gas chromatograph with a flame ionization detector.
The glass column used for the separation was packed with 10% Carbo-
wax 20M and 2% KOH coated on a Chromosorb W support. The attenua-
tion was adjusted such that all of the samples could be run on the
same scale and the corresponding peak areas were determined by an
electronic integrator.

Chemical Shielding Measurements. A JEOL FX-90Q and a Varian XL-200
NMR spectrometer were used to obtain ^{13}C NMR spectra at 22.49 and
50.31 MHz, respectively. Chemical shifts were referenced to TMS
contained in a capillary mounted coaxially within the 10mm NMR tube.
Gated proton-noise decoupling was used to avoid heating the sample
which was thermostated at 27°C. A spectral width of 5000 Hz, acqui-
sition time of 1.6 sec, pulse delay of 6 sec, acquired total of 100
to 3000 transients and flip angle of 45 degrees were typical parameters at 22.5 MHz. The resolution at 22.5 MHz was 0.7 Hz while that
at 50.3 MHz was 0.75 Hz. The measured chemical shifts are accurate
to within 0.05 ppm at both field strengths.

Advantage was taken of the different field geometries in the 22.5 and 50.3 MHz instruments to correct for magnetic susceptibility differences between samples. The 50.3 MHz instrument used a super-conducting magnet with the magnetic field parallel to the sample axis. The 22.5 MHz instrument used an electromagnet whose field was perpendicular to the sample axis. When the same sample was run on both instruments, true chemical shifts, unaffected by susceptibility differences, could be obtained (16). The relationship is

$$\delta_{true} = 1/3 \ [\delta_{sc} + 2\delta_{ic}] \tag{1}$$

where the subscripts refer to superconducting and iron core magnets, respectively.

Experimental Results

The water-rich corner of the phase diagram for the three component system consisting of water, MMA, and SLS contains an extensive iso-tropic region. We have identified this region as an L_1 phase and have determined its extent at 23°C as is shown in Figure 1. The phase diagram is in terms of weight percent. The phase boundary was studied along lines of constant weight ratio of SLS to water and these paths are also depicted in Figure 1. These origins, on the water/SLS axis, are located at 2.5, 5.0, 9.9, 14.7 and 19.4 wt% SLS, respectively.

The effects of adding n-pentanol or n-hexanol to the L_1 phase were studied in the following manner. To systems containing 9.9, 14.7, or 19.4 wt% aqueous SLS was added MMA in 25, 50, or 67% excess of that which would saturate the system (L_1 phase boundary). Alcohol was then added to the resulting two-phase system and the minimum alcohol content necessary to produce a microemulsion determined. The results for these systems are shown in Figure 2. In these phase diagrams, the aqueous SLS solution is considered as a pseudo-compo-nent.

The L_1 phase for the ternary system, water, SLS and MMA has been reported to extend up to equal mole ratios of MMA to SLS (17). Our results indicate that the limiting mole ratio is more like three to one in favor of the MMA. The L_2 and water-in-oil microemulsion regions of the system consisting of water, SLS, MMA and pentanol has recently been investigated (18). Combined with the present work in the L_1 and oil-in-water microemulsion regions, this shows the applicability of the phase diagram approach to microemulsions.

Additional information can be obtained from the phase diagrams if one considers the molar ratios of the structure-forming compo-nents (19). An example of this is Figure 3 in which the MMA/SLS mole ratio for systems on the microemulsion phase boundary is plotted versus the corresponding alcohol/SLS ratio. One notes that the plots form a straight line for a given alcohol type at a fixed sur-factant content. Extrapolated to zero alcohol content, these lines all intercept satisfactorily at a MMA/SLS ratio of 3.0 ± 0.2. Along any given line, the number of moles of SLS is held fixed so that the observed slope is a measure of the increased capacity for the solu-bilization resulting from the addition of the alcohol. The slopes of the lines are positive and are seen to increase with surfactant

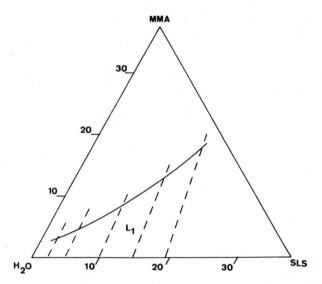

Figure 1. Partial diagram of the system water, MMA and SLS at 23 °C expressed in weight percent. Solid line is the L_1 phase boundary, dashed lines are the composition pathways studied.

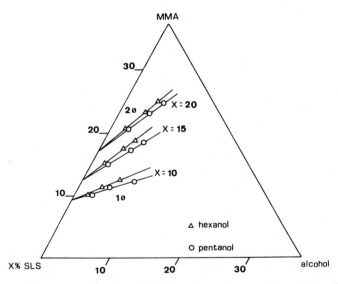

Figure 2. Phase boundaries of the system aqueous X% SLS solution, MMA and pentanol or hexanol at 23 °C expressed in weight percent. At each surfactant concentration, the region above the phase boundary (higher MMA content) consists of two phases while the region just below is a one phase microemulsion.

concentration or with a change from pentanol to hexanol. Hexanol
is thus somewhat more efficient at microemulsifying MMA than is
pentanol.

Head Space Analysis. A plot of peak area as a function of concentration for aqueous solutions of MMA was found to be a straight line as
shown in Figure 4. This plot provides a Henry's Law relationship
between the concentration of MMA in solution and its corresponding
vapor pressure. In addition to these standard systems, the vapor
pressure over samples from the L_1 and microemulsion phases was also
determined for the 14.7 wt% SLS aqueous solution. At low concentrations, the peak area is again linearly related to MMA concentration.
As the saturation point is approached, however, the peak area increases more slowly. For any of the surfactant systems, the concentration of MMA in the continuous aqueous phase can be determined
by constructing a horizontal line from the surfactant curve to the
standard curve and then dropping a vertical line down to the concentration axis. The intercept is the concentration in the continuous phase and the amount of MMA in the micellar phase then follows from mass balance. Figure 4 shows that the concentration of
MMA in the aqueous phase at the L_1 phase boundary and in the microemulsions is approximately 0.15 $\underline{M}$. This is also the solubility
limit of MMA in water.

For low concentrations of MMA, where the surfactant system
areas can be fitted to a straight line, the mole fraction of MMA in
the aqueous phase, X_a, and in the micellar phase, X_m, can be calculated from

$$X_a = C_a \, / \, [C_a + C_w] \tag{2}$$

$$X_m = [C_s - C_a \, \phi] \, / \, [C_s - C_a \, \phi + C_m] \tag{3}$$

where C_s and C_a are the MMA concentrations in the surfactant and
standard aqueous solutions, respectively, while C_m is the concentration of surfactant. Also, the aqueous phase of the surfactant systems is described by its volume fraction ϕ and water concentration
C_w.

Knowing these mole fractions, the distribution constant K_x, for
the partitioning of MMA between the micellar and aqueous environments can be determined from the ratio of X_m to X_a, assuming that
all activity coefficients are unity. Then, the standard free energy
of transfer of MMA from the aqueous to micellar phase can be calculated from

$$\Delta G^\circ = -R \, T \, \ln K_x \tag{4}$$

Table I displays the mole fraction of the MMA which is in the aqueous
phase for the L_1 phase and microemulsion systems studied. These
fractions were found to be reproducible to within ±0.04. As can be
seen, MMA favors the micellar phase by at least a four-to-one ratio.
The free energy of transfer, calculated for systems less than 0.53 $\underline{M}$
in MMA was found to be -14.0 kJ/mole with an uncertainty of 10%.

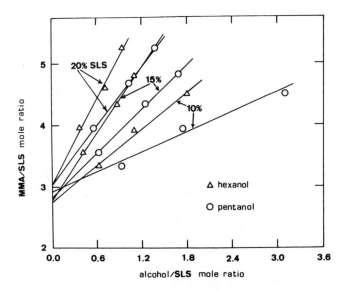

Figure 3. MMA/SLS mole ratio plotted versus the alcohol/SLS mole ratio for systems at the phase boundaries shown in Figure 2.

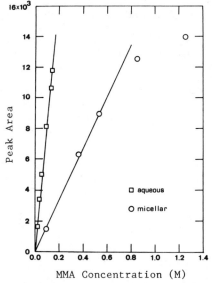

Figure 4. MMA peak area as determined by chromatographic analysis of the head space over aqueous and micellar solutions of MMA. The micellar systems correspond to the L_1 phase at 14.7 wt% SLS.

Table I.   Summary of Data for L_1 Phase and Microemulsions Systems

w% MMA[a]	Fraction[b] MMA in water	X_a[c]	X_i[d]	$\dfrac{SLS}{MMA}$[e]	$\dfrac{Alcohol}{MMA}$[f]	$r(\overset{\circ}{A})$[g]
0.9	0.17	0.66	0.49	11.4	...	24.4
1.8	0.17	0.62	0.45	6.2	...	24.7
3.5	0.17	0.56	0.39	3.6	...	25.4
6.8	0.17	0.45	0.28	2.5	...	28.4
9.9	0.13	0.35	0.22	2.0	...	32.7
12.7	0.09	0.28	0.19	1.9	...	36.7
+25%	0.07	0.18	0.11	2.5	0.6	47.2
+50%	0.06	0.16	0.10	2.3	1.1	52.1
+67%	0.05	0.14	0.09	2.3	1.7	53.7

a – refers to L_1 phase at 14.7% SLS or microemulsion with an X% excess of MMA over that which saturates the L_1 phase.
b – refers to fraction of the MMA which is in the aqueous continuous phase.
c – fraction of MMA in aqueous environment, calc. with eq. (6)
d – fraction of MMA in micelle interface, calculated from $X_i = X_a -$ (fraction in water).
e,f – mole ratio at the interface.
g – radius calculated from $r = 3$ (volume/surface area).

<u>Chemical Shielding Study.</u> ^{13}C NMR spectra were recorded for samples
lying within the L_1 phase and which originated from the 14.7 wt%
aqueous SLS solution. The MMA content was varied up to a composition lying at the phase boundary. Spectra were also recorded for
the microemulsions containing hexanol and the 14.7 wt% SLS solution.
In addition, systems were studied which were models for the distinct
aqueous, interfacial and oily environments found in microemulsions.
These were aqueous solutions of MMA, aqueous micellar solutions of
hexanol and SLS, and MMA solutions of hexanol, respectively.

A typical spectrum, recorded at 22.5 MHz for a microemulsion
containing a 67% excess of MMA, is shown in Figure 5. Except for
the carbons in the middle of the SLS chain, each type of carbon
gave rise to a single, distinct resonance signal. The assignments
shown for the carbons of MMA and hexanol were determined by running
each of the components of the microemulsion separately.

The chemical shifts of two carbon atoms in particular were found
to be particularly sensitive to the composition of the system. These
were the α-carbon of hexanol (assigned as a) and the carbonyl (x)
carbon of MMA.

Figure 6 shows the chemical shift of the α-carbon of hexanol
as a function of composition in three different kinds of solutions.
The first represents hexanol in a 14.7 wt% SLS aqueous solution as
a function of hexanol concentration. The concentration range covered
runs from very dilute in hexanol up to systems in which a fairly
viscous phase results. Next are the microemulsions containing a
25, 50, and 67% excess of MMA over that which saturates the corresponding L_1 phase system. Lastly are water-saturated MMA solutions
of hexanol in which the hexanol/MMA volume ratios bracket those
found in the microemulsion. The α-carbon shifts of hexanol measured
in the micellar solutions are about 0.6 ppm downfield from those
measured in MMA solution. There is only a slight variation in shift
for either the micellar or the MMA solutions with various hexanol
concentrations. The hexanol in the microemulsion systems appears
approximately halfway between that in the other two solutions.

The shift of the carbonyl carbon of MMA was determined in
aqueous solution at a concentration of 0.15 M, as well as in L_1 phase
systems where the wt% of MMA was varied up to 12.7%, which lies
at the phase boundary as shown in Figure 7. In addition, the same
microemulsions and MMA solutions of hexanol were examined as were
used above for the hexanol shift studies. The signal from MMA dissolved in water is downfield by 3.8 ppm from those of MMA/hexanol
mixtures. The latter systems show a small downfield shift with increasing hexanol concentration. Once again, the L_1 phase and microemulsion systems show an intermediate behavior but now there is a
continuous upfield shift with increasing MMA content in the L_1 phase
which is continued in the microemulsions.

Discussion

Our quantitative findings concern the phase behavior of the L_1 and
microemulsion systems are summarized in Figure 3. In the absence of
alcohol, the maximum MMA content of the L_1 phase is fixed by a
limiting MMA/SLS mole ratio of approximately three. Higher mole
ratios will invariably lead to a two-phase system. With alcohol

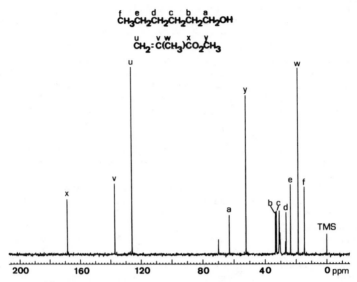

Figure 5. ^{13}C-NMR spectrum at 22.5 MHz recorded for a micro-emulsion containing a 67% excess of MMA over that which saturates the L_1 phase at 14.7 wt% SLS. The reference peak is external tetramethylsilane (TMS) and the peak assignments were made by running hexanol and MMA separately.

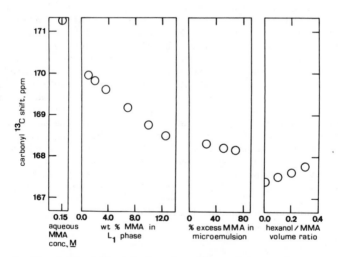

Figure 6. Chemical shift of the α-carbon of hexanol measured with respect to external TMS as a function of environment. The environments studied are: hexanol in aqueous 14.7 wt% SLS solution as a function of hexanol concentration, hexanol in microemulsions and hexanol in MMA solution as a function of hexanol concentration.

present and at any given SLS concentration, the moles of micro-
emulsified MMA are a linearly increasing function of the moles of
alcohol. As Figure 3 demonstrates, there is an increase in slope
on going from pentanol to hexanol at a fixed SLS content, indicating
that the efficiency of the alcohol for microemulsifying MMA increases
with its chain length. From an examination of the slopes of pairs
of lines having the same SLS content but using different alcohols,
we find that 1.6 ± 0.2 moles of pentanol are needed to microemulsify
the same amount of MMA as obtained with one mole of hexanol. The
effect of increasing the SLS concentration is shown in Figure 8
where the slopes of the lines from Figure 3 (the increase in the
number of moles of MMA solubilized per mole of added alcohol) are
plotted versus the original concentration of SLS. One observes an
approximately linear increase in the slope with increasing SLS content. These results indicate a substantial increase in the effectiveness of these alcohols with surfactant content.

The results of the head space analysis are evidence for the applicability of the swollen micelle model to these microemulsion
systems. Despite the solubility of MMA in water, vapor pressure
measurements of L_1 phase systems find that the concentration in the
aqueous phase is small. Thus the majority of the MMA must be bound
in micelles which behave as a distinct phase. Near the saturation
limit in the L_1 phase and in the microemulsions, the measured vapor
pressure is about what one would expect from an aqueous solution
saturated with MMA. This also is consistent with our picture of
distinct aqueous and micellar phases. In fact, the composition of
the continuous phase in water-in-oil microemulsions has been determined in just this way (20).

An additional indication comes from the measured free energy
of transfer of MMA from the aqueous to the micellar phase. Wishnia
measured the free energy for the transfer of alkanes from aqueous
solution to SLS micelles (22). He studied the homologous series,
ethane through pentane, at low concentrations. His result for ethane
was -14.5 kJ/mole.

It has been established that the behavior of a hydrocarbon in
microemulsion formulations can be characterized by its Equivalent
Alkane Carbon Number (EACN) (22). This is the number of methylene
and methyl carbons making up the molecule. The EACN for MMA should
be two and has been determined as such in this laboratory. It is
interesting that this correlation seems to apply also to the measured
free energies of transfer. This suggests that this previously empirical correlation has a theoretical basis in thermodynamics. Further, in the L_1 phase, up to a MMA concentration of at least 0.5 M,
the free energy of transfer is constant, indicating that the micelle
structure is probably not too much different from that at very low
MMA concentrations.

The ^{13}C chemical shielding study, together with the headspace
results, allowed a somewhat more detailed picture of these systems
to be drawn. The chemical shielding experienced by a nucleus can be
directly related to its surrounding electronic environment. In the
absence of susceptibility or anisotropy effects, the measured chemical shift is an accurate expression of this chemical shielding. The
dependence of the chemical shift on intermolecular effects can be
used to sense how a molecule's environment changes when it is placed

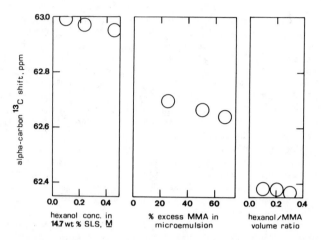

Figure 7. Chemical shift of the carbonyl carbon of MMA measured
with respect to external TMS as a function of environment.   The
environments studied are: MMA at 0.15 $\underline{M}$ in water, MMA in the L_1
phase as a function of MMA content, MMA in microemulsions as a
function of MMA content and MMA solutions as a function of hexanol
concentration.

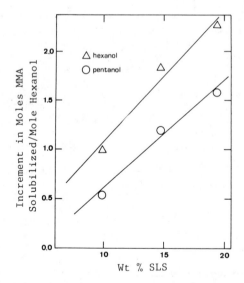

Figure 8. Slopes of the lines from Figure 3 plotted as a function
of the original concentration of SLS in the microemulsions.

in a different solvent. With this knowledge, the kind of environment which a certain nucleus is experiencing can be determined from
its chemical shift. This technique has been used to study micelles
(23) and microemulsions (24).

For ^{13}C NMR studies, the intermolecular effects which can
affect chemical shifts have been documented (25). These are of two
types: hydrogen-bonding effects and conformational or γ effects.
The γ effect is related to the ratio of trans to gauche formations
in a hydrocarbon chain. For the particular carbon atoms of this
study, hydrogen bonding effects are the most important.

We consider first the hexanol shifts of Figure 6. In the absence of MMA, the downfield position of micellar bound hexanol can
be attributed to hydrogen bonding of the hydroxyl group of hexanol
with water. Assuming that the core of the micelle is largely water-
free, this requires that the hydroxyl group be at the micelle surface with the hydrocarbon chain most probably packed along those of
SLS in the micelle. Evidence for the incorporation of pentanol into
the core of SLS micelles when the alcohol concentration is high has
been presented (26). The small upfield trend observed here in the
α-carbon shift as a function of hexanol concentration is probably
due to this incorporation. Thus the shift measured at the lowest
hexanol concentration is considered to be characteristic of micellar
bound hexanol.

The hydrogen bonding of hexanol which is found in the micellar
phase is largely lacking in the water saturated MMA solutions. Hydrogen-bonding of hexanol with MMA can be detected by observing the
MMA carbonyl shift, as will be shown later. However, the extent
of this hydrogen bonding is small with respect to that which occurs
when hexanol encounters water. In the swollen micelle model of
microemulsions, hexanol can be at the micelle interface or in the
hydrocarbon core where it should experience chemical shielding quite
like that measured for micellar or for MMA solutions, respectively.

The microemulsions studied show a single resonance for the
alpha-carbon appearing midway between those of its expected model
environments. The diffusion of cosurfactant within microemulsions
is very rapid on an NMR time scale as has been confirmed in a series
of studies (27,28,29). In this case, the observed chemical shift is
a weighted average of those of the different environments in which
the cosurfactant is located. For our system, the observed shift
could be represented as

$$\delta_{obs} = X_c \, \delta_c + (1-X_c) \, \delta_i \tag{5}$$

where X_c is the mole fraction of hexanol which finds itself (on the
average) inside the hydrocarbon core. δ_c and δ_i are the characteristic shifts of the core and interface, respectively. Applied to our
data, the mole fraction of hexanol located in the core for microemulsions containing a 25, 50, and 67% excess of MMA is 0.5, 0.5,
and 0.6 ± 0.1, respectively. This result compares favorably with the
work on hexadecane in water microemulsions (27). In those systems,
between 20 and 35% of the cosurfactant pentanol was found to be
located within the micelle interior. Even better agreement is found
with the data for paraffin-in-water microemulsions stabilized with
sodium oleate and pentanol (5). The fraction of the pentanol located

in the core was 0.5 - 0.7 in the microemulsion region studied. It
should be recalled that the microemulsions in the present study contain the minimum amount of alcohol necessary to give a transparent
system. This alcohol is then required to split fairly evenly between the micelle core and interface, suggesting that this sort of
partitioning is an important factor in the formation of microemulsions.

In the case of the carbonyl carbon of MMA shown in Figure 7,
the shift measured in a saturated aqueous solution is well downfield,
evidence of the expected strong hydrogen bonding of MMA with water.
The signal from water-saturated MMA solutions is predictably well
upfield. Upon addition of hexanol, a slight downfield shift is observed, increasing in proportion to the amount of hexanol added.
This is due to hydrogen-bonding of hexanol to MMA. The shifts
measured in the L_1 phase and microemulsions are once again intermediate between those of the model systems indicating that the carbonyl's environment is neither all aqueous nor all oily. To interpret these results we have applied our two-site model of microemulsions to MMA. We imagine that MMA may reside in the micelle core,
where its chemical shift should be like that in MMA solution, as
well as at the micelle surface where its shift is like aqueous MMA.
Assuming rapid diffusion of the MMA, the observed shift can again
be written as

$$\delta_{obs} = X_a \delta_a + (1-X_a) \delta_c \tag{6}$$

where now X_a is the fraction of the MMA in an aqueous environment
in which its chemical shift is δ_a. The choice of a value for δ_c is
not critical, since the variation in δ_c with hexanol content is small
compared with the difference between δ_a and δ_c. A value consistent
with the calculated hexanol to MMA ratio in the core was used. The
values of X_a calculated for the L_1 and microemulsion systems are
given in Table I. These X_a values include a contribution from MMA
dissolved in the aqueous phase surrounding the micelles. The amount
of MMA dissolved in this way was determined in the headspace studies
and is also shown in Table I. Subtracting this value from the respective X_a value gives X_i, the fraction of the MMA which is at
the micelle surface. We observe that X_i is fairly large but decreases uniformly with increasing MMA content. With X_i known for
both MMA and hexanol in the microemulsions, the relative ratios of
the moles of interfacial SLS, MMA and hexanol can be calculated. Two
such ratios are given in Table I. The SLS/MMA interfacial molar
ratio (column 5) is particularly interesting, decreasing steadily
with MMA content while in the L_1 phase and becoming fairly constant
in the microemulsions. The data indicate that the amount of interfacial MMA increases as the L_1 phase boundary is approached, even
though the relative fraction of interfacial MMA decreases. The
interfacial alcohol/MMA molar ratio (Column 6) increases steadily
with the MMA content of the microemulsion, as would the corresponding
alcohol/SLS ratio (not shown). These data show that there is no
unique alcohol/MMA or alcohol/SLS molar ratio for the interface of
these systems.

The particular behavior of the SLS/MMA interfacial molar ratio
with increased MMA content suggests an expanding microemulsion

droplet in which the droplet volume grows progressively faster than its surface area. To test this idea, we have assumed a model for a microemulsion consisting of monodisperse spheres of radius r. We have also assumed that all of the surfactant in our system (in excess of the CMC) is involved in forming interface, while the alcohol and MMA are distributed between droplet interface and core according to the ratios determined in the chemical shielding studies. The total interfacial area S in the system can then be written as

$$S = M \sigma_m + X_{in} N \sigma_n + X_{ip} P \sigma_p \qquad (7)$$

where M, N and P are the number of moles of SLS, MMA and hexanol in the microemulsion and the α's are the corresponding surface areas per mole. Further, X_{in} and X_{ip} are the fractions of the MMA and hexanol found at the interface. An analogous expression can be written for the total disperse phase volume V in terms of the appropriate partial molar volumes v as

$$V = M v_m + (1 - X_{an} + X_{in}) N v_n + P v_p \qquad (8)$$

Values for the σ and v were taken from the literature as follows: $\sigma_m = 3.01 \times 10^5$ m^2/mole and $\sigma_p = 1.26 \times 10^5$ m^2/mole (29), $v_m = 2.46 \times 10^{-4}$ m^3/mole (26) and $v_p = 1.23 \times 10^{-4}$ m^3/mole (30). v_n was estimated as 1.07×10^{-4} m^3/mole from the molar volume of the pure liquid and σ_n was calculated to be 2.3×10^5 m^2/mole assuming that MMA is a spherical molecule.

The total interfacial surface area can be simply related to the dispersed phase volume in terms of the microemulsion droplet radius. Calculated radii for the systems studied are shown in Table I. As expected, the droplets increasingly grow in size as the volume of dispersed phase increases. Thus, the observed decrease in X_i is merely a consequence of the decreasing surface-to-volume ratio needing a correspondingly smaller fraction of the MMA to create new interface. The droplet sizes predicted are consistent with the observed transparency of our systems which requires droplet sizes of under 100 Å in radius. Thus, the simple swollen micelle serves as an adequate structural model for this microemulsion system.

From these results, it is obvious that the extensive L$_1$ phase in the water, SLS and MMA phase diagram is a result of MMA possessing a rather favorable water interaction. It is just this interaction that enables MMA to come to the interface and plug the gaps in the expanding surface film. This phenomenon is probably not as important in systems containing less hydrophilic oils.

Acknowledgments

The Varian XL-200 NMR instrument used in this work was located at the AT&T Engineering Research Center, Princeton, NJ. We thank Drs. John A. Emerson and Deni M. Rose for making arrangements to use the instrument and for providing technical advice. We also thank Dr. William R. Anderson, Jr., for advice concerning the susceptibility correction. Support of this work by Lever Research and by IBM through graduate fellowships is gratefully acknowledged.

Literature Cited

1. Hoar, T. P.; Swarup, S. J. Phys. Chem., 1982, 86, 4212.
2. Bowcott, J. E. L.; Schulman, J. H. Z. Electrochem, 1955, 59, 283.
3. Friberg, S. In Microemulsions: Theory and Practice, ed. by L. M. Prince, Academic Press, Inc.: New York, 1977.
4. Schulman, J. H; Friend, J. A. J. Colloid Sci., 1959, 4, 497.
5. Attwood, D.; Currie, L. R. J.; Elworthy, P. H. J. Colloid Interface Sci., 1974, 46, 255.
6. Graciaa, A.; Lachaise, J.; Martinez, A.; Bourrel, M.; Chambu,C. C. R. Acad. Sci., Paris, 1976, 282B, 547.
7. Dvolaitzky, M.; Guyot, M.; Lagues, M.; Le Pesant, J. P.; Ober, R.; Sauterey, C.; Taupin, C. J. Chem. Phys., 1978, 69, 3279.
8. Stoeckenius, W.; Schulman, J. H.; Prince, L. Kolloid-Z., 1960, 169, 170.
9. Biais, J.; Mercier, M.; Bothorel, P.; Clin, B.; Lalanne, P.; Lemanceau, B. J. Microsc., 1981, 121, 169.
10. Roux, A. H.; Roux-Desgranges, G.; Grolier, J. E.; Viallard, A. J. Colloid Interface Sci., 1981, 84, 250.
11. Lindman, B.; Stilbs, P.; Moseley, M. E. J. Colloid Interface Sci., 1981, 83, 569.
12. Atik, S. S.; Thomas, J. K. J. Phys. Chem., 1981, 85, 3921.
13. DeGennes, P. G.; Taupin, C. J. Phys. Chem., 1982, 86, 2294.
14. Hoskins, J. C.; King, Jr., A. D. J. Colloid Interface Sci., 1981, 82, 264.
15. Spink, C. H.; Colgan, S. J. Phys. Chem., 1983, 87, 888.
16. Becconsall, J. K.; Daves, Jr., G. D.; Anderson, Jr., W. R. J. Am. Chem. Soc., 1970, 92, 430.
17. Merlin, A.; Fouassier, J. P. Polymer, 1980, 21, 1363.
18. Stoffer, J. O.; Bone, T. J. Polym. Sci. (Polym. Chem. Ed.), 1980, 18, 2641.
19. Kertes, A. S.; Jernstrom, B.; Friberg, S. J. Colloid Interface Sci., 1975, 52, 122.
20. Biais, J.; Odberg, L.; Stenius, P. J. Colloid Interface Sci., 1982, 86, 350.
21. Wishnia, A. J. Phys. Chem., 1963, 67, 2079.
22. Cayias, J. L.; Schechter, R. S.; Wade, W. H. Soc. Pet. Eng. J., 1976, 16, 3 51.
23. Muller, N.; Birkhahn, R. H. J. Phys. Chem., 1967, 71, 957.
24. Hansen, J. R. J. Phys. Chem., 1974, 78, 256.
25. Rosenholm, J. B.; Drakenberg, T; Lindman, B. J. Colloid Interface Sci., 1981, 84, 250.
26. Lianos, P.; Lang, J.; Strazielle, C.; Zana, R. J. Phys. Chem., 1982, 86, 1019.
27. Tricot, Y.; Kiwi, J.; Niederberger, W.; Gratzel, M. J. Phys. Chem., 1981, 85, 862.
28. Nguyen, T.; Ghaffarie, H. H. J. Chim. Phys. et Phys.-Chem. Bio., 1979, 76, 513.
29. Nguyen, T.; Ghaffarie, H. H. In Magnetic Resonance in Colloid and Interfaces Sciences, ed. by J. P. Fraissard and H. A. Resing, D. Reidel Publishing Co.: New York, 1980.
30. Almgren, M.; Swarup, S. J. Phys. Chem., 1982, 86, 4212.

RECEIVED June 8, 1984

Photochemical Reactions in Microemulsions and Allied Systems

S. ATIK, J. KUCZYNSKI, B. H. MILOSAVLJEVIC, K. CHANDRASEKARAN, and J. K. THOMAS

Department of Chemistry, University of Notre Dame, Notre Dame, IN 46556

Photophysical and photochemical studies of polymerized microemulsions and colloidal semiconductors are investigated. Photochemical studies of pyrene incorporated into polymerized microemulsions indicate that electron or energy transfer occurs via a tunneling or Förster mechanism, respectively. Steady state and pulsed laser excitation techniques have been used to investigate photo-induced processes at the surface of CdS and TiO_2 colloids. The luminescence of CdS and TiO_2 is strongly dependent on either the excitation intensity and the nature of the adsorbed species, or the colloid crystallinity. CdS luminescence is quenched rapidly ($\tau \ll 10^{-9}$ sec) via e^- or hole capture. Photochemical reduction of carbontate to formaldehyde occurs on TiO_2 whereas photoreduction of methyl viologen, MV^{2+}, occurs on both semiconductors, the kinetic decay of MV^+ being sample dependent.

The past twenty years have shown an increased interest in photoreactions in surfactant systems (1,4), and in particular, micelles have played a significant role in catalyzing charge transfer processes in photochemical reactions. At first sight, microemulsions and micelles provide hydrophobic domains in an essentially polar background, so that reactants may be concentrated into these domains, thereby producing locally high concentrations, which lead to increased rates of reaction over those observed for similar concentrations in homogeneous solution. Such effects can be useful. However, a more intriguing situation arises when dealing with photoinduced charge transfer on the surfaces of these assemblies, which lie between the hydrophobic and hydrophillic domains of the system, and provide an intriguing environment to influence significant movement of charge from one species to another. From the point of view of surfaces, one would expect that

0097–6156/85/0272–0303$06.50/0
© 1985 American Chemical Society

micelles would have a greater effect than microemulsions in this type of photochemical reaction (4). In most instances this is indeed found to be true. However, microemulsions themselves, especially when they are used as polymerized microemulsions, can have additional features which are also useful to the photochemist (5-8). This brief survey deals with photoinduced charge separation as an example of a photochemical reaction. Examples of this reaction are discussed in microemulsion systems with comparison to micelles and homogeneous solution. A later part of the work is connected with systems which may be considered to be allied systems to microemulsions, such as colloidal cadmium sulfide, and titanium dioxide, TiO_2. In these latter systems reactions can only occur on the surface of these particles following excitation of the particle bulk.

Experimental

Preparation and Polymerization of (O/W) Cetyltrimethylammonium Bromide Microemulsion (CTAB-µE) (5-7). An oil in water µE composed of 1.0 g of CTAB, 0.5 g of hexanol, and 1.0 g of 50% styrene-divinylbenzene in 50 mL of water was carefully prepared by slowly adding the water to a stirred mixture of the other components to yield a slightly bluish clear solution. A 0.1% solution (w/w) of initiator AIBN (based on monomer) was then solubilized in the system followed by removal of O_2 (by gentle N_2 bubbling for 5 min), and finally the system was heated in an oil bath (50°C) until complete polymerization was achieved as determined spectrophotometrically. Proper dilution with water was then made to give a 0.01 M CTAB-P-µE solution; P-µE indicates polymerized microemulsion.

Preparation of Colloidal CdS. Colloidal CdS samples were prepared by precipitation from an aqueous surfactant solution. Aqueous sodium sulfide was slowly added to a stirred solution of cadmium chloride plus surfactant, which produced a clear, yellow-orange colloidal sample of cadmium sulfide. The particle sizes of the colloids were on the order of 250-300 Å in radius. There was no observable change in the particle radius upon addition of MV^{2+} to the CdS colloids. Furthermore, in the absence of surfactant CdS rapidly precipitates from solution.

Preparation of TiO_2 Colloid Solution. Titanium tetraisopropoxide (Aldrich Chemical Company) at a concentration of 5 ml in 25 ml of isopropanol, was added dropwise to 0.1 M aqueous HCl solution while stirring. After the addition was complete, the solution was stirred for another 10 minutes and then heated slowly to remove the solvent; the residue was dried under vacuum at 118°C. The TiO_2 powder readily peptized in water. Aqueous colloidal solutions tended to precipitate in basic solutions; addition of large amounts of inert salts also precipitated the colloids. Photoplatinization was done as reported earlier (9).

Amorphous TiO_2 was prepared as reported earlier (10). Commerical TiO_2 powder (Alfa Chemical Co.) was crushed in a mortar prior to use.

<u>Instrumentation.</u> Pulsed irradiation studies were carried out with either a 337 nm beam (8 mJ energy, 6 nsec FWHM) from a Lambda Physik EMG-100 laser, or with 490 nm light (0.1 J energy, 120 nsec FWHM) from a Candela SLL-66A dye laser. The short-lived transients produced were monitored by fast spectrophotometry (response $\leqslant$ 1 nsec) and the data was captured by a Tektronix 7912 A digitizer with subsequent processing by a 4051 minicomputer (<u>8</u>). Steady state irradiation studies were performed using a 300 W quartz iodine lamp with cut off filters interposed between the sample and lamp so that only light of wavelength greater than 400 nm reached the sample. Absorption and emission spectra were measured on a Perkin Elmer 552 spectrophotometer and a Perkin Elmer MPF 44B spectrofluorimeter, respectively.

Colloid particle sizes were measured on a Nicomp HN5-90 dynamic light scatter spectrometer. The Nicomp particle analyzer is a light (single mode 6328 Å He-Ne laser) scattering computerized instrument which utilizes the theory of Rayleigh scattering of translational Brownian particles to compute the mean hydrodynamic radius $\bar{R}$ using the Stokes-Einstein relationship for spherical particles, $D = kT/6\pi\eta R$.

A built-in microcomputer system performs rapid quadratic least squares fit to the data, yielding $\bar{D}$, $\bar{R}$, σ (normalized standard deviation of the intensity weighted distribution of diffusion constants) and χ squares goodness of fit. The greater the value of σ the larger the degree of polydispersity present in the particle sizes $-\sigma$ values less than 0.2 are generally considered to correspond to pure monodisperse systems. A typical result obtained for 4 x 10^{-3}M CdS-SDS is: $\bar{D}$= 7.25 x 10^{-8} cm^2/s, $\bar{R}$ = 300 Å, σ = 0.60 and χ^2 = 1.35.

Results and Discussion

Figure 1 shows the photoinduced charge transfer reaction between pyrene and dimethylaniline, DMA. Excitation of the system is usually achieved through the arene leading to a long-lived excited state of pyrene with a lifetime which is several hundred nanoseconds. During this time period, the excited state may interact with dimethylaniline giving rise, in non-polar media to an exciplex, which is an excited complex with established absorbance and emission spectra, and a lifetime in the region of 50 nanoseconds. However, in more polar media, for example in alcohols or acetonitrile, little fluorescence of the exciplex is observed but efficient charge separation is achieved, and ions, such as pyrene anion and dimethylaniline cation, are identified in pulsed laser experiments. The rate of reaction of excited pyrene and dimethylaniline is diffusion controlled in all homogeneous solvents. The lifetime of the ions produced on excitation is short, existing for several microseconds, and depends on the conditions of the systems. The effect of micelles on such systems has been studied; for example, pyrene is located in a micelle such as cetyl trimethyl ammonium bromide, CTAB, close to the surface of the micelle, where it may be excited in the presence of dimethylaniline, which is also contained on the micelle surface. The reactants are separated by a short distance, and in effect a high local reactant

concentration is achieved, which leads to rapid reaction. This is
the concept of the experiment, and the experimental techniques used
to study the photo-events are to observe the quenching of the pyrene
fluorescence, and the development of the ions following pulsed laser
excitation.

It is shown that pyrene fluorescence is rapidly quenched when
dimethylaniline is added to a CTAB solution of these two reactants,
and that ions are very rapidly developed in the pulsed laser
experiments. Figure 2 shows typical spectra observed for the flash
photolysis of pyrene and dimethylaniline in a hydrophobic
homogeneous solvent, such as cyclohexane, in a polar homogeneous
medium such as methanol, and in a CTAB micellar system. It can be
seen that only excited states are developed in the hydrophobic
medium as exemplified by the excited triplet absorption spectrum and
exciplex emission, while in polar media such as methanol,
significant yields of ions develop which replace the excited
states. In CTAB this effect is even further amplified by large
yields of dimethylaniline cation and pyrene anions, which are
identified by their characteristic absorption spectra as indicated
in the figure. The ions are developed rapidly during the laser
pulse, and for a short time period following the pulse, depending on
the concentration of dimethylaniline. As shown by the inserts in
Figure 2, the pyrene anion when formed in CTAB has an extremely long
lifetime and analysis of such data shows that it can live for
several milliseconds, in contrast to a lifetime of microseconds in
homogeneous polar media such as acetonitrile and methanol. An
interesting feature also shown as an insert in Figure 2 is that if
an anionic surfactant is used, such as sodium lauryl sulfate, NaLS,
that the ions again develop rapidly on excitation but that there is
a sharp initial ion decay, followed by a much slower one. This
kinetic feature arises from the fact that the dimethylaniline cation
cannot escape from the anionic micelle where the pyrene anion is
located, and rapid recombination is observed that is even faster
than that in homogeneous polar solutions.

However, our concern is with the cationic surface which
promotes a rapid exchange of an electron from dimethylaniline to
pyrene, and thereafter maintains a long-lived ion which can react
with further solutes added to the system. The concept of the
experiment is, that dimethylaniline transfers the electron rapidly
to pyrene via a diffusion controlled reaction, which occurs by
movement of the reactants on the surface of the micelle until they
encounter each other. Electron transfer then occurs, and the back
reaction of the two ions is prevented by the surface of the micelle,
which holds the reactants in an unsuitable configuration for back
reaction to occur. However, the repulsive positive force of the
micelle on the dimethylaniline cation rapidly drives it away from
the micelle, and effective and efficient charge separation is
achieved, with a quantum yield Q of unity for the process of charge
separation.

There is some question as to whether such a system actually
transfers electrons via an encounter, or whether the electron is
actually ejected from DMA to pyrene over a distance, and before
encounter occurs. This latter statement is shown not to be correct
(5) as the kinetics follow that of a diffusion controlled reaction

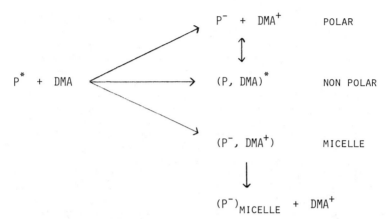

Figure 1. Electron transfer processes of pyrene (P) with dimethylaniline (DMA).

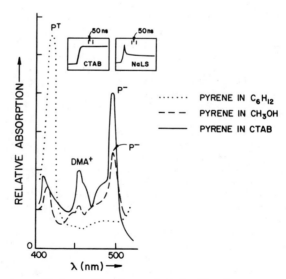

Figure 2. Spectra of short-lived species in the laser photolysis of pyrene and dimethylaniline in C_6H_{12}, CH_3OH and CTAB micelles. Insert: oscilloscope time traces of pyrene anion P^- in CTAB and NaLS micelles. (P^T is the pyrene triplet, P^- the pyrene anion and DMA^+ the dimethylaniline cation).

at the surface of the micelle, and compare favorably with several
other reactions which have been previously determined to occur via
encounters, for example pyrene excimer formation. Also, if such a
system is frozen to 77 K so that diffusion cannot occur, then it is
impossible to promote a reaction even when the DMA and the pyrene
are in close proximity. To sum up, no reaction occurs unless the
reactants encounter each other.

Table I shows the effect of various systems such as micelles,
swollen micelles (achieved by adding hexanol to CTAB), microemulsion
systems, vesicles formed from a double-chain CTAB surfactant, and
reversed micelles with water cores formed with benzyl
dimethylcetylammonium bromide in benzene. The active chromophore
exists either as pyrene, pyrene sulfonic acid or pyrene
tetrasulfonic acid. Essentially the concept here is that the polar
derivatives of pyrene will always locate pyrene at the surface of
the micelle as these anionic species of pyrene complex with the
positively charged surface. Dimethylaniline is used as an electron
donor in each case. It can be seen that for pyrene, a continual
decrease in the yield of the pyrene anion (ion yield of unity in the
micelle) is observed on going from micelle to swollen micelle, to
microemulsion, and no yield of ions is observed in a reversed
micelle system. With pyrene tetrasulfonic acid the yield of ions
over the different systems is fairly constant, even across to the
reverse micellar system. However, the lifetime of the ions is
extremely short in the reversed micellar system. An explanation for
such behavior can be given as follows: as we transverse across the
various assemblies, from micelle to microemulsion through swollen
micelles, it is found that pyrene is gradually located in a more
hydrophobic environment and further away from the particle
surface. Thus, reaction of DMA and pyrene occurs under various
geometric arrangements, first of all via a randomly organised
encounter on the surface. Back electron transfer of the products is
restricted due to their random geometry and rearrangement into a
more suitable complex is inhibited by the micelle surface. A
geometric arrangement which gives rise to rapid back reaction is
readily seen in the reversed micelle system, where pyrene and DMA
originally reside in the benzene phase, and form an exciplex after
excitation. The exciplex is quenched at a diffusion controlled rate
at the reversed micellar water pools, and no ions are formed.
Hence, the exiplex has a sandwich structure which promotes efficient
back e^- transfer at the water pool, and the ion yield is very
small. However, a sandwich reactant pair of this sort is not formed
on a micelle surface and back reaction is slower than the escape of
the cation from the surface. The swollen micelle and microemulsion
systems lead to both randomly organised ionic products and sandwich
pairs, to varying extents, which are reflected in the observed yield
of ions. With polar derivatives of pyrene, e.g. pyrene sulfonic
acid, etc., the reactants are kept on the assembly surface where
reaction occurs, giving rise to ions from a non-sandwiched type of
configuration. In the reverse micellar system, these ions although
they are formed, nevertheless have a short lifetime, as they cannot
escape to any great distance in the small water pool. Thus,
micelles are far superior to microemulsions in various aspects of

promoting and extending yields of ions in photoinduced electron
transfer reactions.

Table II shows various reactions, where electrons are
transferred from pyrene anion to various entities such as carbon
dioxide and methyl viologen. The reactions with some of these
solutes, in particular CO_2, are very slow in homogeneous solution in
micelles. Nevertheless, the long lifetime of the pyrene anion in
micellar systems allows sufficient time for complete electron
transfer to the added solute to produce reduced methyl viologen,
MV^+, or CO_2^-.

Polymerized Microemulsion Systems. A microemulsion of styrene and
divinylbenzene with CTAB + hexanol may readily be made, and
subsequently polymerized to form a polymerized microemulsion
(5,6,7). This system exhibits two sites of solubilisation for
photosystems such as pyrene, one in the surfactant skin layer, and
the other in the polymerized styrene-divinylbenzene core.
Photochemical reactions induced in the surfactant skin are very
similar to those observed in micelles and are not immediately of
concern to us at this stage. However, photochemical reactions
induced in the rigid polymerized core are of interest, as they
essentially confine reactants to a small region of space where
movement is restricted as compared to a fluid non-polymerised
microemulsion or a micelle. Thus, diffusion is minimised, and it
may be possible to investigate reactions which occur over a distance
rather than reactions which occur by diffusion. In order to
eliminate reactions in the surfactant skin a microemulsion can be
constructed which contains cetyl pyridinium chloride in place of
CTAB. The pyrene that resides in the surfactant skin layer is
immediately quenched by the pyridinium group following excitation.
In the systems discussed here, 20% of the pyrene is located in the
skin region and is immediately quenched, while about 80% of the
pyrene is in the polymerised core where it carries out reactions,
fluoresces, forms triplet states, etc. Reactions such as energy
transfer from excited pyrene to perylene occur quite readily in this
system, and it has been shown that the mechanism is via transfer
over distances of 20-50 Å in the polymerized core of the
microemulsion. However, it is not possible to carry out quenching
of excited pyrene by dimethylaniline in the polymerized core of the
microemulsion. This again fits in with the concept that this
reaction requires diffusion to form encounter complexes. However,
an electron transfer reaction has been observed between excited
pyrene and nitrobenzene to give the pyrene cation and the
nitrobenzene anion. This also occurs efficiently in homogeneous
polar media, in micelles, and in non-polymerised microemulsions. In
polymerized microemulsions, nitrobenzene quenching of the pyrene
fluorescence is quite inefficient compared to that observed in
microemulsions or in micellar media. However, reaction does occur
at high nitrobenzene concentrations (10^{-2}M), and typical data are
shown in Figure 3 for the quenching of pyrene fluorescence. The
ragged line is that due to the experimental data, while the smooth
curve is the data calculated from a tunnelling type process of
electron transfer from pyrene to nitrobenzene. The agreement of
theoretical and experimental data is excellent and indicates that in

Table I. Relative Yield of P^- (493 nm)[a]

	CTAB	A	B	C	D
P	1.0(0.40)	0.50(0.32)	0.24(0.12)	0.85(0.67)	no ions
PBA	1.0	0.90	0.77	0.80	short lived ions
PSA	1.0	0.87	0.66	0.96	short lived ions
PTS	1.0	0.95	0.90		short lived ions

[a] 5.0×10^{-5}M P(PBA, PSA, PTS)/(5-10) $\times 10^{-3}$ M DMA. Better than
95% of the fluorescence is quenched: A, 0.05 M CTAB + 0.1 M
hexanol; B, 0.05 M CTAB + 0.25 M hexanol + 0.3 M dodecane
(1.83% (w/w) CTAB, 2.5% hexanol, 5.0% dodecane); C, 0.01 M
DDAB; D, reversed micelle system. Values in parentheses are
obtained for DBA (0.005 M), dibutylaniline.

Table II. Reaction Rate Constants for
Pyrene Anions, P^-, with Acceptors

Solute	Homogeneous Soln. $M^{-1} s^{-1}$	Cationic Micelle, $M^{-1}s^{-1}$
O_2	2.0×10^{10}	5.0×10^9
Eu^{3+}	2.7×10^9	1.3×10^7
		(10^{-2}M CTAB)
MV^{2+}	2.6×10^{10}	3.3×10^7
		(6×10^{-2} M CTAB)
		1.8×10^7
		(10^{-2}M CTAB)
Cetylpyridinium Chloride	2.6×10^{10}	10^7 (s^{-1}) [1$\underline{st}$ order decay]
CO_2	∿10^7	∿10^7

this system, due to the rigid nature of the assembly that holds the
pyrene and the nitrobenzene, where it is not possible to achieve
efficient diffusion, that electron transfer occurs over distances of
the order of 10-20 Å. Polymerized microemulsions provide assemblies
where it is possible to investigate whether reactions can occur
while the reactants are separated--the reaction may be either energy
transfer or electron transfer reactions. This experimental
arrangement is not possible with micelles, as it is not possible to
polymerize small entities such as micelles in order to give rigid
assemblies.

Photochemistry in Colloidal Cadmium Sulfide.   Figure 4 shows an
illustration of a typical cadmium sulfide particle prepared by a
precipitation technique (11-13). A small, roughly spherical cadmium
sulfide particle is pictured, coated with adsorbed surfactant, such
as SDS or CTAB, in order to stabilize the system and keep it
suspended for long periods of time. Excitation of this particle
leads to electron-hole pairs in the bulk of the semiconductor which
migrate to the CdS surface. On the surface the pairs are trapped by
surface defects. Recombination of electrons with trapped holes on
the surface leads to emission which is observed in the red part of
the spectrum. However, recombination of electron-hole pairs in the
bulk of the semiconductor leads to an emission which is in the green
and at higher energy, as observed on excitation of single crystals
of cadmium sulfide. The electrons may be transferred from the
particle to suitable acceptors, such as methyl viologen, which are
adsorbed on the particle surface. This may be encouraged by the use
of a negatively charged surfactant such as sodium lauryl sulfate.
The net result is that shining light on the system gives rise to
electron hole pairs, and the electron is transferred out to a
suitable acceptor, where it is stored for extended periods of time
(hours).
     Figure 5 shows typical emission and absorption spectra for a
dispersion of cadmium sulfide in water, and for two colloids, one
stabilized with sodium lauryl sulfate, and one stabilized with the
surfactant cetyldimethylbenzylammoniumchloride, CDBAC. At low
excitation intensities the emission spectra of dispersions and
colloids are red shifted from that observed for single crystals,
which normally exhibit a maximum emission peak at 520 nanometers.
However, excitation at high intensity, e.g. with a laser, produces a
blue shift in the emission spectra and in the case of a dispersion,
an emission spectrum is observed which is very similar to that of a
single crystal. In the colloidal systems a progressive blue shift
of the emission spectra is observed at higher excitation
intensities. This is associated with a saturation of the surface
trapping sites such that the majority of electron hole-pairs are
produced in the bulk of the colloidal semiconductor, where they emit
light characteristic of bulk cadmium sulfide. Surface adsorption of
ions such as cupric ions modify the surface, and at low excitation
intensities these systems exhibit a red-shifted emission spectra.
Copper forms cupric sulfide on the surface of the particle by
exchanging with surface cadmium. The copper site traps holes which
upon recombination with surface electrons gives rise to a red-
shifted emission.

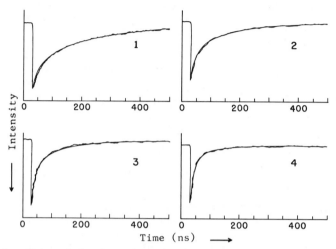

Figure 3. Rate of decay of pyrene fluorescence in CPC-P-µE in
the presence of nitrobenzene; [pyrene] = 10^{-5}M;
[nitrobenzene]: (1) 10^{-3} M; (2) 2.5×10^{-3}M; (3) 5×10^{-3} M;
and (4) 10^{-2} M. Coarse curve, experimental data; smooth curve,
calculated curve according to Eq. 2 from Reference 7.

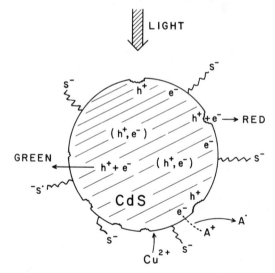

Figure 4. Illustration of a colloidal CdS particle stabilized
via adsorbed surfactant. S^- represents the polar head group of
the surfactant; A^+, the oxidized form of an acceptor; A, the
reduced form of the electron acceptor; e^-, H^+ represents the
electron/hole pairs formed upon excitation of the CdS particle
which may either recombine in the bulk and emit green light or
migrate to the surface and emit red light upon recombination.

Figure 6 shows the effect of methyl viologen on the emission
intensity of cadmium sulfide, as well as the appearance of reduced
methyl viologen, which results from transfer of electrons from the
excited particle to adsorbed methyl viologen. These data indicate
that the emission quenching occurs via electron transfer from the
CdS surface to adsorbed acceptors. Figure 7 shows Stern-Volmer and
Perrin plots for the quenching of CdS emission in a sodium lauryl
sulfate colloid and a CDBAC colloid. With the sodium lauryl sulfate
colloid, the quenching is quite marked at low concentrations of
cupric ions, methyl viologen, and propyl viologen sulfonate, while a
negatively charged quencher, such as iodide ion, has no effect on
the system. In the case of the positively charged CDBAC colloid,
methyl viologen quenches inefficiently, propyl viologen still
quenches to some extent, but copper is still very efficient. The
Stern-Volmer plots of the data are not linear: in the case of
methyl viologen there is a downwards curvature in the plot while
with cupric chloride there is an upwards curvature with increasing
concentration. If the copper data is plotted as a Perrin plot of
$\log {}^{Io}/I$, verses the concentration of adsorbed copper on the surface
of the particle, then a linear plot is obtained indicating static
quenching of the CdS luminescence by copper. In the case of methyl
viologen the downwards curvature is mainly the result of saturation
of the CdS surface by methyl viologen. If the actual concentration
of methyl viologen adsorbed on the CdS surface is plotted, rather
than bulk MV^{2+} concentration, then a linear Stern-Volmer plot is
obtained. These data are explained as follows: the hole and
electron migrate to the CdS surface where the hole is captured by
copper in the form of CuS. However, the hole on reaching the
surface is no longer mobile, and if it meets up with the cooper
sulfide on the surface then it is captured leading to a reduction in
CdS emission intensity. This gives rise to static quenching.
However, if it is not immediately captured by copper then no
quenching occurs as it combines with electrons in the normal
fashion. The kinetic processes occur in a very short period of
time, much shorter than the emission lifetime which in these
colloids is less than 10^{-9} sec.

In the case of methyl viologen it is the electron that is
captured at the CdS surface. The electron is still very mobile at
the CdS surface where it may be captured by adsorbed MV^{2+} during its
diffusive motion over the surface. This gives rise to a Stern-
Volmer type of quenching even though it is over a very short period
of time in much less than 10^{-9} sec. These data also show that it is
essential for the reactive entities to be on the surface of the CdS
particle in order for reaction to occur.

Figure 8 shows typical pulsed laser data for the production of
methyl viologen on the surface of a CdS particle stabilized with
sodium lauryl sulfate. The reduced methyl viologen is formed
immediately in the laser pulse but shows a subsequent decay due to
back reaction at the particle surface. This figure also shows the
effect of oxygen on the quenching of reduced methyl viologen, MV^+.
The rate constant for this quenching is extremely low in the case of
methyl viologen on cadmium sulfide. It is known that the reaction
of reduced methyl viologen with oxygen is environment-dependent
(14,15) and the present data indicate that reduced methyl viologen

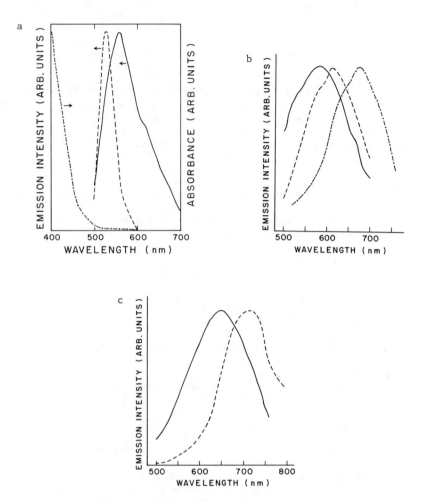

Figure 5.   a) Absorption(-.-) spectrum of a CdS dispersion and
a surfactant stabilized CdS colloid; emission spectra of a CdS
dispersion upon low (-) and high (--) excitation intensity.
b) Emission spectra of SDS stabilized CdS colloids at low (--)
and high (-) excitation intensity; emission spectrum of CdS/SDS
+ 10^{-5} M Cu^{2+} (-.-).   c) Emission spectrum of CDBAC stabilized
CdS colloid (-); effect of 1.5 x 10^{-5}M Cu^{2+} (--).

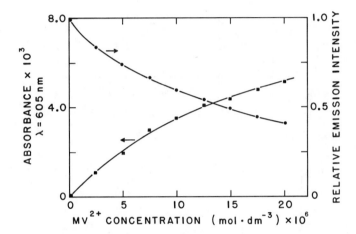

Figure 6. Yield of $MV^{+\cdot}$ and relative intensity of CdS emission in the photolysis of a CdS/SDS colloid upon addition of MV^{2+}.

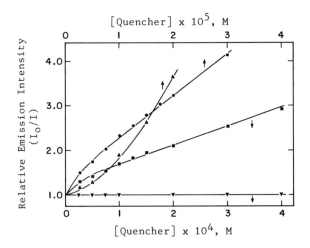

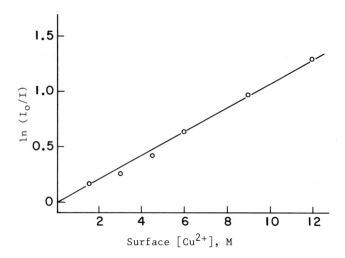

Figure 7a. Effect of quenchers on the CdS emission intensity.
CdS/SDS$_2$ colloid: Stern-Volmer plots for quenching by MV^{2+}
($\bullet$), Cu^{2+} ($\blacktriangle$), PVS ($\blacksquare$), and I$^-$ ($\blacktriangledown$); top: Perrin plot of
Cu^{2+} data.

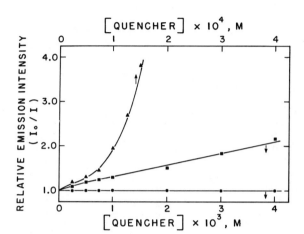

Figure 7b.   Effect of quenchers on the CdS emission intensity. CdS/CDBAC colloid: Stern-Volmer plots for quenching by Cu^{2+} (▲), PVS (■), and MV^{2+}, I^- (●).

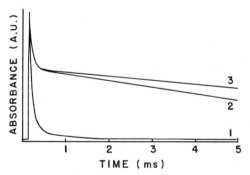

Figure 8.   Pulsed laser studies of $MV^{+\cdot}$ formation and decay in CdS/SDS colloids: Effect of O_2 (1) and pH on $MV^{+\cdot}$ decay (2 = pH 8.2, 3 = pH 11.2).

remains on the particle surface for several hundred microseconds. This explains the rapid back reaction of reduced methyl viologen with the hole on the surface of the particle. It can be shown that reduced methyl viologen migrates around the surface of the particle, as adsorbed copper causes an increased rate of decay of reduced methyl viologen on the surface. The net result of the above processes is that the efficiency of electron transfer from excited cadmium sulfide to acceptors is low, and the quantum yield is less than 10^{-3}. This is mainly due to the short lived nature of the electron-hole pair which does not give ample opportunity for transfer to acceptors, and also due to the fact that the reaction product or reduced methyl viologen does not escape from the surface, and hence back reaction leads to low yields.

Two features have been utilized to correct this problem and to increase the yield of photoreduced products. In the first instance a positively charged colloid was employed using either CTAB or CDBAC. This colloid does not adsorb methyl viologen. However, if EDTA is placed in the system, a complex of methyl viologen and EDTA with a resultant negative charge is formed, and this is adsorbed at the positive surface of the colloid. The methyl viologen EDTA complex is adsorbed at the surface of the colloid and participates in electron transfer from excited CdS and forms reduced methyl viologen. The resulting MV^+-EDTA complex breaks up and the positively charged MV^+ is repelled from the surface of the colloid. The photo yield now increases up to about 0.01, at least a factor of a hundred increase in quantum yield.

A further feature may be obtained by producing a less structured red form of cadmium sulfide. The red CdS colloid exhibits a red shifted emission spectrum compared to that of normal cadmium sulfide, and it also has a much longer lifetime. This gives a much longer lifetime for the electron-hole pair and larger yields of reduced product are expected from such systems.

Photochemistry of Titanium Dioxide Colloids. Another semiconductor colloid used in our studies is titanium dioxide which has a band gap of 3.2 eV. As in the case of cadmium sulfide, excitation of aqueous suspensions of this particle leads to electron-hole pair separation which can be intercepted with suitable redox reagents. In the absence of externally added solutes, the photogenerated electron-hole pair recombines to give the starting material and the light energy is dissipated to the medium as heat. Two types of TiO_2 samples are used in this study. TiO_2 prepared at high temperature (80°C) which behaves very similarly to commercial samples, and TiO_2 prepared at low temperature (35°C) which has a particle size of 300 ± 100 Å radius and shows different properties.

Figure 9 shows the emission spectrum of crystalline TiO_2, $(TiO_2)_c$, when excited at 320 nm. The emission intensity increases sharply after 340 nm and reaches a maximuma at 375 nm then tails up to 700 nm. The emission maximum corresponds to the band gap of the semiconductor. Single crystal semiconductors show a sharp emission maximum corresponding to the band gap energy. The broad emission is typical of semiconductor particle suspensions in water. The low energy tail may by due to the electron exchange leading to Ti^{4+}. The emission quantum yield was calculated to be ~10^{-3} when compared to emission from quinine sulfate in 0.5 M aqueous sulfuric acid. Addition of small amounts of MV^{++} or I^- (10^{-3} M) do not quench the emission. However, quenching behavior was observed at high concentrations (>10 mM) of these quenchers.

Amorphous TiO_2, $(TiO_2)_A$, prepared at 35°C does not show any emission at room temperature.

Photoreduction of MV^{++} on TiO_2 Colloids. Pulse photolysis of solutions containing $(TiO_2)_c$ and MV^{2+} leads to instantaneous formation of MV^+, (Fig. 10), which is characterized by its absorption spectrum with maxima at 395 nm and 605 nm. These data can be explained by direct electron transfer from excited $(TiO_2)_c$ to adsorbed MV^{2+}. Benzylviologen, a derivative of MV^{2+} is not adsorbed on the surface, and hence is not reduced on photoexcitation.

The charge separated energy-rich products do not live long enough so that any stored energy can be utilized subsequently. The photoreduced MV^{2+} is still adsorbed on the $(TiO_2)_c$ surface and readily reacts with the hole giving back the starting materials. MV^+ adsorbed on $(TiO_2)_c$ shows an initial fast decay ($\tau_{1/2} \sim 60$ ns) followed by a slower decay with a lifetime of 1 μs. The fraction of the initial decay depends on the pH of the medium. In acidic medium (pH 1) the initial fast decay accounts for all of the process, whereas in basic medium only about 60% of the signal decays by the fast process, while the remaining MV^+ is stable. It should be noted that more MV^{2+} is adsorbed on $(TiO_2)_c$ in basic media than in acidic media.

Amorphous TiO_2/MV^{2+} systems behave quite differently upon photo-excitation. MV^{2+} is not adsorbed on amorphous TiO_2 and so instantaneous formation of MV^+ is not observed. However, in the presence of electron donors such as polyvinyl alcohol (PVA), or halide ions, a slow growth of MV^+ formation is observed over a period of several μs (Fig. 10). In the case of PVA, a permanent reduction of MV^{2+} is observed as reported by Grätzel et al. (16), but in the case of Cl^- no permanent reduction is observed. Pulse photolysis studies show large yields of initially formed MV^+, but rapid back reaction of MV^+ and Cl_2^- yield the starting materials. In 0.1 M $HClO_4$ the reduction of MV^{2+} is not efficient, as electron transfer from ClO_4^- to the TiO_2 hole is not efficient.

The results can be explained as follows:

$$(TiO_2)_a \xrightarrow{h\nu} TiO_2 \ (e^-, h^+)$$

$$e^- + MV^{++} \longrightarrow MV^+$$

$$h^+ + RCH_2-OH \longrightarrow R\overset{\bullet}{C}H-OH + H^+$$

$$MV^{2+} + R\overset{\bullet}{C}H-OH \longrightarrow MV^+ + RCHO + H^+$$

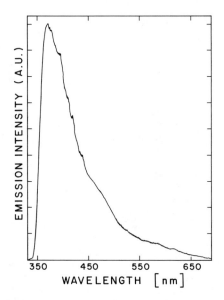

Figure 9. Emission spectrum of TiO$_2$ particulate suspension in water. Excitation wavelength 320 nm.

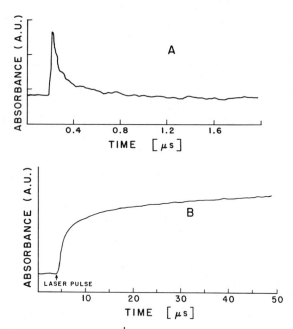

Figure 10. a) Decay of MV$^+$ upon laser excitation of crystalline TiO$_2$ suspensions in water monitored at 395 nm. b) Growth of MV$^+$ upon laser excitation of an Amorphous TiO$_2$ colloid in 2 mM NaOH containing 0.1% w/w polyvinylalcohol monitored at 605 nm. [MV^{2+}] = 1 mM.

The slower growth of MV^+ on excitation of $(TiO_2)_a/MV^{2+}$ suggests that the photoproduced electrons have a longer lifetime in these systems than on the $(TiO_2)_c$.

Photooxidation of Halide Ions. The photogenerated hole can be intercepted by various inorganic and organic ions, such as $S_2O_8^{2-}$, acetate, and I^-, some of which have been used in this study (17). In the case of photolysis of $S_2O_8^{2-}$ on $(TiO_2)_c$, a transient spectrum with a maximum at 455 nm is observed which is attributed to the $SO_4^{\cdot-}$ radical. Similar studies with iodide give a transient with a maximum at 390 nm, which has been assigned to the I_2^- radical (18). These transitory species appear within the pulse duration (6 $\times 10^{-9}$s) even at 10^{-4} M anion concentrations. It is reasonable to assume that only the ions adsorbed on the surface are reactive. In homogeneous solution the halide anion radicals disporportionate to give neutral molecules and halide ions, and the decay kinetics follow a second-order process (18). However, the transients generated on the particle surface of $(TiO_2)_c$ colloid in water decay by an apparent first-order process. This indicates that the transients are formed on the surface. Before the radical escapes from the surface of the particle to the solution bulk, they react with the photogenerated electrons on the surface, and the decay appears to be first-order.

Iodide ions are adsorbed on amorphous TiO_2 particles. Flash photolysis of amorphous TiO_2 and iodide results in the instantaneous formation of I_2^- as the adsorbed iodide transfers an electron to the photogenerated hole and the iodine atom combines with the excess iodide to give I_2^- ion. Similar to crystalline TiO_2, the conduction band electron and I_2^- back react to give the starting materials. Since the electron lives longer on the amorphous TiO_2 as seen from the reaction with MV^{2+}, I_2^- is expected to live longer. Indeed the half-life of the I_2^- decay is ~35 µs compared to 14 µs for crystalline TiO_2 particle suspensions.

Photochemical Reduction of Carbonate to Formaldehyde.
Carbonate is adsorbed on colloidal TiO_2, and flash photolysis of these solutions exhibits an instantaneous formation of carbonate anion radical (19). As with other systems, carbonate anion radicals back react with conduction band electrons giving the starting materials. Carbonate anion radicals decay faster in aerated solution ($\tau_{1/2}$ = 0.9 ± 0.1 µs) than in deaereated solution ($\tau_{1/2}$ = 2.9 ± 0.2 µs) indicating that oxygen reacts with one of the photoproducts (19).

Illumination of aerated or deaerated solutions containing TiO_2/Pt and carbonate (1 M), results in the formation of formaldehyde (Figure 11). Decreasing the concentration of carbonate by 100 fold does not appreciably affect the formation of the final product, as the surface is saturated with CO_3^{2-} under these conditions. However, the formaldehyde yield decreases at concentrations less than 1 mM carbonate. Illumination of the solution for longer periods does not result in an increased yield of formaldehyde. In fact, the yield decreases at longer irradiation times. Since larger amounts of formaldehyde are formed in the initial stages, it is reasonable to assume that formaldehyde

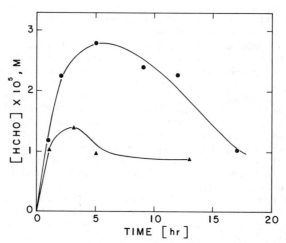

Figure 11.   Steady state yield of formaldehyde as a function of irradiation time.   (●) 1 M sodium carbonate   (▲) 1 x 10^{-3} sodium carbonate.

undergoes secondary reactions. Since methanol was not observed as
one of the products at longer irradiation times, oxidation of
formaldehyde to formate is proposed to be the secondary reaction.

$$TiO_2 \xrightarrow{h\nu} TiO_2 \ (e^-, \ h^+)$$

$$CO_3^{2-} + h^+ \longrightarrow CO_3^- \ ; \ O_2 + e^- \longrightarrow O_2^-$$

$$CO_3^- + O_2^- \longrightarrow CO_3^{2-} + O_2$$

$$CO_3^- + e^- \longrightarrow CO_3^{2-}$$

$$CO_3^- + O_2 \longrightarrow CO_3 + O_2^-$$

$$CO_3 \longrightarrow CO + O_2$$

$$CO + 2e^- \xrightarrow{2H^+} HCHO$$

$$\text{Net reaction} \quad CO_3^{2-} \xrightarrow[H^+]{e^-, \ h^+} HCHO + O_2$$

The important aspect of the reaction is that both hole and electron
are used to drive the net chemical reaction. The free energy stored
in this reaction is 138 kcals/mole.

Conclusions

The foregoing studies show the unique features introduced into
photochemical systems by the use of microemulsions which assemble
reactants and promote reaction features of interest. The use of
large surface areas, as with semiconductor colloids, also leads to
unique photochemical reactions which are quite different to bulk
reactions. Finally, colloidal systems provide excellent vehicles
for studies in interfacial photochemistry.

Acknowledgments
The authors wish to thank the Army Research Office, the National
Science Foundation and the Petroleum Research Foundation of the
American Chemical Society, for support of this work.

Literature Cited

1.   Fendler, J. H. "Membrane Mimetic System". Wiley, N. Y.
     (1983).
2.   Turro, N.; Braun, A; Gratzel, M. Angew. Chem. 1980, 19, 675.
3.   Thomas, J. K. Chem. Rev. 1980, 80, 283.
4.   Atik, S.; Thomas, J. K. J. Am. Chem. Soc. 1981, 103, 3550.
5.   Atik, S.; Thomas, J. K. J. Am. Chem. Soc. 1981, 103, 4279.
6.   Atik, S.; Thomas, J. K. J. Am. Chem. Soc. 1982, 104, 5868.
7.   Atik, S.; Thomas, J. K. J. Am. Chem. Soc. 1983, 105, 4515.
8.   Atik, S.; Thomas, J. K. J. Am. Chem. Soc. 1981, 103, 4367.

9.   Krautler, B.; Bard, A. J. J. Am. Chem. Soc. **1978**, <u>100</u>, 4317.
10. Duonghong, D.; Borgarello, E.; Grätzel, M. J. Am. Chem. Soc. **1981**, <u>103</u>, 4685.
11. Kuczynski, J.; Thomas, J. K. Chem. Phys. Letts. **1982**, <u>88</u>, 445.
12. Kuczynski, J.; Milosavljevic, B. H.; Thomas, J. K. J. Phys. Chem., in press.
13. Kuczynski, J. P.; Milosavljevic, B. H.; Thomas, J. K. J. Phys. Chem. **1983**, <u>87</u>, 3368.
14. Farrington, J. A.; Ebert, M.; Land, E. J.; Fletcher, K. B.B.A. **1973**, <u>314</u>, 372.
15. Patterson, L. K.; Small, R. D.; Scaiano, J. C. Rad. Res. **1977**, <u>72</u>, 218.
16. Duonghong, D.; Ramsden, J.; Grätzel, M. J. Am. Chem. Soc. **1982**, <u>104</u>, 2977.
17. Henglein, A. Ber. Bunenges. Phys. Chem. **1982**, <u>86</u>, 241.
18. Grossweiner, L. S.; Matheson, M. S. J. Phys. Chem. **1957**, <u>61</u>, 1089.
19. Chandrasekaran, K.; Thomas, J. K. Chem. Phys. Lett. **1983**, <u>99</u>, 7.

RECEIVED June 8, 1984

Reaction Kinetics as a Probe for the Dynamic Structure of Microemulsions

R. LEUNG, M. J. HOU, C. MANOHAR[1], D. O. SHAH[2], and P. W. CHUN[3]

Departments of Chemical Engineering and Anesthesiology, University of Florida, Gainesville, FL 32611

The coagulation of the hydrophobic AgCl sols has been investigated by stopped-flow method employing microemulsions composed of SDS, IPA, water and benzene as reaction media. Two enhancement peaks of the coagulation rate have been observed. In order to correlate the enhancements with the dynamic structures of reaction media, the physico-chemical properties of the microemulsions have been measured using various techniques including conductance, viscosity, light scattering, ultracentrifuge, ultrasonic absorption and pressure-jump relaxation. Subregions consisting of different microstructures within a clear single microemulsion phase have thus been delineated. Accordingly, the broad enhancement peak of the coagulation of 0.56 water mass fraction is associated with surfactant aggregates in the alcohol-rich solvent. The sharp enhancement peak at 0.855 water mass fraction has been attributed to the fast coagulation of normal micelles leading to micellar growth. It has been found by the titration method that a minimum of eight water molecules per surfactant molecule are required to hydrate the sulfate group for the dissolution of SDS into IPA.

A microemulsion is a thermodynamically stable isotropic dispersion of two relatively immiscible liquids, consisting of microdomains of one or both liquids stabilized by a interfacial film of surface-active molecules. In practice, one often identifies the microemulsion by the formation of a clear isotropic mixture of the two immiscible liquids in the presence of appropriate emulsifiers. In a phase diagram, such region is referred as the microemulsion phase. It has been shown that microemulsion regions consist of different microstructures (1,2), e.g., water-in-oil (W/O), oil-in-water (O/W),

[1] Current address: Chemistry Division, Bhabha Atomic Research Center, Bombay-400085, India
[2] To whom correspondence should be addressed.
[3] Current address: Department of Biochemistry and Molecular Biology, College of Medicine, University of Florida, Gainesville, FL 32611

etc. A number of different techniques such as light scattering, neutron-scattering, ultrasonic absorption, ultracentrifugation, as well as relaxation measurements have been used to characterize different microstructures within a single phase microemulsion region (1-12). The presence of different microstructures in reaction media can affect the mechanism and rate of a chemical reaction. Therefore, kinetics of a chemical reaction can be used as a probe for investigating the transition of the microstructure within a single phase microemulsion region. As a matter of fact, the microemulsions, being transparent, are very suitable systems for monitoring the chemical reactions spectrophotometrically.

Silver chloride precipitation is one of the most extensively investigated reactions due to its importance in photographic industry (13-22). Aqueous suspension of silver halide in the presence of protectice colloid such as gelatin, surfactants, is usually called emulsions in photography. The turbidity measurement is conventionally used to monitor the precipitation process. It is well-known that the turbidity of the aqueous solution increases rapidly upon mixing silver nitrate with alkali halid in the absence of protective colloids. Matijevic and Ottewill (13,14) have attributed such turbidity enhancement to the fast coagulation and precipitation of silver halide sols. They investigated the effect of cationic detergents on the stability of negatively-charged silver halide sols. Periodic "sensitized" coagulation and stabilization regions of the sols were observed upon increasing the detergent concentration. When the detergent concentration approached its critical micelle concentration, the detergents acted as protective colloids and the coagulation rate of silver halide sols was drastically reduced. In another set of studies on the stability of positively charged silver halide sols in the presence of anionic surface active agents, Ottewill and Watanabe (15) have shown that the stability of the sols decreased initially (coagulated faster) and then increased again up to a limiting value due to the adsorption of surface active molecules on the sols. A theory has also been proposed to account for these experimental findings (14,15).

In this preliminary study, we have investigated the coagulation rate of silver chloride sols in microemulsion media. The results are intimately related to structural properties of the microemulsions.

Experimental

Materials and Methods. The sodium dodecyl sulfate (SDS) was of purity higher than 99% from BDH. $AgNO_3$, NaCl were ACS certified grade from Fisher Scientific Company. Isopropyl alcohol (IPA) and benzene were of 99% purity from Fisher Scientific Company. All chemicals were used as received without further purification.

The viscosity was measured by Cannon-Fenske viscometer (#100). Light scattering was monitored by Duophotometer Model 5200 (Wood Mfg. Co.). The ultrasonic absorption was measured using Matec Pulse Modulator and Receiver model 6600. Pressure-jump studies were performed using DIA-LOG system with conductivity detection. The stopped-flow experiments were carried out using Durrum Model D-115 systems. The ultracentrifuge studies were carried using a Beckman Model E analytical ultracentrufuge with Schlieren optics. All the measurements were carried out at 25°C.

<u>Preparation of AgCl Sols</u>. Two stock aqueous solutions of 5 mM AgNO$_3$
and of 5 mM NaCl were first prepared separately. Microemulsions
were then prepared by mixing specific amount of SDS, IPA, benzene
and either one of the two aqueous stock solutions so that two identical microemulsion samples at a desired composition were formed
except one contained AgNO$_3$ and the other contained NaCl. Using
stopped-flow apparatus, the AgCl sols were formed upon a rapid injection of these two identical microemulsion samples into a mixing
chamber. The turbidity development through the coagulation of the
AgCl sols was then followed by transmittance measurement.

<u>Coagulation Rate Measurements</u>. As shown by Matijevic and Ottewill
(13), the turbidity τ resulting from the formation of the solid
phase in a solution can be defined by the relation:

$$I = I_0 e^{-\tau l} \tag{1}$$

where I_0 and I are the intensities of the incident and transmitted
radiation respectively, and l is the optical path length of the cell
employed. For small particles (r < $\lambda/20$) in the absence of consump-
tive light absorption, τ is related to the number of particles
per milliliter N_p, and their individual volume V_p by Rayleigh equation,
viz.,

$$\tau = A\, N_p V_p^2 \tag{2}$$

here A is an optical constant given by:

$$A = \frac{24\pi^3 n_0^4}{\lambda^4}\left(\frac{n^2 - n_0^2}{n^2 + 2n_0^2}\right)2 \tag{3}$$

where n_0 is the refractive index of the solvent while n is the refractive index of the particles. λ is the wavelength of the light
used (in vacuo).
     For a coagulation process, the change of turbidity with time is
given by the equation,

$$\tau = A\, N_p V_p^2 (1+kt) \tag{4}$$

where k is a rate constant. When the particles are small enough to
obey Rayleigh's equation (2), a linear relationship between turbidity
and time is obtained (13).
     For all samples studied, the use of initial coagulation rate (the
first 2 seconds in a overall process longer than 100 seconds) does
provide a linear (or pseudo-linear) rate constant k for comparative
purpose. Moreover, the use of initial coagulation rate for comparison
is justified in view of the fact that the concentration product of
AgCl in our samples is considerably greater than the solubility product (1.765 x 10^{-10} at 25°C) and hence, upon mixing, the nuclei of
AgCl are formed spontaneously without significant induction-time delay. However, due to the possible complication involved in the multicomponent systems, we report all our initial coagulation rates in
terms of a relative rate constant k_{rel} using pure water as a reference,

$$k_{rel} = \frac{k \text{ in microemulsion media}}{k \text{ in pure water}} \tag{5}$$

Assuming the changes of A and V_p on various microemulsion media are negligible, and N_p is proportional to the volume fraction of water in a microemulsion sample, the following equation is then used to obtain a relative rate constant for a microemulsion sample,

$$k_{rel} = \frac{\left[\dfrac{1}{N_p} \dfrac{d\ln(I/I_0)}{dt}\right]_{microemulsion}}{\left[\dfrac{1}{N_p} \dfrac{d\ln(I/I_0)}{dt}\right]_{water}} \tag{6}$$

Results and Discussions

<u>Coagulation of Hydrophobic AgCl Sols</u>. The experimental line along which all the measurements were made in a microemulsion phase is shown in Figure 1. The phase diagram for IPA + SDS + benzene + water system was obtained by Clausse (23). The mass ratio of SDS/IPA is 0.5 throughout the phase diagram, and the mass ratio of benzene/SDS is 0.33 for all the samples studied. The structure of the microemulsions employed as reaction media may differ markedly from alcohol-rich (point A) to water-rich (point B) region in the phase diagram. Figure 1 shows the variation of relative rate constant as a function of mass fraction of water in microemulsions. The enhancements in the coagulation rate at specific water mass fractions of 0.56 and 0.855 are quite striking. Of the two enhancement peaks observed, the highest peak at 0.855 mass fraction of water is very narrow and intense, while the other peak at 0.56 is broad but less intense. The former does not seem to be related to the formation of micelles because the surfactant concentration is about 15 times higher than the critical micelle concentration. The broadness of the latter peak is indicative of a certain structural property that may persist in a wide range of alcohol-rich region of the microemulsion phase.

It is of interest to note the resemblance of our data to that reported by Friberg et al. (24). They have investigated the rate of hydrolysis of p-nitrophenol laurate in the microemulsion system consisting of ceryltrimethyl ammonium bromide, butanol and water. Two pronounced and broad peaks of the reaction rate were observed. The enhancements have been ascribed to the conventional micellar catalysis effect in which the micellar surface charge density plays a dominant role. However, this seems unlikely to be the reasons for the enhancements observed in our studies in view of the sharpness of the peak at 0.855 as compared to that reported by Friberg et al. (24).

We will discuss first the reaction kinetics by which the turbidity increases. As shown by Ottewill and Watanabe (15) in the case of sol formation, a complex series of consecutive and simultaneous reactions occur. These can be schematically presented as follows (15):

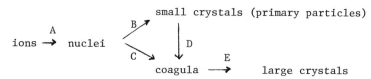

It involves nucleus formation (step A), crystal growth (step B) and coagulation (step C and D) of primary particles. The turbidity development observed in our experiments can not be related to the reaction A because the nucleus formation is usually very fast at reasonably high supersaturation and therefore will terminate before experimental observations are recorded. Another supporting evidence is that the nuclei of AgCl sols may only consist of about five ions as reported by Klein et al. (25). The change in opalescence resulted from such small nuclei is probably not perceivable. Thus, the reactions mostly observed in the experiments would appear to be B, C and D.

In many classical coagulation studies using <u>performed sols</u> (14, 15) with the addition of coagulatory agent (such as surface active molecules), a simplified analysis is feasible in which the growth of nuclei to primary particles is very rapid, and hence the reaction observed is predominantly the coagulation of the primary particles, i.e. the reaction D. This reaction has been found to depend strongly on the surface potential of the sol particles. The adsorption of surface active agents on sols may modify the surface potential and consequently alter the coagulation rate. In our studies, we have chosen a different approach, namely by monitoring the formation of hydrophobic sols <u>in statu nascendi</u> (13) using microemulsions as reaction media. The analysis thus appears complicated due to the possible influences of the microemulsions on the nucleation and crystal growth. Some of these influences are even not clear at present. It is our opinion, however, that the surface potential of the sols and the nature of the media are principally the controlling factors in the crystal growth and coagulation process.

<u>Physico-Chemical Properties of the Microemulsions</u>. In order to delineate the correlation between the coagulation rate and the dynamic structure of microemulsions, we have performed a number of measurements to determine various physico-chemical properties of the microemulsion phase along the experimental line. Figure 2 shows the conductance of the microemulsions. The maximum conductance around 0.56 water mass fraction corresponds to the peak of the coagulation rate in Figure 1. The increasing conductance with the addition of water in the alcohol-rich region may be attributed to the increased ionization of SDS molecules. The decreasing conductance beyond the maximum is presumably due to the dilution effect on the conductance by additional quantity of water.

Figure 3 represents the plot of viscosity vs. water content. It shows that as the amount of water decreases, the viscosity increases up to a water mass fraction of 0.56, and with further reduction in water content there is no significant change in viscosity. The decrease in viscosity in the region of water mass fraction greater than 0.56 is consistent with the dilution effect shown in the electrical conductivity measurements (Figure 2).

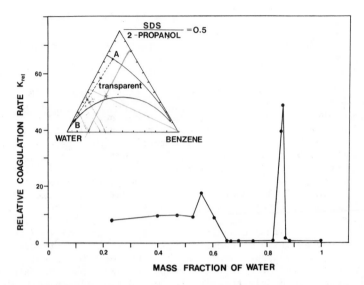

Figure 1. Variation of the relative coagulation rate constant k_{rel}
of AgCl sols as a function of water content along
the experimental line AB in the phase diagram of SDS–IPA–
benzene–water microemulsions.

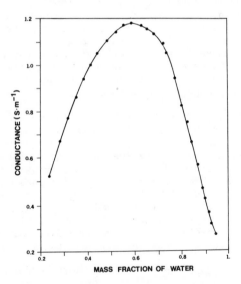

Figure 2. Electrical conductance as a function of water mass fraction along the line AB (Figure 1) of SDS–IPA–Benzene–Water
microemulsion system.

In order to further understand the association structure of the surfactant in the solution, we have measured the light scattering as shown in Figure 4. Interestingly, the results exhibit two peaks corresponding to the peaks of the coagulation rate in Figure 1. The broad peak in Figure 4 also bears a strong resemblance to the results reported by Friberg et al. (24). It can be stated from these results that starting from the alcohol-rich region, the addition of water in microemulsions results in certain association structure which enhances the light scattering from the solution. Similar conclusions have also been drawn by Sjoblom and Friberg (10) for water, pentanol, potassium oleate and hydrocarbon oil microemulsion systems. The association structure of the surfactant in the alcohol-rich region may resemble the water-in-oil (W/O) microemulsions (10).

It should be noted that a concentration fluctuation can also increase the light scattering intensity by orders of magnitude near the vicinity of a critical point. It is well established from recent studies (26-30) that a critical-like behavior has been observed near the percolation threshold in a W/O microemulsion system where a strong concentration fluctuation occurs due to the long range van der Waals interaction force between the microemulsion droplets. However, at present we do not have any experimental evidence to distinguish between the micellar growth and concentration fluctuation mechanism to explain the light scattering data. It is to be noted that a sharp peak in light scattering around 0.855 water mass fraction indeed is observed. A broad and smooth transition zone has been observed in the transition of spherical to rod (cylindrical) shape of micelles using viscosity, light scattering and magnetic field measurement (31,32). It is likely, therefore, that this sharp peak may not be related to a structural transition, but a concentration fluctuation which we will discuss in more details later.

Besides the light scattering data, the ultracentrifugation results further confirm the existence of surfactant aggregates in this system (Figure 5). For the samples with 0.855 water mass fraction or greater, no sedimentation peak was observed. We had expected that normal SDS micelles with solubilized benzene may exist in this region. The absence of the sedimentation peak is probably due to the electric repulsion force between the micelles. This is consistent with the observation that no sedimentation peak was observed in a pure 0.5 M SDS aqueous solution containing normal micellar aggregates. For the solutions with water fraction 0.8 down to 0.25, sedimentation peaks were observed and the sedimentation coefficients shown in Figure 5 were calculated from the velocity of the sedimenting peaks (33-35). As seen in Figure 5, the change of sedimentation coefficients with rotor speed was unexpected. It suggests that the aggregates of the surfactant are rather fragile and sensitive to the centrifugal force. The increase in the value of sedimentation coefficient as the amount of water decreases does not necessarily indicate the growth of aggregates. It may also be attributed in part to the decreasing buoyancy of the solvent due to the continuous addition of IPA into solutions. Further attempts to obtain the particle size from the sedimentation coefficient is thus thwarted due to the constant variation in the composition of the continuous media. For the solutions with water fraction less than 0.25, no sedimentation peak was observed.

Figure 6 represent various Schlieren patterns of the samples of different compositions. The Schlieren peak appears upward if the re-

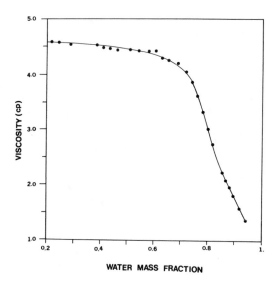

Figure 3. Viscosity as a function of water mass fraction along the
           line AB (Figure 1) of the SDS-IPA-Benzene Water micro-
           emulsion system.

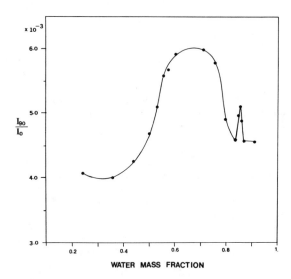

Figure 4. Ratio of light scattering intensity I_{90}/I_0 as a function
           of water mass fraction along the line AB (Figure 1) of
           the SDS-IPA-Benzene-Water microemulsion system.

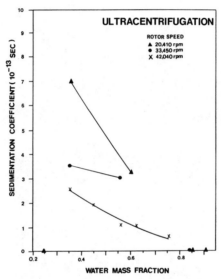

Figure 5.   Sedimentation coefficients at various rotor speeds as a function of water mass fraction along the line AB (Figure 1) of the SDS-IPA-Benzene-Water microemulsion system.

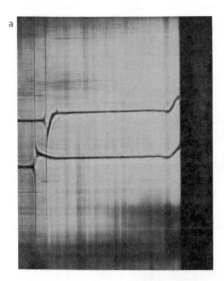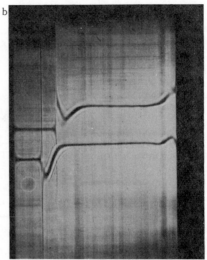

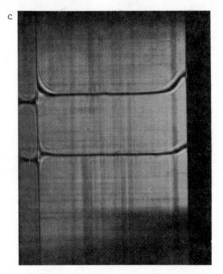

Figure 6. Ultracentrifuge Schlieren patterns of the microemulsions along the line AB in Figure 1. Note the left hand side is the meniscus, and the right hand side corresponds to the bottom of the cell.
(a) Upper curve corresponds to the sample at 0.8 water mass fraction, lower curve to the sample at 0.853 water mass fraction; 1651 sec elapsed after reaching the speed of 33,350 rpm.
(b) Upper curve corresponds to the sample at 0.35 water mass fraction, lower curve to the sample at 0.567 water mass fraction; 2595 sec elapsed after reaching the speed of 42,040 rpm.
(c) Upper curve corresponds to the sample at 0.24 water mass fraction, lower curve to the sample at 0.908 water mass fraction; 1377 sec elapsed after reaching the speed of 20,410 rpm.

fractive index increment is positive, dn/dc>0; and downward if negative, dn/dc<0, where n denotes the local refractive index of the solution and c is the local solute concentration in the solution (33-35).
Figure 6(a) shows an upward meniscus boundary for the sample of 0.853
water mass fraction (lower curve) and a downward peak for the sample
of 0.8 water mass fraction (upper curve). It is evident that an inversion of the refractive index increment dn/dc occurs in the region
of water fraction between 0.85 and 0.8, indicating a change in the
solution properties. Such inversion is also accompanied with the
onset of the sedimentation of aggregates. Figure 6(b) indicates
the sedimentation peaks observed at 0.567 (lower curve) and 0.35
(upper curve) water fraction. Both peaks are downward. It is noteworthy that there exists another upward peak near the bottom of the
cell for both samples. This peak appears to float up against the
centrifugal field during the course of centrifugation. The reason
for this floating peak is not clear at present. However, it can not
be attributed to the flotation of the aggregates because a flotation
peak usually appears downward (with negative refractive index increment). We propose that this peak may be related to the compressibility of the system for which the aggregates under compression tend to
relax back through a back-diffusion of the solutes against the centrifugal field (36). Figure 6(c) represent the Schlieren patterns
of the sample at 0.24 water fraction. No sedimentation peak was
observed and the meniscus boundary was inverted from downward to
upward again, indicating the occurrence of a structural transition
(phase separation occurs at composition A, Figure 1).

The ultrasonic absorption of the solutions at 5MHz also corroborates
the picture that emerges from the ultracentrifugation results; Fig. 7.
The maximum absorption observed at 0.56 water fraction corresponds
to the peak of light scattering and coagulation rate of AgCl, and is
indicative of the ease of perturbation of the structures by the ultrasonic pressure. The processes to which the ultrasonic relaxation of
surfactant solutions can be attributed are: (1) the exchange of alcohols between the mixed micelles and surrounding solution; (2) the
exchange of surfactants between the micelles and surrounding solution;
(3) the ion association-dissociation equilibrium of the electrolytes;
and (4) concentration fluctuation of the solution. Zana et al. (28)
have investigated the ultrasonic absorption behavior in many W/O
microemulsion systems. It was found that the high ultrasonic absorption could only be detected if large concentration fluctuations
occurred in the system. Hence, the first and second process can be
excluded. The third process is also unlikely to be the cause for
the maximum absorption in view of the results reported by Friberg
et al. (24) that a continuous increase of [81]Br line width occurs as
the water content decreases. The broadening of [81]Br line width indicates the increasing strength in the counter-ion binding. The
ultrasonic absorption, if any, will then appear as a monotonic function, instead of a maximum. We shall therefore only consider the concentration fluctuation as the probable cause of absorption maximum.

The concentration fluctuations in our system can possibly further be subdivided into a solute (surfactant aggregates) concentration
fluctuation and a solvent concentration fluctuation. The solute concentration fluctuation is similar to that of critical-like behavior
observed in many W/O microemulsion systems (28), while the solvent

concentration fluctuation may result from the mixed solvent of IPA and water. It has been reported (37) that a maximum ultrasonic absorption occurs at 0.84 mass fraction of water (mole fraction of IPA is 0.057) in a IPA + water mixture. A shift in the composition of this maximum absorption may occur upon addition of other additives. Therefore, we can not rule out the solvent concentration fluctuation as a possible cause of the maximum absorption.

We have also performed a pressure-jump relaxation study on the system in an attempt to directly probe the microstructures in our system. It is now well established that the existence of normal micellar structures gives rise to two well separated relaxation processes (38-40). The fast relaxation process with relaxation time τ_1 in micro-seconds range is related to the fast exchange of surfactant monomers between the micelles and the surrounding solution. The slow relaxation process with relaxation time τ_2 in milli-seconds range is associated with the micellar formation-dissolution equilibrium. No experimental data have yet been reported regarding the pressure-jump relaxation in a reverse micellar system. In the water-rich region (water mass fraction greater than 0.75) of our microemulsion system, the relaxation spectra resemble to that of normal micelles except that there exists an extra slow relaxation process (referred as τ_{slow}). τ_1 is too fast to measure by pressure-jump apparatus. The value of τ_2 as plotted in Figure 8 is smaller than that of pure SDS micelles (800 milliseconds to 5 seconds) in the same SDS concentration range of 100 to 200 mM by Kahlweit (38). This may be attributed to the presence of short chain alcohol (39). We have also shown in a recent study (41) that the addition of a short chain alcohol (propanol) increases the rate of micelle formation-dissolution significantly. The amplitude of τ_{slow} process in the water-rich region is very small and hence the resolution of relaxation time τ_{slow} is poor. Within the range of experimental accuracy, τ_{slow} was found to be independent of the microemulsion composition.

As the amount of IPA and SDS continuously increases in the solution, the relaxation spectra seem to undergo a smooth transition. The amplitude of τ_{slow} process is gradually increasing, while the amplitude of τ_2 process is diminishing. We only reported τ_2's for the water rich region in Figure 8 due to the poor resolution of τ_2's with increasing alcohol concentration. However, the value of τ_2 appears to increase with alcohol concentration and approach a plateau value around 0.8 sec at alcohol-rich corner. We should note that τ_{slow} is probably not related to micelle formation. We have attributed these slow processes to the compressibility of the IPA + water solvent in our recent study (41). We have shown (41) that a 10 mM potassium chloride in a mixed IPA + water solvent gives rise to a similar relaxation signal (consisting of three relaxation processes) and relaxation times to that of the sample at 0.24 water mass fraction. It is obvious that the signal resulted from the relaxation of normal micelles dominates the relaxation spectra in the water-rich region and hence the amplitude of τ_{slow} process is small. But as the alcohol concentration increases, the normal micellar aggregates disappear gradually and a structural transition takes place. The relaxation spectra will then be overshadowed by the relaxation of the mixed solvent of IPA + water. The mechanisms corresponding to this mixed solvent relaxation are not established at present.

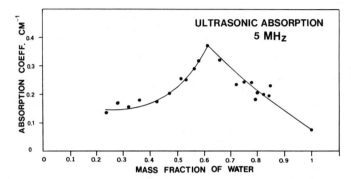

Figure 7. Ultrasonic absorption as a function of water mass fraction along the line AB (Figure 1) of the SDS–IPA–Benzene–Water microemulsion system.

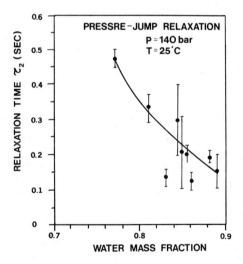

Figure 8. Variation of relaxation time τ_2 as a function of water mass fraction.

Piecing together all the experimental data we have thus far presented, we can attain the following sketch regarding the structural properties of our microemulsion system. The sedimentation studies confirm the existence of association structures of surfactants in the alcohol-rich region. Based upon the light scattering data (Figure 4) we can state that the addition of water induces the association of the surfactants starting at about 0.4 water mass fraction. Between 0.56 to 0.7 water fraction, the surfactant aggregates and concentration fluctuation may coexist in the system. The association structures over the alcohol-rich region should resemble the inverted micelles or W/O microemulsions. Beyond 0.7 water mass fraction, a structural transition from inverted to normal micellar structures occurs and hence the light scattering decreases. If the normal micellar structure persistently exists over the alcohol-rich region, we would have expected that the value of τ_2 decreases continuously instead of increases according to the relaxation study of high concentration SDS solution reported by Kahlweit (38).

The structural studies reported by Bellocq et al. (5,6) on the microemulsion system composed of SDS, butanol, water and toluene can be referred to support the types of association structures mentioned above. Three subregions consisting of different microstructures in a single microemulsion phase region have been identified using quasielastic Rayleigh scattering. In view of the striking resemblance of the phase diagram reported by Bellocq et al. (5,6) to ours, similar conclusions can also be drawn for our system. The formation of spherical inverted structures are responsible for the light scattering enhancement starting at 0.4 water fraction as shown in Figure 4. It is noteworthy that the light scattering intensity remains constant upon further increasing the water fraction from 0.56 to 0.7. This suggests that the additional water does not induce further growth of inverted micelles, but partition in the continuous medium. This not only explains the decrease of the conductance beyond the 0.56 water mass fraction, but also indicates that large micellar structures are not formed in the IPA + water solvent. Hence, the concentration fluctuation may play an important role in light scattering. This can explain the fact that the association structures in the alcohol-rich region are not rigid. They are fragile and easy to perturb as concluded from the ultracentrifuge study.

The structural transition from inverted to normal micellar structure occurs around 0.8 to 0.7 water mass fraction. It is obviously a progressive transition. The microstructure in this region is not yet well established. The ultracentrifuge study has shown that the inverted micellar structure may exist persistently down to 0.8 water mass fraction. But the quasielastic light scattering has detected the trace of normal micellar structure as low as 0.7 water mass fraction (5).

Lastly, we would like to point out that the head group of the ionic surfactant have to be hydrated by a minimum amount of water in order to dissolve into a low polarity solvent (e.g. short chain alcohols). In the hydrocarbon oil rich corner of a microemulsion phase diagram, micellization occurs as long as the minimum water required to hydrate the ionic head group is added (5). Hence the minimum water to surfactant molar ratio required for such hydration can be determined by light scattering measurement. The ratio has been found to be 10 for sulfate surfactants in toluene and 8 for carboxy-

late surfactants in dodecane (5).   But in the alcohol (short chain)
rich corner, the micellization does not occur upon the addition of
necessary minimum amount of water.   Therefore the light scattering
measurement cannot be used to determine the minimum water required.
Instead, we have developed a titration method for this purpose.
Starting from the surfactant-rich region of our system, the added
water is expected to partition in both surfactant phase (hydration)
and continuous medium (IPA).   Then the total number of water molecules
N_w added to the solution is:

$$N_w = N_w^s + N_w^a \tag{7}$$

where N_w^s denotes the number of water molecules that hydrate the surfactant head groups, and N_w^a is the number of water molecules partitioning in the IPA.   Assuming h is the minimum number of water molecules per surfactant molecule required for hydration, we then have:

$$N_w^s = hN_s \tag{8}$$

where the N_s is the total number of surfactant molecules in the solution.   We can also express the N_w^a as:

$$N_w^a = kN_a \tag{9}$$

where k is the proportional constant of water molecules partitioning
in the IPA, and N_a is the total number of alcohol molecules in the
solution.   Combining the equation (7), (8) and (9) gives:

$$\frac{N_w}{N_s} = k\,\frac{N_a}{N_s} + h \tag{10}$$

We took a clear microemulsion sample of 0.22 water mass fraction
near the phase boundary, first titrated with IPA till the sample
just became turbid, then titrated with water till the sample became
clear again.   Repeating these procedures many times and plotting
the ratio of N_w/N_s versus N_a/N_s, we then obtained a straight line
as shown in Figure 9.   The slope yields the constant k and the
intercept on y-axis corresponds to the minimum number of water
molecules per surfactant molecule required for dissolution.   It
was concluded that minimum 8 water molecules are needed to hydrate
each sulfate group for dissolution of SDS into IPA.   It should be
noted that this titration method can only be used in the miscibility
range of the short chain alcohol with water.

Interrelationship Between the Reaction Kinetics and the Dynamic
Structure of Microemulsions. We now come to our final goal to correlate the reaction kinetics with the dynamic structures of the microemulsions.   The coagulation rate of AgCl sols depends on the surface charge of the sols.   High surface charge density prevents the
collisions of the preliminary particles of AgCl crystals, and consequently results in the slow coagulation rate and small size of
precipitates.   The enhancement of the caogulation in Figure 1 from
water mass fraction of 0.22 to 0.65 corresponds to the region where
the nature of the continuous phase of the microemulsions is dominated

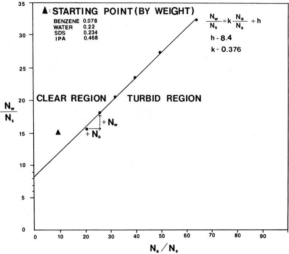

Figure 9.   Minimum number of water molecules, h, per surfactant
molecule needed to hydrate the sulfate group for the dissolution of SDS into IPA.

by alcohol (low polarity). The association structures in this region are probably inverted micelles.

It has been shown that the stability of colloidal suspensions
can also be influenced by a pure alcohol-water mixture, without
the addition of any surface active agent. In a study of the flocculation of polystyrene emulsions in ethanol-water mixtures (42),
the concentration of sodium chloride required to produce rapid flocculation increases with increasing ethanol concentration up to 0.09
molar fraction, beyond this composition, the concentration of sodium
chloride required for flocculation decreases rapidly. It will be very
informative, therefore, to compare our coagulation rate obtained in
the microemulsion media to that in pure IPA + water mixture. The
results can be used to further delineate the role of inverted micellar structure on the enhancement of coagulation.

The reasons for the sharp peak of coagulation rate at 0.855
water fraction can be explained as follows. The sharp increase
in the coagulation rate is presumably due to the fast mutual coagulation of the normal micelles as a result of increasing concentration
of SDS. Many facts substantiate this conjecture. Using density and
heat capacity measurements, Roux et al. (7,8) have found that the micellar growth starts at about 0.8 water mass fraction along a dilution
line (by water), and at about 0.85 water mass fraction along the
lower demixing line in a microemulsion system consisting of SDS,
butanol, water and toluene. The discrepancy in these two water fractions appears to result from the micelles swollen by the solubilized
toluene along the lower demixing line. In view of the striking similarity in the phase diagram of our microemulsion system with that of
Roux et al., it is plausible to propose that the micellar growth occurs
at 0.85 water mass fraction in our systems.

We would like to point out that this micellar growth is probably
not the type of transition from spherical to cylindrical micelles
as usually observed in a concentrated surfactant solution (31,32).
A tighter packing micelle is usually expected as a result of this
transition. But in contrast, Roux et al. (7,8) have found that a
less structured micelle results after the micellar growth. Two
counter-acting forces may exist during the course of this micellar
growth; one force tends to increase the micellar size due to the
continuous increase of SDS concentration, while the other tends
to break down the micelles due to the increasing concentration of alcohol (43-45). It is likely due to these counter-balancing forces,
that the micellar growth in these microemulsion systems differs from
that of sphere to cylinder transition.

It is generally considered that the micelle formation represents
a step-wise association process (40). For an ionic surfactant system
at low concentration, the mutual coagulation between the micelles is
forbidden due to the electrostatic repulsion force. Hence, the micelles grow only by incorporation of monomers only. However, a recent review paper by Kahlweit (38) indicates that at high counter
ion concentrations (i.e. high surfactant concentration), the micelles can grow through a bypass of a reversible coagulation of sub-
micellar aggregates. For SDS micellar solution, such bypass coagulation occurs at the concentration of about 200 mM which corresponds
to the minimum of $1/\tau_2$. The concentration of SDS at 0.855 water mass
fraction of our microemulsion system is about 150mM. We propose that
a   mutual coagulation of micelles may occur due to the micellar

growth at 0.855 water mass fraction, if taking into account that the presence of IPA may decrease the surface charge density of the micelles. This can explain the occurrence of the sharp light scattering peak at 0.855 water mass fraction in Figure 4. This light scattering peak is due to the concentration fluctuation resulting from the mutual coagulation of micelles.

The sharp enhancement in the coagulation of AgCl sols at 0.855 water mass fraction is the consequence of this mutual coagulation of the micelles. It is likely that the primary particles of AgCl sols are located in the micelles (14) or adsorbed at micelle surface. The electrostatic repulsion force prohibits the coagulation of the micelles, hence the growth of the particles is slow and the size of the precipitate is small in the water-rich region as compared to alcohol-rich region. Our preliminary study on the particle size using scanning electron micrographs indeed confirms that the particles precipitated in the water-rich region is much smaller than in the alcohol-rich region. However, at the composition of micellar growth, the rate of particle growth is enhanced through a fast coagulation of the micelles. The sharpness of the coagulation peak suggests that the micellar growth is only limited to a very finite composition range.

Conclusions

The coagulation of the hydrophobic AgCl sols has been investigated using microemulsions as reaction media. Both equilibrium and dynamic studies have been carried out to delineate the microstructures in a single microemulsion phase. The existence of different microstructures in the single phase region has been established. For the microemulsions with water mass fraction of 0.4 to 0.56, inverted micellar structures are formed. The concentration fluctuation of these inverted micelles may also play an important role in the region from 0.4 to 0.7 water mass fraction. For the microemulsions with water fractions greater than 0.8, the existence of normal SDS micelles have been indicated. It has been shown that the kinetics of the chemical reaction is intimately correlated with the structures and the nature of the medium in the microemulsion phase. Hence, the chemical reaction can serve as a useful approach for probing the dynamic structures in microemulsions. Besides the kinetics of reactions, the morphology of the products from the chemical reaction is also influenced by the microemulsion. This study is thus relevant to various technological applications such as the manufacturing of photographic films (46), catalysis (47) and in general, fine powder technology.

Acknowledgments

We are grateful to the National Science Foundation (Grant No. NSF-CPE 8005851) and the American Chemical Society Petroleum Research Fund (Grant No. PRF-14718-AC5) for supporting this research. We thank Professor M. Clausse at the University of Pau, France, for many stimulating discussions and suggestions.

Literature Cited

1. Clausse, M., Peyrelasse, J., Heil, J., Boned, C., Lagourette, B., Nature, 293, 636 (1981)
2. Boned, C., Clausse, M., Lagourette, B., Peyrelasse, J., McClean, V.E.R., Sheppard, R.J., J. Phys. Chem., 84, 1520 (1980).
3. DeGennes, P.G., Taupin, C., J. Phys. Chem., 86, 2294 (1982).
4. Dvolaitzky, M., Guyot, M., Lagues, M., Le Pesant, J.P., Ober, R., Sauterey, C., and Taupin, C., J. Phys. Chem., 69, 1 (1978).
5. Bellocq, A.M. Fourche, G., J. Colloid Interf. Sci., 78, 275 (1980).
6. Bellocq, A.M., Biais, J., Clin, B., Lalanne, P., and Lemanceau, B., J. Colloid Interf. Sci., 70, 524 (1979).
7. Roux, A.H., Roux-Desgranges, G., Grolier, J.P.E., and Viallard, A., J. Colloid Interf. Sci., 84, 250 (1981).
8. Roux-Desgranges, G., Roux, A.H., Grolier, J.P.E., and Viallard, A., J. Colloid Interf. Sci., 84, 536 (1981).
9. Roux-Desgranges, G., Roux, A.H., Grolier, J.P.E., and Viallard, A., J. Solution Chem., 11, 357 (1982).
10. Sjoblom, E., and Friberg, S., J. Colloid Interf. Sci., 67, 16 (1978).
11. Lindman, B., Stilbs, P., and Moseley, M.E., J. Colloid Interf. Sci., 83, 569 (1981).
12. Tondre, C. and Zana, R., J. Dispersion Sci. Techn., 1, 179 (1980).
13. Matijevic, E. and Ottewill, R.H., J. Colloid Interf. Sci., 13, 242 (1958).
14. Ottewill, R.H., Rastogi, M.C., and Watanabe, A., Trans. Faraday Soc., 56, 855-891 (1960).
15. Ottewill, R.H., and Watanabe, A., Kolloid Z., 170, 38 (1960); Kolloid Z., 170, 132 (1960); Kolloid Z., 171, 33 (1960); Kolloid Z., 173, 7 (1960).
16. Horne, R.W., Matijevic, E., Ottewill, R.H., and Weymouth, J.W., Kolloid Z., 161, 50 (1958).
17. Tezak, B., Matijevic, E., and Schulz, K., J. Colloid Interf. Sci., 55, 1557-1576 (1951).
18. Davies, C.W. and Jones, A.L., Faraday Soc. Discussions, 5, 103 (1949).
19. Davies, C.W. and Jones, A.L., Trans. Faraday Soc., 51, 812-829 (1955).
20. Howard, J.R., Nancollas, G.H., and Purdie, N., Trans. Faraday Soc. 56, 278 (1960).
21. Ohyama, Y. and Futaki, K., Bull. Chem. Soc. Jpn., 28, 243 (1955).
22. Ohyama, Y. and Futaki, K., Bull. Chem. Soc. Jpn., 31, 10 (1958).
23. Clausse, M., private communication.
24. Friberg, S., Rydhag, L., and Lindblom, G., J. Phys. Chem., 77, 1280 (1973).
25. Klein, D.H., Gordon, L., and Walnut, T.H., Talanta, 3, 187 (1959).
26. Cazabat, A.M., Langevin, D., Meunier, J., and Pouchelon, A., Adv. Colloid Interf. Sci., 16, 175-199 (1982).
27. Fourche, G., Bellocq, A.M., Brunetti, S., J. Colloid Interf. Sci., 88, 302 (1982).
28. Zana, R., Lang, J., Sorba, O., Cazabat, A.M., and Langevin, D., J. Physique-Lett., 43, L-829 (1982).
29. Cazabat, A.M., Langevin, D., and Sorba, O., J. Physique-Lett., 43, L-505 (1982).
30. Lagues, M. and Sauterey, C., J. Phys. Chem., 84, 3503 (1980).
31. Porte, G. and Poggi, Y., Phys. Rev. Lett., 41, 1481 (1978).

32. Ekwall, P., Mandell, L., and Solyom, P., J. Colloid Interf. Sci.,
    35, 519 (1971).
33. Schachman, H.K., "Ultracentrifugation in Biochemistry", Acad.
    Press, 1959.
34. Freifelder, D., "Physical Biochemistry, Application to Biochemis-
    try and Molecular Biology", W.H. Freeman and Company.
35. Bowen, T.J., "An Introduction to Ultracentrifugation", John
    Wiley & Sons, (1970).
36. Chun, P.W., Richar, A.J., Herschler, W.P., Krista, M.L., Bio-
    polymers, 12, 1931 (1973).
37. Frank, F., "Water: A Comprehensive Treatise", V. 2, P. 551, Plenum,
    N.Y. 1973
38. Kahlweit, M., Pure & Appl. Chem., 53, 2069 (1981).
39. Yiv, S. Zana, R., Ulbricht, W., and Hoffmann, H., J. Colloid
    Interf. Sci., 80, 224 (1981).
40. Aniansson, E.A.G., Wall, S.N., Almgren, M., Hoffmann, H., Kielmann,
    I., Ubricht, W., Zana, R., Lang, J. and Tondre, C., J. Phys.
    Chem., 80, 905 (1976).
41. Leung, R. and Shah, D.O., Submitted to J. Colloid Interf. Sci.
42. Franks, F., "Physico-Chemical Processes in Mixed Aqueous Solvents",
    P. 63, American Elsevier Pub., N.Y. (1967).
43. Birdi, K.S., Backlund, S., Sorensen, K., Krag, T., and Dalsager,
    S., J. Colloid Interf. Sci., 66, 118 (1978).
44. Oakenfull, D., J. Colloid Interf. Sci., 88, 562 (1982).
45. Stilbs, P., J. Colloid Interf. Sci., 89, 547 (1982).
46. Dvolaitzky, M., Ober, R., Taupin, C., Anthore, R., Auvray, X.,
    and Petipas, C., J. Dispersion Sci. & Tech., 4(1), 29 (1983).
47. Boutonnet, M., Kizling, J., Stenius, P., Colloids and Surfaces,
    5, 209 (1982).

RECEIVED January 23, 1985

Interfacial Phenomena of Miniemulsions

C. D. LACK, M. S. EL-AASSER, J. W. VANDERHOFF, and F. M. FOWKES

Emulsion Polymers Institute, Departments of Chemical Engineering and Chemistry, Lehigh University, Bethlehem, PA 18015

The interfacial characteristics between an oil drop and aqueous mixed emulsifier solutions were studied with a spinning drop interfacial tensiometer. An interfacial layer was observed at the oil/aqueous phase interface, as evidenced by the formation of "tails" on the rotating drop. The length of these "tails" increased with spinning time and rotation speed. The interfacial tensions between styrene and aqueous mixed emulsifier solutions were unexpectedly high, 5 to 13 dynes/cm, whereas tensions in the range of 10^{-2} dynes/cm were measured between the "tails" and the aqueous solution.

Static, equilibration studies also indicated that a molecular association forms at the oil/water interface in the presence of mixed emulsifiers. Spinning drop experiments with pre-equilibrated oil and aqueous phases suggested that the presence of oil in association with the mixed emulsifier molecules in the aqueous phase affects the formation of an interfacial layer.

Emulsions are generally characterized by the droplet size even though there may be other equally significant differences. Oil-in-water miniemulsions are prepared using a combination of a surfactant and a cosurfactant (which is not a surface active agent) and have droplet diameters in the range of 100 to 400nm. Typically, a mixture of an ionic surfactant, such as sodium lauryl sulfate (SLS) or hexadecyltrimethylammoniumbromide (HTAB), and a long chain fatty alcohol is used in concentrations of 0.5 to 3% by weight based on the oil phase. In contrast, microemulsions contain 10 to 100nm diameter droplets and are prepared using mixed emulsifiers in concentrations of 15 to 30% by weight based on the oil phase.

In addition to mixed emulsifier concentration and the droplet size, micro- and miniemulsions differ in the fatty alcohol chain length used. Stable miniemulsions can only be prepared with a fatty alcohol chain length of at least 12 carbon atoms compared to the shorter chain lengths used for most microemulsions. Also, the order

0097–6156/85/0272–0345$06.00/0

of mixing of ingredients is different in the two systems. Successful emulsification of miniemulsions requires an aqueous solution of ionic surfactant and fatty alcohol be pre-emulsified for 0.5 to 1.0 hour, at a temperature above the melting point of the fatty alcohol, before the oil phase is added. In contrast, microemulsions can be prepared by dissolving the fatty alcohol in the oil phase prior to mixing with the water phase containing the ionic surfactant.

Mixed emulsifier systems in the same low concentrations used to prepare miniemulsions (1 to 3% wt.) have also been used successfully to prepare polymer latexes of 100-400nm diameter particles by direct emulsification of polymer solutions (1-7). The resulting latexes showed excellent shelf-stability for more than one year. The significance of this recent finding lies in the ability of mixed emulsifiers, in such low concentrations, to form, by direct emulsification, stable polymer particles that are in the same size range as latexes prepared by emulsion polymerization. Based on this development, many organic solvent based polymer coatings can now alternately be applied from aqueous dispersion systems, thereby avoiding the numerous economic, logistic, and environmental constraints associated with organic solvents.

The formation and stabilization of O/W emulsions prepared with mixed emulsifier systems has been extensively investigated. However, the mechanisms proposed differ greatly. One of the primary hypotheses attributes the enhanced stability to the formation of a molecular "complex" or layer at the oil/water interface (8-11). The mixture of emulsifier types increases the packing density of the adsorbed interfacial film. Several investigators have shown that more closely packed complexes produce more stable emulsions (9,12-14). Friberg, et al. (15-17) have attributed the enhanced stability of mixed emulsifier emulsions to the formation of liquid crystals at the oil/water interface, which reduce the van der Waals attractive forces.

Another explanation of the good stability of micro- and miniemulsions in the presence of mixed emulsifiers is based on Higuchi and Misra's work (18). According to this concept, the degradation by diffusion is greatly retarded in the presence of a less water soluble component at the interface, or in the oil phase. Ugelstad, et al. (19-21) have shown that an emulsion can be stabilized against diffusion by decreasing the oil phase water solubility, which is accomplished through the solubilization of long chain alkanes or fatty alcohols in the oil phase.

In the present research on miniemulsions, the maximum stability was found at ionic surfactant to fatty alcohol molar ratios of between 1:1 and 1:3, when the alcohol and surfactant had near equal chain lengths (22,23).

The overall objective of this research program is to determine the effect of mixed emulsifiers on interfacial properties in relation to miniemulsion formation and stabilization. Based on the previously discussed affects, the immediate objective of this phase of the research program was aimed at understanding the interactions that occur between mixed emulsifier molecules and miniemulsion oil droplets.

Interfacial Tension Theory

The spinning drop method was first proposed by Vonnegut in 1942 and

is based on the elongation of an immiscible lighter phase drop in a
more dense, continuous phase as the two are subjected to centrifugal
forces (24). Often the two fluids are placed in a capillary tube
which rotates about its longitudinal axis at high speed. The lighter
phase drop is forced to the center of the capillary tube where it
will tend to elongate. The particular shape of the drop depends on
the interfacial tension and centrifugal pressure forces. The interfacial tension between the two phases acts to resist elongation, and
the pressure which results from the centrifugal forces enhances
elongation. The centrifugal forces are a function of the rotation
speed and the densities of the two phases. Because these forces are
balanced at equilibrium, the interfacial tension can be determined
from the centrifugal forces.

Materials and Methods

Experimental Materials. The emulsifiers used were sodium lauryl sulfate, SLS (Stepan Chemical Company) and hexadecyltrimethylammoniumbromide, HTAB (Fisher Scientific Company). The SLS was purified by
recrystallization from absolute ethanol and dried at 20°C under
vacuum. The SLS was then Soxhlet-extracted with diethyl ether for
48 hours, dried under vacuum at 20°C, and stored under vacuum at 5°C
until use.

The following fatty alcohols were used as received from Conoco
Chemical Company: n-decanol $[C_{10}]$, lauryl alcohol, LA $[C_{12}]$, tetra-
decanol $[C_{14}]$, cetyl alcohol, CA $[C_{16}]$, and octadecanol $[C_{18}]$.

The styrene monomer (certified grade, Fisher Scientific Company)
was cleaned before use. The inhibitor was removed by washing with
10% wt. aqueous sodium hydroxide solution. The monomer was then
washed with distilled deionized water, dried overnight (at 5°C) with
anhydrous sodium sulfate, vacuum distilled under dry nitrogen, and
then stored at 5°C until use.

Toluene (Fisher Scientific Company) was used as received.
Double distilled deionized (DDI) water was used in preparation of
aqueous solutions.

Interfacial Tension Measurements. A Site Model LP-10 Spinning Drop
Interfacial Tensiometer was used in this study. A description of
the instrument and its capabilities have been reported (25-28). The
apparatus consists of four major components: the spinning cell, the
tilting base, the electronic components, and the microscope. The
spinning cell contains the rotating glass capillary, mounts, bearings
and seals as well as a direct-contact thermal oil chamber and inlet
and outlet chambers for the surfactant solution. It operates at
speeds up to 10,000 rpm and temperatures up to 100°C. The instrument
is suitable for interfacial tensions from 10^{-5} to 10^2 dyne/cm (29).

In order to simulate the conditions of the actual emulsfication
process, all interfacial tension measurements were made at 65°C, in
the following manner. Each aqueous mixed emulsifier solution was
pre-emulsified, by mixing for 1.0 hour at 65°C, and then loaded into
the capillary tube of the tensiometer. The oil drop was injected into the capillary and the rotation speed increased to the desired
level. After allowing time for equilibration, the dimensions of the
oil drop were measured in two perpendicular directions using a micrometer in the microscope eyepiece. Multiple oil drops were measured
with each surfactant solution by repeating the above procedure.

The diameter measured is the apparent diameter (D_{app}) and should be corrected for the refraction of the light beam as it passes through the various mediums to the optical microscope. An empirical calibration equation was developed earlier to convert D_{app} to the actual droplet diameter (D_{act}) (30):

$$D_{act} = D_{app}/[2.084 - 1.37 \text{x} RI_o + 0.629 \text{x} RI_c] \tag{1}$$

where RI_o and RI_c are the refractive indices of the thermal oil and continuous phase respectively, at the temperature of interest.

The interfacial tension, γ, was calculated knowing the actual drop diameter, D_{act}, and length, L, the angular velocity, ω, and the density difference between the continuous and oil phases, $\Delta\rho$, using the following expression (29):

$$\gamma = (1/32) \text{ x } (\Delta\rho) \text{ x } \omega^2 D_{act}^3 (\phi+1) \tag{2}$$

where ϕ is a correction for insufficiently elongated drops, L/D<4.

The effect of equilibration on interfacial layer formation was studied by photographing a drop when it was first injected into the rotating capillary and at regular time intervals until an equilibrium configuration was reached. All photographic studies were conducted at 3000 rpm in order to slow the droplet deformation. A Nikon 35mm SLR camera body was mounted in place of the Ziess microscope eye-piece using a Nikon T-mount and a 10X ocular. A diffused, incandescent light source was used to illuminate the rotating capillary from behind.

Results and Discussion

Mixed emulsifiers are commonly used in combination with electrolytes to attain oil/water interfacial tensions substantially less than 1 dyne/cm, eg. 10^{-1} to 10^{-4} dynes/cm (31). The stability of the resulting microemulsions is usually attributed to the formation of an interfacial film (32,33). Even though the mechanism of stabilization has not yet been resolved, the excess surfactant used in microemulsions usually assures good stability. However, due to the very low mixed emulsifier concentrations used in miniemulsions, an understanding of the interactions between mixed emulsifier molecules at oil/water interfaces should greatly facilitate the development of miniemulsion and mini-latex formulations to achieve good stability.

The objective of this research program was to investigate the characteristics of the interfacial films observed in our miniemulsion systems. This study of oil/aqueous mixed emulsifier solution interfacial properties included the effects of mixed emulsifier molar ratio and concentration, fatty alcohol initial location and chain length, and oil phase water solubility. The effect of equilibration on the formation of interfacial layers was also studied.

Interfacial Layer Visualization. One of the key results of extensive spinning drop experiments between aqueous mixed emulsifier solutions and styrene was visual evidence for the formation of mixed emulsifier interfacial films. This interfacial layer is depicted in Figure 1 by the formation of "tails" as a function of time on the rotating styrene drop in an aqueous solution of 1:1 SLS/CA, based on 10mM SLS.

One possible explanation for these "tails" is that the SLS and fatty alcohol adsorb from solution onto the surface of the styrene drop in some type of molecular association. This surface layer, together with some penetration of the styrene, forms an interfacial film. The diffusion of styrene into the interfacial film is suggested by the decreasing oil drop volume shown in Figure 1. The centrifugal force in the rotating capillary then forces the less dense surface film to migrate to the ends of the drop aligned with the axis of rotation. As new surface area is exposed, additional emulsifier is adsorbed. As this cycle repeats itself, the "tails" continue to grow. These "tails" were observed to have viscoelastic properties as evidenced by the viscous fluid-like deformation that occurs when a neighboring droplet comes in contact with a "tail". The formation of such films between the components of similar mixed emulsifier systems has been suggested by Blakey, et al. (13) and Tadros (34).

The results of experiments run with both styrene and toluene in a 1:1 SLS/lauryl alcohol solution showed that identical "tails" formed on the styrene and toluene droplets, indicating that this phenomenon was not the result of thermally initiated polymerization of the styrene. Also, no "tails" were observed on air drops in the same mixed emulsifier solution or on styrene droplets in an aqueous solution of SLS alone.

The effect of initial fatty alcohol location on the formation of an interfacial layer is depicted in Figure 2. With the fatty alcohol solubilized in the oil phase, no "tails" formed at the interface between the oil-alcohol droplet and an aqueous solution of 10mM SLS. Since microemulsions are usually prepared with the fatty alcohol solubilized in the oil phase and miniemulsions are not, the mechanism of interfacial layer formation and hence stabilization for micro- and miniemulsions appears to be different.

The effect of oil phase water solubility on the formation of interfacial layers was studied in a series of spinning drop experiments run in 1:1 SLS/CA (based on 10mM SLS) using as oil phases, methyl methacrylate (MMA), n-butyl acrylate (BA), and 2-ethylhexyl acrylate (EHA) monomers. Despite the large range of water solubilities, 1.6%, 0.2%, and 0.01%, respectively, indentical, narrow "tails" were formed on the different spinning drops. These narrow "tails" are shown for EHA in Figure 3.

The chemical structure of the oil phase itself has a pronounced effect on the formation of an interfacial layer. Spinning drop experiments with oil phases of different chemical structure and the same water solubilities illustrated this effect. Comparison of Figures 1 and 3 shows that interfacial layers having significantly different characteristics form between the same 1:1 SLS/CA solution and different oil phases (styrene and EHA, respectively) with approximately the same water solubilities. This may be the result of different types of specific interactions between the various components of the mixed emulsifier system and either one of the two types of oils. NMR studies will be conducted in order to investigate this point.

Interfacial Tension Values. The results for the effect of ionic surfactant to fatty alcohol molar ratio and concentration on interfacial tensions with styrene are shown in Figure 4. Maximum interfacial

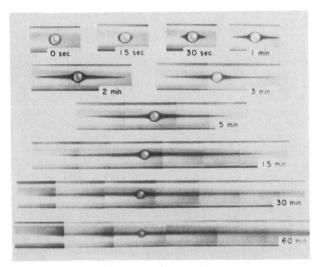

Figure 1. Spinning drop photographs of a styrene droplet in 1:1 SLS/cetyl alcohol solution, based on 10mM SLS, at 65°C.

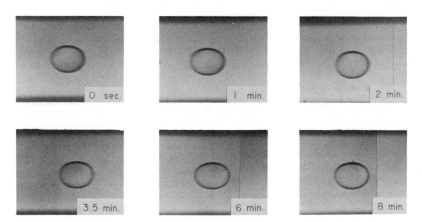

Figure 2. Spinning drop photographs of a styrene droplet containing cetyl alcohol in an aqueous 10mM SLS solution at 65°C.

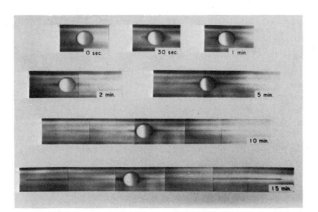

Figure 3. Spinning drop photographs of 2-ethylhexyl acrylate in 1:1 SLS/cetyl alcohol solution, based on 10mM SLS, at 65°C.

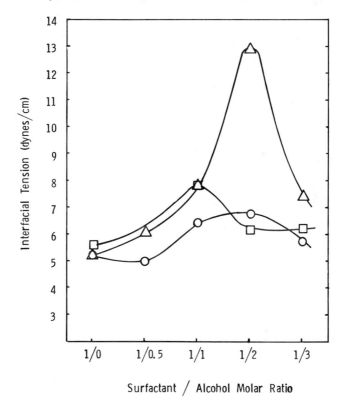

Surfactant / Alcohol Molar Ratio

Figure 4. Effect of molar ratio on interfacial tension between styrene and aqueous solutions of 10mM SLS/lauryl alcohol [-□-], 16.7mM SLS/lauryl alcohol [-O-] and 16.7mM SLS/cetyl alcohol [-Δ-].

tensions with styrene were observed at a molar ratio of 1:2 for both
the SLS/LA and SLS/CA solutions that were based on 16.7mM SLS. The
values of these maxima, 6.7 and 12.9 dynes/cm, respectively, were
significantly higher than that measured for the 16.7mM SLS solution
in the absence of fatty alcohol.

The results from a similar experiment with SLS/LA based on 10mM
SLS are also shown in Figure 4. As expected, the styrene/aqueous
phase interfacial tension was increased by this decrease in emulsifier concentration. The molar ratio corresponding to the maximum
interfacial tension shifted from 1:2 for the higher emulsifier concentration to 1:1 for the lower emulsifier concentration.

Earlier conductivity measurements have indicated that the most
stable miniemulsions are produced with mixed emulsifier molar ratios
between 1:1 and 1:3 (22,23). This correlation agrees with a theoretical analysis of mixed emulsifier adsorption onto oil droplets by
Lucassen-Reynders (35), who have determined the optimum stability to
occur at molar ratios near 1:1. However, the maximum interfacial
tensions at these molar ratios were unexpected because, minimum interfacial tensions are usually associated with maximum emulsion stability. In fact, minima values substantially less than 1 dyne/cm
have been reported for several oil/mixed emulsifier systems (31,33,
36,37).

The results for the effect on interfacial tension of the fatty
alcohol chain length are given in Table I for 1:1 molar ratio mixed
emulsifier solutions.

Table I. The Effect of Fatty Alcohol Chain Length on Interfacial
          Tension at 65°C between Styrene and Aqueous Mixed Emul-
          sifier Solutions, based on 10mM SLS and a Molar Ratio
          of 1:1

Continuous Phase	Molar Ratio	Avg. γ (dyne/cm)	Std. Deviation (dyne/cm)
SLS/Decanol	1:1	4.1	0.6
SLS/Lauryl Alcohol	1:1	5.8	1.9
SLS/Tetradecanol	1:1	6.8	1.2
SLS/Cetyl Alcohol	1:1	8.0	3.8
SLS/Octadecanol	1:1	8.3	1.1

The interfacial tension values increase from 4.1 dynes/cm for SLS/
decanol to 8.3 dynes/cm for SLS/octadecanol. Conductometric titration results have indicated that all of these mixed emulsifier systems, except the one with decanol, should give a relatively stable
emulsion (22,23). Interestingly, the SLS/decanol mixed emulsifier
solution was the only case in which the presence of the fatty alcohol
reduced the interfacial tension with styrene to below the value measured for SLS alone. Studies are in progress to investigate this
phenomenon and to determine the effect of alcohol chain length on
miniemulsion stability.

The relatively large interfacial tension values given in Table I
and depicted in Figure 4 may actually be an indication of the interfacial tension between the oil droplet and the mixed emulsifier interfacial layer. This hypothesis is supported by the low interfacial
tensions measured for "tails" which have detached themselves from the
oil drop. The interfacial tension between these detached, free-

floating "tails" and the aqueous mixed emulsifier solution was less than 10^{-2}dynes/cm. This interfacial tension is of the same order of magnitude as the values usually cited for interfacial tensions of oil measured in aqueous, mixed emulsifier systems (36,38-40).

Equilibration Studies. Because the formation of an interfacial layer is a dynamic phenomenon, experiments were conducted to study the effect of oil--water phase equilibria on interfacial properties. Two experiments were carried out where aqueous solutions of 1:1 SLS/LA were equilibrated with both styrene and toluene in sealed containers, without agitation for five weeks. Several important observations were made: (a) Despite the presence of mixed emulsifiers, which typically yield very low interfacial tensions, spontaneous emulsification did not occur (36,41-43). (b) An association of the emulsifiers formed as indicated by a milky, cloudy layer floating at the oil/water interface. However, a third phase did not form, as often occurs in microemulsion systems containing electrolytes. (c) The birefringence was a maximum near the oil/water interface and dropped off to non-birefringent in the bulk aqueous phase. In contrast, a control sample of 1:1 SLS/LA was uniformly birefringent.

The results of spinning drop experiments with the equilibrated oil (styrene) and aqueous (1:1 SLS/LA) phases are shown in Figure 5. No "tails" were formed on droplets of the pre-equilibrated styrene when injected into the capillary tube containing the pre-equilibrated aqueous phase. Thus, the formation of an interfacial layer in the spinning drop tensiometer is a non-equilibrium affect.

Similar experiments with a fresh styrene drop, which had not been pre-equilibrated, in the same pre-equilibrated, 1:1 SLS/LA mixed emulsifier solution, did not yield any visible interfacial layer. Therefore, the diffusion of the mixed emulsifiers into the oil phase evidently does not effect the formation of an interfacial layer. Therefore, the controlling factor appears to be the diffusion of the mixed emulsifiers in association with oil molecules into the aqueous phase.

Conclusions

Static, equilibration studies indicated that a molecular association forms at the styrene/water interface in the presence of mixed emulsifiers. Interfacial layers were also observed in spinning drop experiments between various oil phases and aqueous mixed emulsifier solutions. The formation of these interfacial layers as a function of time was found to be a non-equilibrium effect that depended primarily on the chemical structure of the oil phase. Oil phase water solubility had little effect.

Interfacial tension values between styrene and several mixed emulsifier solutions were relatively high, 5-13 dynes/cm, while the apparent interfacial tensions between the aqueous phase and the resulting interfacial layer were substantially less than 1 dyne/cm. The maximum interfacial tension occurred at an SLS/fatty alcohol molar ratio of 1:1 to 1:2.

The interfacial tensions between styrene and mixed emulsifier solutions increased with increasing fatty alcohol chain length at a constant molar ratio.

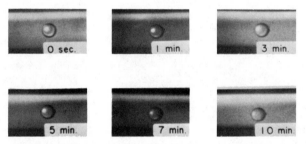

Figure 5. Spinning drop photographs of a pre-equilibrated styrene
droplet in pre-equilibrated, 1:1 SLS/lauryl alcohol solution,
based on 10mM SLS, at 65°C.

Acknowledgments

The financial support of this research by the National Science
Foundation, under Grant No. CPE-8119223, is greatly appreciated.

Literature Cited

1.   El-Aasser, M.S.; Hoffman, J.D.; Kiefer, C.; Leidheiser, Jr., H.;
     Manson, J.A.; Poehlein, G.W.; Stoisits, R.; and Vanderhoff,
     J.W., Final Report AFML-TR-74-208, "Water-Base Coatings, Part
     I," July 1973 - Aug. 1974 (dated Nov. 1974).
2.   El-Aasser, M.S.; Misra, S.C.; Vanderhoff, J.W.; and Manson,
     J.A., J. Coatings Tech., 1977, 49, 71.
3.   El-Aasser, M.S.; Vanderhoff, J.W.; and Poehlein, G.W., Preprints,
     A.C.S. Div. Org. Coatings Plastic Chem., 1977, 37, 92.
4.   Chou, Y.J.; Confer, L.M.; Earhart, K.A.; El-Aasser, M.S.;
     Hoffman, J.D.; Manson, J.A.; Misra, S.C.; Poehlein, G.W.;
     Scolere, J.P.; and Vanderhoff, J.W., Final Report AFML-TR-74-
     208, "Water-Base Coatings, Part II," Feb. 1975 - Nov. 1975
     (dated Aug. 1976).
5.   Vanderhoff, J.W.; El-Aasser, M.S.; and Hoffman, J.D., U.S.
     Patent 4,070,323 (to Lehigh University), Jan. 1978.
6.   Vanderhoff, J.W.; El-Aasser, M.S.; and Ugelstad, J., U.S.
     Patent 4,177,177 (to Lehigh University), Dec. 1979.
7.   Misra, S.C.; Manson, J.A.; and Vanderhoff, J.W., Preprints, ACS
     Div. Org. Coatings Plastics Chem., 1978, 38, 213.
8.   Pithayanukul, P.; and Pilpel, N., J. Coll. Int. Sci., 1982, 89
     (2), 494.
9.   Schulman, J.H.; and Cockbain, E.G., Trans. Faraday Soc., 1940,
     36, 651.
10. Biswas, B.; and Haydon, D.A., Kolloid-Z. Polymere, 1962, 185(1),
     31.
11. Hallworth, G.W.; and Carless, J.E., J. Pharm. Pharmac., 1972,
     24, 71p.
12. Izmailova, V.N.; Tulovskaya, Z.D.; Al'-shimi, A.F.; Nadel, L.G.;
     and Alekseeva, I.G., Daklady Akademii Nauk SSSR, 1970, 191(5),
     1081.
13. Blakey, B.C.; and Lawerence, A.S.C., Disc. Faraday Soc., 1954,
     18, 268.
14. Elworthy, P.H.; Florence, A.T.; and Rogers, J.A., J. Coll. Int.
     Sci., 1971, 35(1), 34.
15. Friberg, S.; Mandell, L.; and Ekwall, P., Kolloid-Z. Polymere,
     1969, 233, 955.
16. Friberg, S.; Mandell, L.; and Larsson, M., J. Coll. Int. Sci.,
     1969, 29, 155.
17. Friberg, S.; and Rydhag, L., Kolloid-Z. Polymere, 1971, 244, 233.
18. Higuchi, W.I.; and Misra, J., J. Pharm. Sci., 1962, 51(5), 459.
19. Ugelstad, J.; El-Aasser, M.S.; and Vanderhoff, J.W., Polym. Sci.
     Polym. Letters Ed., 1973, 11, 503.
20. Ugelstad, J., Makromol. Chem., 1978, 179, 815.
21. Ugelstad, J.; and Mork, P.C., Adv. Coll. and Interface Sci.,
     1980, 13, 101.
22. Grimm, W.L., M.S. Thesis, Lehigh University, Bethlehem, Pa.,
     1982.

23. Chou, Y.J., Ph.D. Thesis, Lehigh University, Bethlehem, Pa., 1978.
24. Vonnegut, B., Rev. Sci. Instr., 1942, 13, 6.
25. Burkowsky, M.; and Marx, C., Oil-Gas Eur. Mag., 1977, 4, 33.
26. Burkowsky, M.; and Marx, C., Tenside Detergents, 1978, 15, 247.
27. Burkowsky, M.; and Marx, C., Oil-Gas Eur. Mag., 1979, 5, 36.
28. Marx, C.; Murtada, H.; and Burkowsky, M., Erdoel-Erdgas-Zeit-schrift, Jg., 1977, 93, S303.
29. EOR Inc., Houston, Tx., "Operating Manual for SITE Model LP-10 Spinning Drop Interfacial Tensiometers," 1980.
30. Lack. C.D.; El-Aasser, M.S.; and Vanderhoff, J.W., "Calibration of SITE Spinning Drop Interfacial Tensiometers", submitted for publication, 1983.
31. Shah, D.O.; Bansal, V.K.; Chan, K.; and Hsieh, W.C., "Improved Oil Recovery by Surfactant and Polymer Flooding", Academic: New York, 1977; pp. 293-337.
32. Sumner, C.G., J. Appl. Chem., 1957, 7, 504.
33. Prince, L.M., J. Coll. Int. Sci., 1967, 23, 165.
34. Tadros, T.F., Colloids and Surfaces, 1980, 1, 3.
35. Lucassen-Reynders, E.H., J. Coll. Int. Sci., 1982, 85(1), 178.
36. "Microemulsions: Theory and Practice", Prince, L.M., Ed.; Academic: New York, 1977; pp. 1-19 and 91-141.
37. Gerbacia, W.; and Rosano, H.L., J. Coll. Int. Sci., 1973, 44(2), 242.
38. Cayias, J.L.; Schecter, R.S.; and Wade, W.H., "The Measurement of Low Interfacial Tensions via the Spinning Drop Technique", Departments of Chemistry and Chemical Engineering, University of Texas at Austin, 1975.
39. Hsieh, W.C.; and Shah, D.O., "The Effect of Chain Length of Oil and Alcohol as well as Surfactant Alcohol Ratio on the Solubil-ity, Phase Behavior and Interfacial Tension of Oil/Brine/Surfac-tant/Alcohol Systems," Soc. of Pet. Eng. of AIME Paper No. 6594, 1976.
40. Rubin, E.; and Radke, C.J., Chem. Eng. Sci., 1980, 35, 1129.
41. Benton, W.J.; Miller, C.A.; and Fort Jr., T., J. Disp. Sci. Tech., 1982, 3(1), 1.
42. Bellocq, A.M.; Bourbon, D.; and Lemanceau, B., J. Coll. Int. Sci., 1981, 79(2), 419.
43. Cayias, J.L.; Schecter, R.S.; and Wade, W.H., in "Adsorption at Interfaces", Mittal, K.L., Ed.; ACS SYMPOSIUM SERIES No. 8, American Chemical Society: Washington, D.C., 1976; pp. 234-247.

RECEIVED January 8, 1985

MACROEMULSIONS

Stability and Stabilization of Water-in-Oil-in-Water Multiple Emulsions

ALEXANDER T. FLORENCE and DOUGLAS WHITEHILL[1]

Department of Pharmacy, University of Strathclyde, Glasgow G1 1XW, Scotland

Water-in-oil-in-water (w/o/w) and oil-in-water-in-oil (o/w/o) multiple emulsions are inherently unstable dispersions because of the several possible routes to physical breakdown, some of which are peculiar to these 'double' emulsions. In w/o/w systems coalescence of the oil droplets carrying the dispersion of smaller water droplets occurs with the water droplets influencing only marginally the process; coalescence of the internal water droplets may also occur, diffusion of surfactant from the original w/o interface to the interface of the oil droplets with the bulk aqueous phase complicating formulation and stabilization. The most difficult feature to control is the osmotic flux of water molecules across the oil lamellae in w/o/w systems. Electrolyte and other additives including drugs can affect the stability of the systems by both osmotic and interfacial effects.

Strategems to overcome some of these basic problems include (i) the use of a high viscosity oil phase in w/o/w emulsions to prevent or decrease diffusion of individual surfactant molecules and water molecules, (ii) the polymerisation of interfacially adsorbed surfactant molecules, and (iii) the gelation of the oily or aqueous phases of the emulsions.

In spite of the extensive literature on the theoretical and practical aspects of conventional o/w and w/o emulsions, little

[1] Current address: Department of Pharmacy, Robert Gordon's Institute of Technology, Aberdeen AB9 1FR, Scotland

attention has been paid to multiple emulsions*. These complex
systems have shown promise, particularly in pharmaceutics and in
separation science. Their potential biopharmaceutical
applications (1,2), a consequence of the presence of a 'reservoir'
phase inside droplets of another, include use as adjuvant vaccines
(3) as prolonged drug delivery systems (4,5,6,7,8), as sorbent
reservoirs in drug overdosage treatment (9,10) and in immobilization
of enzymes (11). The use of multiple emulsions in the separation
field has included the separation of hydrocarbons (12) and the
removal of toxic materials from waste water (13). Multiple emulsions
have been used as the basis of liposome-like lipid vesicles (14)
and microcapsules (15).

Seifriz described in 1925 (16) oil drops of an o/w emulsion
containing droplets of water, noting small oil globules inside the
water drops of a w/o emulsion; Seifriz termed these 'bimultiple'
systems, and also found that more complex 'trimultiple', 'quatre-
multiple' and even 'quinquemultiple' systems existed. Although
multiple systems have been known for such a long period of time,
it is only in the past 15 years that they have been purposely
prepared and studied. The two major types of multiple emulsion
are water-in-oil-in-water (w/o/w) emulsions in which internal and
external aqueous phases are separated by an oil layer, and oil-in-
water-in-oil (o/w/o) emulsions where an aqueous phase separates
internal and external oil phases.

The formulation and stability of multiple emulsion systems
has recently been reviewed by the present authors (17).

Here we use the term "multiple drop" to describe the oil
droplets in w/o/w emulsions containing dispersed aqueous droplets,
"primary" surfactant the stabilizer for the w/o emulsion and
"secondary" surfactant to denote the more hydrophilic surfactant
used to stabilize the o/w component. The "internal" phase is the
dispersed aqueous phase, the "external" phase is the continuous
aqueous phase and the "middle" phase the carrier oil droplets.

In emulsions with high disperse phase volumes (>0.74) the
drops may assume a complex structure and often multiple drops will
be observed (18). Many workers have noted the appearance of multiple
globules during emulsion inversion (16,19,20). This appears to
result from partitioning of the surfactants between the two phases.
Lin et al (20) noted that when a hydrophilic surfactant was initially
placed in the oil phase a transient w/o/w emulsion was formed in
the process of phase inversion from w/o to o/w emulsion type. A
portion of the aqueous phase, added during the emulsification
process, will be emulsified in the oil phase forming a primary w/o
emulsion. If the conditions favour o/w emulsification, i.e. if the
HLB of the surfactant is high this initial w/o emulsion will be
unstable and on further agitation the w/o emulsion is mixed into
the excess water to form a w/o/w emulsion. As surfactant migrates

* In some disciplines, certain multiple emulsions have been
termed 'liquid membrane' systems, as the liquid film which separates
the other liquid phases acts as a thin semi-permeable film through
which solute must diffuse moving from one phase to another. There
are, therefore many potential practical applications of multiple
emulsions.

to the outer aqueous phase, the unstable, larger globules of water readily coalesce with smaller droplets to form a final o/w emulsion. This is not found if the hydrophilic surfactant is initially placed in the aqueous phase. Becher (21) reported the inversion of o/w and w/o emulsions to multiple systems followed later by reversal to the opposite type. It appears that, under certain conditions, when inversion takes place, some of the original structure becomes trapped in the final emulsion.

One of the main drawbacks to the commercial development of multiple emulsions is their inherent instability. The intention of this paper is to review studies on the stability and mechanism of breakdown of multiple systems and attempts to minimise such instability, for example, by appropriate choice of surfactant, polymerisable surfactants or gelation of the aqueous or oily phases.

Multiple emulsions which will possess some degree of stability may be prepared by using pairs of surfactants, one of which will stabilize a w/o emulsion (lipophilic) and this forms the basis for the intentional preparation of multiple systems.

Multiple emulsion types

The structure of w/o/w multiple globules is dependent on the nature of the secondary surfactant. We (22,23) have shown that it is possible to prepare 3 different types of w/o/w emulsion, so called type A, type B and type C. Three different water-isopropyl myristate-water emulsions were prepared using various non-ionic surfactants, each system consisting of a primary water-isopropyl myristate emulsion stabilized with 2.5% sorbitan mono-oleate and containing 50% water. The multiple emulsions were then prepared by re-dispersing the primary w/o system in an equal volume of water containing 2% hydrophilic surfactants; Brij 30 (Type A), Triton X-165 (Type B) and Span 80: Tween 80 (3:1) Type C. Their particle size distributions are shown in Fig. 1. Type A emulsions were composed principally of small multiple drops (mean diameter 8.6 μm) 82% of which contained only one internal aqueous droplet (mean diameter 3.3 μm). Type B consisted of larger multiple drops (mean diameter 19 μm) containing smaller, but more numerous internal aqueous droplets (mean diameter 2.2 μm), whereas type C emulsions were composed of very large multiple drops (mean diameter 25 μm) which contained very large numbers of internal droplets which were difficult to resolve. In type C emulsions, it is thought that the primary (encapsulated) system is a flocculated w/o emulsion. The structure of the multiple drop probably depends on the efficiency of the second emulsification step and perhaps also on the rapidity of transfer of surfactants between the interfaces. (Figure 1)

In each system, however, simple oil drops are also present. All of these structures may coexist to some extent, although the emulsions we have described consist predominantly of either type A, B or C drops.

Recently Di Stefano and coworkers (24) have found with o/w/o systems that the nature of the internal droplets depended on the rate of agitation used in preparation, agitation rates of 212 rpm forming types A or B systems and 425 rpm a type C emulsion. These systems were unusual in that the multiple droplets were extremely small, with diameters in the range 0.05-0.20 μm and the internal droplets in the range 0.02-0.05 μm.

Effect of composition of multiple emulsion systems

Oils

As with conventional emulsions the nature of the oil can affect the behaviour of the system. For pharmaceutical uses, oils used include the refined hydrocarbon oils such as light liquid paraffin and esters of long-chain fatty acids including vegetable oils, for example, ethyl oleate and isopropyl myristate, olive oil and sesame oil. Frankenfeld et al (9) used mixtures of 'Solvent 100 Neutral' (an isoparaffinic, dewaxed oil of high viscosity) and 'Norpar 13' (a non-viscous, normal paraffinic solvent) to vary the viscosity of the oil phase in attempts to control the transfer of solutes across the oil membrane.

The relative proportions of each oil determines the overall viscosity of the oil (membrane) phase. The higher the concentration of 'Norpar 13', the lower the viscosity of the membrane, and the greater the rate of transfer of material from the external to the internal phase. The stability of the membrane toward rupture and leakage of entrapped materials, however, decreased with decreasing viscosity. In fact, formulations containing greater than 50% 'Norpar 13' were found to be too unstable for practical use. However, multiple emulsions of liquid paraffin (222 cP) prepared by Panchal et al (25) were less stable than those prepared with kerosene (viscosity 15 cP), stability being measured by loss of internal aqueous phase.

Surfactants

It is necessary to use at least two surfactants, one for the primary emulsion and the other for the dispersion of this emulsion to form the multiple system. The optimum surfactant to emulsify a given oil can be determined by use of the hydrophile-lipophile balance (HLB) approach. The present authors have carried out an investigation into the optimal HLB required for both primary and secondary emulsification steps in the formulation of a water/isopropyl myristate/water emulsion. W/o emulsions containing 47.5% isopropyl myristate and 2.5% surfactant had an optimal HLB of 4.5.

Apart from the fact that the use of the HLB system is limited as it is based on the observation of creaming or separation of the emulsions, as an index of instability the HLB system also neglects the effects of surfactant concentration on stability (26) and of course it is irrelevant to the particular problems with multiple emulsion systems. Nevertheless, it provides a useful approach to the choice of optimal surfactant system. In general, in a w/o/w emulsion, the optimal HLB value of the primary surfactant will be in the range 2-7 and in the range 6-16 for the secondary surfactant. Equilibration of the systems after mixing will undoubtedly result in the transfer of surfactant between the aqueous and nonaqueous components. Saturation of the phases with the two surfactants used should prevent instability during this equilibration.

There will be an optimal concentration of surfactant required to stabilize the system; low concentrations may not be sufficient to stabilize the emulsions and may result in the rapid degradation of the emulsion. High concentrations may serve to increase the viscosity of the system. The general range is between 1 and 10%, but Matsumoto et al (27) suggest that concentrations of greater than 30% by weight of Span 80 in the oil phase are required to obtain yields of more than 90% of multiple drops. They also found that yields of w/o/w emulsion fell markedly when the concentration of secondary emulsifier (Tween 20) in the external phase was increased. Matsumoto et al considered that the ratio of concentration of primary to secondary surfactant was significant in this particular system, and they suggested that more than 10 times as much Span 80 as Tween 20 was required to obtain 90% or higher yields of w/o/w emulsion. This is thought to be due to the solubilization of molecules of the primary (lipophilic) surfactant in the outer aqueous phase when the concentration of the secondary (hydrophilic) surfactant exceeds the critical micelle concentration. As the concentration of secondary surfactant increases, more of the primary surfactant may be incorporated into the secondary surfactant micelles, causing the concentration of primary surfactant at the interface to fall, and leading to the rupture of the oil layer, which results in the loss of the internal aqueous drops. This would explain the significance of the ratio of surfactant concentration on the initial yield of multiple drops.

Inversion of multiple w/o/w emulsions to o/w emulsions has been found to occur (28) only when the oil droplet size is reduced below a critical size or if the HLB of the emulsifiers approaches the 'required' HLB of the oil phase. When these droplets were reduced in size below about 5 μm they no longer could accommodate an inner aqueous phase. Droplet size reduces with increasing concentrations of secondary surfactant (Fig. 2) which might, as Magdassi and co workers (28) point out, explain the results of Matsumoto et al (27).

Reducing the size of the external droplet for a given w/o (Fig. 2) primary system will lead to an increased opportunity for internal droplet coalescence with the continuous phase, because of the denser packing of the system and because of the greater attraction between internal water droplets and the continuous aqueous phase than between water droplets (Fig. 3). Coalescence of the external droplets will thus lead to greater stability of the internal drop-lets (if this reasoning is correct).

The HLB shift caused by emulsifier migration to the external oil surface has been estimated by Magdassi et al (28). Increasing concentrations of the primary (w/o) emulsifier causes a significant increase in the optimal HLB e.g. by increasing its concentration from 5 to 30%, a shift of 2 units in optimal HLB is achieved. At a fixed concentration of the primary emulsifier (10%) decreasing the concentration of the secondary emulsifier from 5 to 1% causes a similar increase in the optimal HLB (see Fig. 4); as the hydrophilic surfactant diffuses to the external droplet surface, there inter-acting with the stabilizing layer, a more hydrophilic mix of surf-actants is required to maintain the optimal HLB (29). (Fig. 4)

The thickness of the membrane phase and the type of surfactant used may also be important in determining stability and transport rates. Li (30) found that ionic surfactants gave a much higher transport rate of toluene through an aqueous membrane. The diffusion rate of toluene also increased with increasing hydrophilic

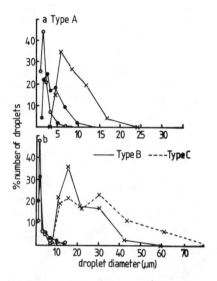

Figure 1. Particle size distribution of a) type A, b) type B
and type C w/o/w emulsions (where O=isopropyl myristate) just
after preparation as described in the text. Key: x=multiple oil
drops; o=simple or empty oil drops; and ●=internal aqueous
droplets.

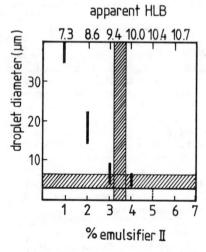

Figure 2. Change in droplet diameter of multiple emulsions
as a function of the concentration of secondary emulsifier (II)
and of the calculated weighted or apparent HLB of the surfactant
system. Hatched regions represent boundaries for inversion.
Reprinted with permission from Ref. 28. Copyright 1979, Academic
Press.

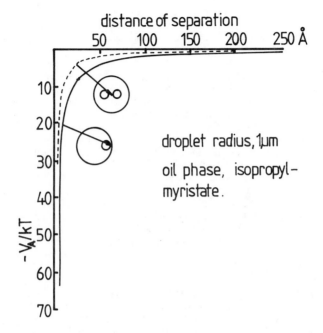

Figure 3. Forces of attraction (V_A/kT) as a function of
distance of separation of the aqueous phases in an isopropyl
myristate/water system. Inner droplet 1 μm radius.

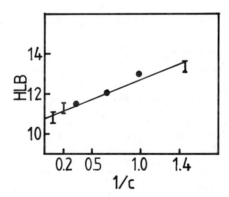

Figure 4. The change in the optimal HLB of secondary surfactants
used to prepare multiple w/o/w emulsions in which the primary
w/o emulsion has been stabilized by 10% Brij 92. Reproduced
with permission from Ref. 28. Copyright 1979, J. Colloid
Interface Sci.

chain length of alkyl ether non-ionic surfactants. The concentration
of emulsifier may also be an important factor. Collings (5)
investigated the effect of the concentration of primary surfactant on
the release of sodium chloride from the internal aqueous phase of
w/o/w emulsions.

The presence of liquid crystal structures at both the w-o and
o-w interfaces in multiple emulsions has been investigated by
Kavaliunas and Frank (31). Microscopic examination of w/o/w
emulsions between crossed polarizers revealed the presence of liquid
crystal phases at both inner (w-o) and outer (o-w) interfaces in a
w/o/w system composed of water, p-xylene and nonylphenol diethylene
glycol ether. Liquid crystalline phases were also detected in
o/w/o emulsions at both interfaces. The presence of these liquid
crystal structures was found to improve the stability of the
emulsions markedly. Matsumoto (32, 33) have concluded that the oil
layers in w/o/w systems are likely to be composed of or contain, at
least in proximity to the aqueous phase, multilamellar layers of the
lipophilic surfactant used in the formulation; this is postulated in
part to explain the rate of volume flux of water through the oily
layer.

Phase volumes

Matsumoto et al (27) found that internal phase volume ($\emptyset_{w/o}$)
had no significant effect on the yield of w/o/w emulsion under
the experimental conditions studied. It would appear that w/o/w
emulsions can be prepared using a wide range of internal phase
volumes. Collings (5) quotes a range of 5-75% water-in-oil, but
the optimal range is 25-50%. He found, however, that the internal
phase volume influenced the release of materials from the internal
aqueous phase. Interesting results by Matsumoto et al (27)
suggested that the secondary phase volume ($\emptyset_{w/o/w}$) influences the
yield of multiple drops over a range of low volume fractions.
When $\emptyset_{w/o/w}$ exceeded about 0.4 there was no significant effect.

Nature of entrapped materials

The nature of entrapped materials may have a bearing on the
stability of the system. Due to the nature of the multiple emulsion,
the middle phase may act as an osmotic reservoir, thus virtually
all additions to this phase will set up osmotic gradients. This
might include high concentrations of surfactant. To this end
polymeric microspheres have been used as the internal reservoir
when osmotic transfer of water will not compromise stability.

Release of methotrexate, metoclopramide and sodium chloride
from type A, B and C w/isopropyl myristate/w emulsions have been
compared (Fig. 5 a,b,c). In all cases, release from the type C
emulsion is not prolonged, which may be a reflection of stability
or structure or a combination of these two parameters. In the
case of methotrexate, variation of the concentration of secondary
surfactant (polysorbate 80) from 0.5 to 20% had no significant
effect on the rate of drug release from the system.

Stability and release mechanisms

The ability of multiple emulsions to entrap materials is one
of their most useful assets and so the passage of materials from the

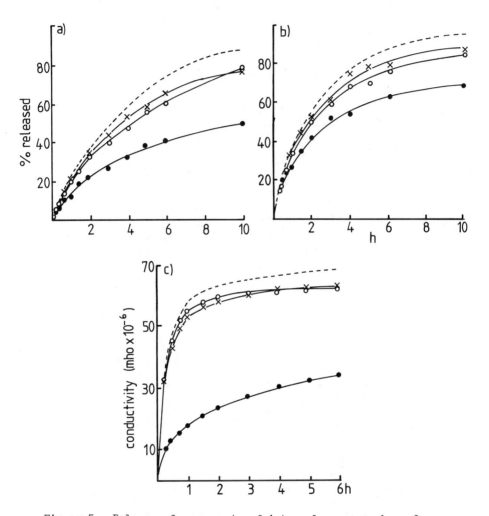

Figure 5. Release of components of internal aqueous phase from multiple water-isopropyl myristate-water emulsions placed in dialysis sacs at 37 °C, compared with release rates of equivalent concentrations of solutions of the same substances (------).
a) Methotrexate. Key: -----, transport of 0.1% MTX; o, type A; x, type B; and ●, type C systems. Ø w/o = 0.5, Ø w/o/w = 0.5.
b) Metoclopromide. Dotted line represents transport of 0.1% metoclopromide HCL release. Symbols as for a).
C) Sodium chloride. Systems as for a).

internal to external phase, across the middle phase, is important.
This is particularly so in pharmaceutical systems where multiple
emulsions are envisaged as possible controlled-release drug delivery
systems. The stability and release characteristics of emulsion
systems are influenced by a number of factors such as the composition of the emulsion, droplet size, viscosity, phase volumes, pH etc.
This becomes even more complex in multiple systems because there are
two dispersed phases, two phase volumes, at least two surfactants,
and 3 different droplet size distributions.

The mechanisms of instability in multiple emulsions are complex
and difficult to study. We earlier (23) attempted an analysis of
the possible mechanisms of instability. w/o/w multiple drops, for
example, may coalesce with other oil drops (single or multiple) or
they may lose their internal droplets by rupture of the oil layer
on the surface of the internal droplets, leaving simple oil drops.
Under the influence of an osmotic gradient, the oily lamellae of the
multiple drops act as 'semi-permeable' membranes resulting in the
passage of water across the oil phase. This leads to either
swelling or shrinkage of the internal droplets, depending on the
direction of the osmotic gradient. Another possible breakdown
mechanism may be coalescence of the internal aqueous droplets within the oil phase. A combination of these mechanisms may take
place; the likelihood of events taking place may be predicted by
analysis of the van der Waals' attractive forces and free-energy
changes in these systems (23).

Specific factors controlling the stability of multiple emulsion

Effect of electrolytes
Electrolyte presence appears to be one of the most important
factors in determining the stability and release of materials from
the internal droplets. The effects of electrolytes are two-fold:
(a) osmotic; and (b) interfacial, the former being only observed in
multiple systems. The effects of electrolytes on electrical double
layers, etc., will not be considered here as they are not specific
to multiple systems.

(a) Osmotic effects
If the osmotic pressure is higher in the internal aqueous
phase, water will pass into this phase resulting in swelling of the
internal droplets which eventually burst, releasing their contents
into the external phase. Transfer of water from the internal to
external aqueous phase causing shrinkage of the internal droplets
occurs if the reverse gradient exists (Fig. 6). If the osmotic
pressure difference across the oil layer is extreme, then passage
of water is so rapid that almost immediate rupture of the oil
drops occurs with expulsion of the internal droplets. When the oil
layer ruptures the inner aqueous phase disappears instantaneously
and is followed by mixing of the internal aqueous phase with the
external aqueous medium, leaving a simple oil drop. This appears
to occur frequently where the oil layer is thin, for example in
Type A and type C drops, but also occurs to some extent in type B
drops. Materials other than electrolytes, such as proteins and
sugars and of course drugs, in either aqueous phase can also exert
this effect. Collings (5) partially solved the problem in relation

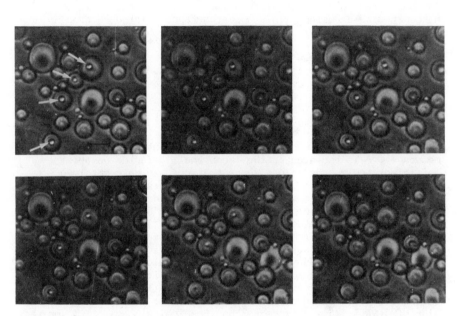

Figure 6. Sequence from cinematographic film of the shrinkage of internal droplets (arrowed) in a type A w/o/w system.

to administration of these systems by the addition of small amounts
of sodium chloride to the internal aqueous phase so that it was
isotonic with the final external phase but this approach can lead
to inequality of pressure on storage which is unsatisfactory.
Osmolarity may also be adjusted by the addition of other materials
such as glucose or glycerol.

Davis and Burbage (34) while investigating the effects of
sodium chloride on the size of multiple drops found that a threshold concentration of electrolyte existed below which little or no
change in size occurred. This appeared to be related to drop
diameter - larger multiple drops required a higher concentration of
sodium chloride to effect shrinkage.

Matsumoto and Khoda (35) measured the water permeation
coefficient of the oil layer indirectly by measuring the change in
viscosity of the w/o/w emulsion to determine change in globule
diameter and using the following equation, φ_w being the flux of
water in moles per unit time:

$$\varphi_w = -P_oA(g_2c_2 - g_1c_1)$$

where P_o is the osmotic permeability coefficient of the oil
'membrane', A is the area of the 'membrane', g is the osmotic
coefficient and c, the solute concentration. We have obtained
estimates of P_o using a more direct cinemicrography technique (23)
by measuring the change in volume of internal droplets with time,
i.e. $d\varphi_w/dt$. Values of P_o in the range -0.018×10^{-4} to
-0.582×10^{-4} (mean value $-0.116 \pm 0.071 \times 10^{-4}$)cm.s^{-1} were found
for passage of water from the internal to external aqueous phase in
an osmotic gradient created by 3M sodium chloride, and in the range
-0.042×10^{-4} to -0.438×10^{-4} cm.s^{-1} (mean value $-0.162 \pm 0.15 \times
10^{-4}$ cm.s^{-1} for the flux of water in the opposite direction under
the same conditions (36). The osmotic permeability can be used to
calculate the diffusion coefficient for water in the oil layer; a
value of $3.94 \pm 1.5 \times 10^{-3}$ cm.s^{-1} has been calculated.   The
diffusion coefficient of water in bulk hydrocarbon is about 1×10^{-5}
cm.s^{-1} and appears to be the same in biological membranes (36)
leading to the assumption that water transport in the multiple
emulsion oil globules is by carriage in inverse micelles of
surfactant (17). The translational diffusion coefficient of proteins
and lipids in biological membranes is in the range 1×10^{-9} to
1×10^{-8} cm^{2}s^{-1} lending support to the hypothesis that water transport is achieved by attachment to large (micellar) units.

(b) Interfacial effects

The release of the narcotic antagonist, naltrexone hydrochloride, from the internal aqueous phase of a w/o/w emulsion was
found (7) to decrease with a value of D   of 1×10^{-5} cm^{2}s^{-1} with
increasing sodium chloride concentration in the internal aqueous
phase.   A decrease in the diffusion coefficient of the drug of 73%
was obtained with 9% W/v sodium chloride dissolved in the internal
aqueous phase. Sorbitol also caused a decrease in the diffusion
coefficient but at an equivalent sodium chloride concentration of
about 6% W/v. These results indicate that factors other than
osmotic gradients are affecting passage of the drug. Brodin et al
(7) suggested that sodium chloride competes with surfactant for

water molecules at the inner w-o interface, which would result in
a rigid interfacial layer which might be a more effective mechanical
barrier to drug transfer.

Properties of the primary emulsion

Instability of the primary emulsion may result in coalescence
of the internal droplets within the membrane phase, their osmotic
growth leading to rupture of the external droplet due to breakdown of the thin oil lamella. Loss of internal droplets of course
leads to rapid appearance of drug in the outer phase, circumventing
diffusional transfer process. The aqueous internal droplets may
be individually expelled or may coalesce before being expelled;
or water may pass by diffusion through the oil phase gradually
resulting in shrinkage of the internal droplet. Whether all these
mechanisms occur in all systems is not clear; neither is the
relative importance of each mechanism in different w/o/w systems.
On the other hand, a combination of the above events may take place.
Comparison of release of methotrexate, metoclopramide and sodium
chloride from type A, B and C multiple emulsions   (Fig. 5) is
not necessarily reflective of stability per se, but may also reflect
differences in the initial states of the emulsions. In all three
cases, however, release is slowest from type C systems.

Calculation of the change in the free-energy of the system
associated with each step indicates for example that coalescence
of the multiple drops would result in a relatively large change in
the free-energy, brought about by the reduction in w-o interfacial
area. Coalescence between the small internal aqueous droplets, on
the other hand, would not be expected to be a major route of
breakdown. These predictions generally agree with experimental
evidence (23).

Some progress toward an understanding of these systems is also
possible by considering the influence of the presence of water
within the oil drops on the interaction between the oil drops and
by consideration of the influence of the size of the internal water
droplets on their internal stability and on the possibility of
coalescence with the external aqueous phase. It is premature to
consider all this in detail as the application of colloid stability
theory to simpler emulsions has not been particularly successful
(37). For type A w/o/w emulsions, the approach of Vold (38) may
perhaps be used if the oil layer is thought of as the homogeneous
'adsorbed' layer.

The nett interaction for two particles covered with an adsorbed
layer is given by :

$$V = V_f - 2V_D$$

where V_f is the sum of the interaction of two solvated particles
in contact and the interaction energy of two imaginary particles
with the composition of the medium, and V_D is the interaction of
an imaginary particle and a real solvated particle in contact.

Modifying Vold's approach to type A multiple drops, the total
interaction energy controlling flocculation is given by the
equation

$$-12V = \left\{ (A_{w_e}^{\frac{1}{2}} - A_o^{\frac{1}{2}})^2 H_o + (A_o^{\frac{1}{2}} - A_{w_i}^{\frac{1}{2}})^2 H_w + \right.$$

$$+2(A_{w_e}^{\frac{1}{2}} - A_o^{\frac{1}{2}})(A_o^{\frac{1}{2}} - A_{w_i}^{\frac{1}{2}})H_{wo}\bigg\}$$

where the subscripts w_i, o and w_e refer to the internal aqueous phase, oil phase and external phase respectively. Since the internal and external phases are both water, this simplifies to

$$V = -\frac{1}{12}\bigg\{(A_w^{\frac{1}{2}} - A_o^{\frac{1}{2}})^2 H_o + (A_o^{\frac{1}{2}} - A_w^{\frac{1}{2}})^2 H_w$$

$$+ 2 (A_w^{\frac{1}{2}} - A_o^{\frac{1}{2}})(A_o^{\frac{1}{2}} - A_w^{\frac{1}{2}}) H_{wo}\bigg\}$$

The symbol H_o represents two drops of radius $(r + \delta)$ where δ is the thickness of the oil layer and separation Δ. H_{wo} represents the H function for a sphere of radius r and one of radius $(r + \delta)$ separated by a distance $(\delta + \Delta)$. H_w represents the function for two spheres of radius r separated by a distance $(\Delta + 2\delta)$.

If the Hamaker constant for the oil phase is close to that of water (as is the case) for the water/isopropyl myristate/water systems investigated by the authors (23) the effect of replacing part of the oil drop with water is not great; in fact in a typical type A drop the effect of large internal droplet on attractive energies is found to be insignificant. Only when the internal droplet almost fills the whole diameter of the multiple drop is the influence of the internal phase noticeable.

A more significant influence on van der Waals' forces of attraction between multiple drops appears to reside in the reduction in size which follows from expulsion of the internal droplets. The resultant reduction in diameter leads to a reduction in the force of attraction directly as V_A is related to globule radius, r, by

$$V_A = -\frac{(A_o^{\frac{1}{2}} - A_w^{\frac{1}{2}})^2 r}{\Delta}$$

A full analysis of interactions in multiple emulsions would obviously have to take account of forces of repulsion. The systems are too complex to (Fig. 7) allow any reasonable estimate of repulsive forces at this stage, although simplified models are being developed to allow an approach along this route. One complication resides in the possible lamellar nature of interfacial films or liquid crystalline structures, discussed earlier.

Attempts to improve the stability of multiple emulsions

It will be apparent that multiple emulsions - whilst having many potential uses - are complex, inherently unstable systems. They are unlikely to be commercially acceptable until problems with their stability in vitro and in vivo are solved. Despite this, there are few reports in the literature, regarding attempts to improve their stability. We have concentrated our efforts on the potential of polymerisable surfactants or other monomers to enhance stability following early attempts to use an oil phase which would solidify at room temperature. Release profiles of methotrexate from water-octadecane-water emulsions are shown in Fig. 8; the photomicrograph insert shows the structure of the emulsion.

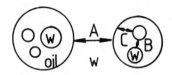

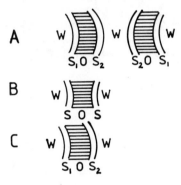

Figure 7.   Three possible interactions (attractive and repulsive) between phases in multiple emulsions are shown in the upper diagram, while below those arrangements of the aqueous (W), surfactant (s_1 and s_2) and oil phases are shown which must be taken into account in calculation of attractive and repulsive forces in these interactions.

a

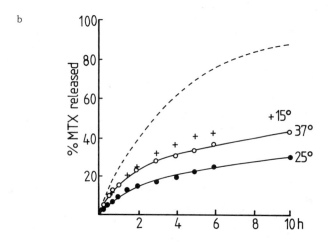

b

Figure 8. a) Photomicrograph of a water–octadecane–water system.
b) Release from a dialysis bag of MTX from an 0.1% solution of
MTX at 37 °C and from the emulsion at 15, 25, and 37 °C. The oil
phase of the emulsion is solid at 25 °C and liquid at 37 °C; at
15 °C the system tends to break up.

Because the preparation of a w/o/w system is a two-stage
procedure, it is possible to modify either the primary aqueous
phase (which becomes the internal aqueous phase of the multiple
system) or the secondary aqueous phase, which subsequently becomes
the continuous aqueous phase. Each aqueous phase can be gelled by
in situ polymerization reactions. Polyacrylamide has been used to
demonstrate the technique, but less toxic systems are required for
pharmaceutical use.

Al-Saden et al (39,40) reported the gelation of aqueous solutions
of poloxamer surfactants by the action of γ-irradiation.
Poloxamers* are relatively non-toxic poly(oxyethylene)-poly(oxypropylene) block copolymers with the general formula

$$HO[C_2H_4O]_a[C_2H_6O]_b[C_2H_4O]_aH$$

Crosslinking of the surfactant molecules may be induced by simultaneous activation of two neighbouring molecules with the net result
that the molecular weight of the polymer increases until a three
dimensional gel network is formed. As the hydrophilic poloxamers
are surface-active, promoting o/w emulsification, oil-in-water
emulsions may be prepared which contain the poloxamer in the
continuous aqueous phase. After emulsification, the surfactant
molecules can be crosslinked at the oil-in-water interface and in
the continuous phase by γ-irradiation, forming a network of
surfactant molecules which link the dispersed oil globules.
Similarly, the poloxamers can be used in the second emulsification
step in the preparation of the multiple emulsion (41).

As the hydrophilic nature of the poloxamer surfactants prevented
their use as stabilizers of the primary w/o emulsion, and as the
more lipophilic members of the series degrade on irradiation, an
alternative approach - a modified emulsion polymerization method
based on the technique of Ekman and Sjoholm (42) has been used to
gel the internal aqueous phase.

Initial experiments, in which a w/o/w emulsion containing the
monomer and crosslinking agent initially in the internal aqueous
phase was irradiated, were unsuccessful. It was found that release
of the monomer and crosslinking agent into the external aqueous
phase caused this phase to gel.

The problem of gelling the internal phase itself was solved
by irradiating the primary w/o emulsion containing the monomer and
crosslinking agent in the aqueous phase. The resulting poly(acrylamide)-in-oil dispersion was then redispersed in an aqueous phase
containing hydrophilic surfactant to produce a w/o/w system
containing a cross-linked poly(acrylamide) gel in the internal
aqueous phase. This system has similarities to the gelatin micro-
sphere/oil/water system described by Yoshioka et al (43). Preliminary release rate experiments on these systems have been disappointing,
bo significant difference in the release of an iodide marker being
found when release was compared to that from ungelled systems. It
may be that other conditions are required to produce a more rigid
gel. Stability of this system has not been assessed, although it

Footnote * Poloxamers - sold under the trade name Pluronic,
               registered trade name of Wyandotte.

is obvious that gelation of the internal phase blocks, internal droplet coalescence and osmotic growth and shrinkage of the internal droplets.

The first method used to gel the external continuous phase was the production of a crosslinked poly(acrylamide) gel. Apart from varying the ratio and concentration of monomer and crosslinking agent, little control could be achieved over the reaction - rigid gels were produced resulting in immobile systems.

The poly(oxyethylene)-poly(oxypropylene)-poly(oxyethylene) block copolymers were also used to gel the continuous aqueous phase. Poloxamers may be used as the secondary hydrophilic surfactant in the preparation of the w/o/w system, and the finished emulsion is then irradiated. The polymerisation reaction can be monitored by cone-and-plate viscometry. Fig. 9 shows the flow curve obtained for a water/isopropyl myristate/water emulsion as a function of the radiation dose. As the dose of γ-irradiation is increased, the viscosity of the w/o/w emulsion increased up to a 'gel-point'. The 'gel-point' of the emulsion is dependent on the type and concentration of poloxamer. In the example shown, prepared using a mixture of 5% (w/v) Pluronic F87 and 5% (w/v) Pluronic F88 in the external phase, the 'gel-point' was reached at 4.2 (Fig. 9). Fig. 10 shows the changes in the properties of irradiated systems on storage.

The main disadvantage of the use of γ-irradiation is that the drug has to be incorporated at the primary emulsification step and is therefore exposed to the γ-irradiation. To overcome this difficulty Law et al (44) have synthesised a series of acryolyl derivatives of certain poloxamer surfactants. Bredimas et al (45) have shown that it is possible to stabilise an unstable o/w emulsion by polymerisation of diacrylate surfactant molecules in surface crosslinking reactions. We have confirmed this (44) and have succeeded in stabilizing w/o/w emulsions, the primary emulsion comprising isopropyl myristate with 5% Span 80 and 0.9% of the diacryloyl derivative of Pluronic L44 in the aqueous phase ($\emptyset$ = 0.5) with 4% $(NH_4)_2S_2O_8$ as initiator. Stirred at 2000 rpm at 37º for 60 minutes the Pluronic cross-links at the interface; the multiple system is prepared with e.g. an equal solution of aqueous Pluronic P123 (0.8%) and Triton X165 (0.4% or 0.6%). Stable inner droplets were seen to be released intact from the oil droplets under certain conditions but normally samples retained their inner droplets for at least 2 weeks.

Conclusions

Knowledge of surfactant equilibration and interactions will probably lead to improved formulations of multiple emulsions. Failing this the use of polymerisable surfactants can lead to obvious strengthening of interfacial barriers and allow control of stability and drug release. Nonetheless further detailed work on both w/o/w and o/w/o systems is justified.

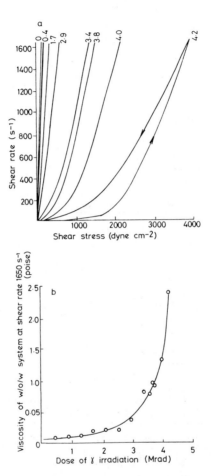

Figure 9. a) Flow curve using a cone and plate viscometer for a water-isopropyl myristate-water emulsion as a function of dose of gamma-irradiation shown in Mrad. b)Viscosity of a w/o/w emulsion at shear rate of 1650 s^{-1} stabilized with Pluronic surfactants (5% F85/5% F88) in the external phase as a function of radiation dose. The increase in viscosity shown by the flow curves indicates structure build-up in the external phase. The hysteresis loop exhibited by the gelled sample indicates that the structure of the external gel phase is broken down during shearing. Reproduced with permission from Ref. 41. Copyright 1982, J. Pharm. Pharmacol.

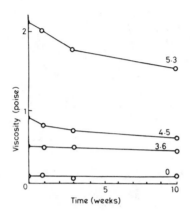

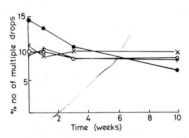

Figure 10. a) Viscosity of w/o/w emulsions (at shear rate 1650 s^{-1}) exposed to different doses of gamma-irradiation and its change with time. b) Percentage number of multiple drops in w/o/w emulsions exposed to o, 3.6 Mrad; x, 4.8 Mrad; +, 5.3 Mrad (control •) as function of time. The decrease in viscosity with time is most likely to be due to changes in the gel structure, as the model system chosen for study contained a low percentage of 'multiple' drops, making a decrease in viscosity due to the loss of water from the internal aqueous phase unlikely. Photomicrographic studies showed that there was little change in the size of the multiple and simple oil drops or of the internal aqueous droplets of each system, although there was some evidence of multiple drop coalescence in the control system, not evident in the irradiated systems. The irradiated systems also appeared to be more stable to multiple drop rupture.

Literature Cited

1. Davis, S.S., J.Clin.Pharm., 1, 11, (1976).
2. Davis, S.S., Chem.Ind., 684 (1981).
3. Taylor, P.J., Miller, C.L., Pollock, T.M., Perkins, F.T. and Westwood, M.A., J.Hyg., Camb., 67, 485 (1969).
4. Elson, L.A., Mitchley, B.C.V., Collings, A.J. and Schneider, R., Rev.Europ.Etudes Clin.Biol., 15, 87 (1970).
5. Collings, A.J. British Patent 1235667 (1971).
6. Benoy, C.H., Elson, L.A. and Schneider, R., Br.J.Pharmacol., 45, 135P.
7. Brodin, A.F., Kavaliunas, D.R. and Frank, S.G., Acta.Pharm. Suec., 15, 1 (1978).
8. Brodin, A.F. and Frank, S.G., Acta.Pharm.Suec., 15, 111, (1978).
9. Frankenfeld, J.W., Fuller, G.C. and Rhodes, C.T., Drug Develop. Commun., 2, 405 (1976).
10. Chiang, C., Fuller, G.C., Frankenfeld, J.W. and Rhodes, C.T., J.Pharm.Sci., 67, 63 (1978).
11. May, S.W. and Li, N.N., Enzyme Eng., 2, 77 (1974).
12. Li, N.N., U.S. Patent 3410794 (1968).
13. Li, N.N. and Shrier, A.L., Recent Develop.Separation Sci., 1, 163 (1972).
14. Matsumoto, S., Kohda, N. and Murata, S. J.Colloid Interface Sci., 62, 149 (1977).
15. Yoshida, H., Uesegi, T. and Noro, S., Yakugaku Zasshi, 100, 1203 (1980).
16. Seifriz, W., J.Phys.Chem., 29, 738 (1925).
17. Florence, A.T. and Whitehill, D., Int.J.Pharmaceutics, 11, 277 (1982b).
18. Sherman, P. in P. Sherman (Ed.) Emulsion Science, Academic Press, London, 206-207 (1968).
19. Becher, P., In: Emulsions, Theory and Practice, 2nd Edn., Reinhold, New York, p.2 (1965).
20. Lin, T.J., Kurihara, H. and Ohta, H., J.Soc.Cosmet.Chem., 26, 121 (1975).
21. Becher, P., J.Soc.Cosmet.Chem., 9, 141 (1958).
22. Whitehill, D. and Florence, A.T., J.Pharm.Pharmacol., 31 (Suppl) 3P (1979).
23. Florence, A.T. and Whitehill, D., J.Colloid Interface Sci., 79, 243 (1981).
24. Di Stefano, F.V., Shaffer, O.M., El-Aasser, M.S., and Vanderhoff, J.W., J.Colloid Interface Sci., 92, 269 (1983).
25. Panchal, C.J., Zajic, J.E. and Gerson, D.F., J.Colloid Interface Sci., 68, 195 (1979).
26. Florence, A.T. and Rogers, J.A., J.Pharm.Pharmacol., 23, 233 (1971b).
27. Matsumoto, S., Kita, Y. and Yonezawa, D., J.Colloid Interface Sci., 57, 353 (1976).
28. Magdassi, S., Frenkel, M., Garti, N. and Casan, R., J.Colloid Interface Sci., 97, 374 (1984).
29. Florence, A.T., Madsen, F. and Puisieux, F., J.Pharm.Pharmacol., 27 385 (1975).
30. Li, N.N., A.I.Chem.J., 17, 459 (1971).
31. Kavaliunas, D.R. and Frank, S.G., J.Colloid Interface Sci., 66, 586 (1978).

32. Matsumoto, S., Kohda, N. and Murata, S. J.Colloid Interface
     Sci., $\underline{77}$, 555 (1980a).
33. Matsumoto, S., Inoue, T., Kohda, M. and Ohta, T., J.Colloid
     Interface Sci., $\underline{77}$, 564 (1980b).
34. Davis, S.S. and Burbage, A.S. J.Colloid Interface Sci., $\underline{62}$,
     361 (1977); Davis, S.S. and Burbage, A.S. in M.J. Groves (Ed.)
     Particle Size Analysis, Heyden, London, 395 (1978).
35. Matsumoto, S. and M. Kohda, J.Colloid Interface Sci., $\underline{73}$,
     13 (1980).
36. Whitehill, D., Ph.D. Thesis, University of Strathclyde, (1981).
37. Florence, A.T. and Rogers, J.A., J.Pharm.Pharmacol., $\underline{23}$, 153
     (1971).
38. Vold, M.J., J.Colloid Sci., $\underline{16}$, 1 (1961).
39. Al-Saden, A.A., Florence, A.T. and Whateley, T.L., Int.J.
     Pharmaceutics, $\underline{5}$, 317 (1980a).
40. Al-Saden, A.A., Florence, A.T. and Whateley, T.L., J.Pharm.
     Pharmacol., Suppl., $\underline{32}$, 5P (1980b).
41. Florence, A.T. and Whitehill, D., J.Pharm.Pharmacol., $\underline{34}$,
     687 (1982).
42. Ekman, B. and Sjoholm, I., J.Pharm.Sci., $\underline{67}$, 693 (1978).
43. Yoshioka, T., Ikeuchi, K., Hashida, M., Muranishi, S. and
     Sezaki, H., Chem.Pharm.Bull., $\underline{30}$, 140B (1982).
44. Law, T.K., Florence, A.T. and Whateley, T.L., Unpublished
     results, (1983).
45. Bredimas, M., Veyssie, M. and Strezlecki, L., Colloid and
     Polymer Sci., $\underline{255}$, 975 (1977).

RECEIVED December 27, 1984

Emulsion Breaking in Electrical Fields

A. KRIECHBAUMER and R. MARR

Institute of Chemical Engineering, Technical University, Graz, Austria

The stability of disperse systems is controlled by
interparticulary attraction and repulsion forces,
and by the range of the free energy of the particle surface. Pure water/oil-systems are not stable.
Addition of surfactants sometimes leads to high
stability. Stability parameters and coalescence
processes, especially when using electrical fields
for emulsion breaking, are discussed in theory and
compared to practical experimental data. The influence of parameters like applied field strength,
contact time, surfactant concentration and type of
surfactant, ionic strength of inner water phase,
and droplet diameter on the breaking efficiency
have been investigated. Advantages and technical
applications of electrical emulsion breaking are
discussed.

The use of emulsions and their range of practical application has
been expanded enormously. As a result, the field of the theory of
emulsions and technical emulsion science, as a part of classical
colloid chemistry, can use a lot of theory developed there.
    One of the greatest concerns for emulsions is the question of
their stability. A very typical example of the different requirements on the stability of an emulsion is their application in
Liquid-Membrane-Permeation (Figure 1) (1,2). In this process, a
water-in-oil emulsion is dispersed by stirring in a bulk water phase
containing metal-ions. Under certain conditions these ions will
permeate through the oil-phase of the emulsion into the inner water
phase of the emulsion. During this time, the emulsion should be
very stable but after the permeation, the emulsion is to be
separated from the bulk water and has to be broken; that mean that
at this step the emulsion is required to be unstable.
    The stability of dispersed systems is influenced by an enormous
number of parameters. Part of them can be predicted by considering
the theory of stability, others can be obtained by experiments, but
the influence of some parameters could not be explained until now.

0097–6156/85/0272–0381$06.00/0
© 1985 American Chemical Society

There are some important advantages of splitting emulsions by means of an electrical field in contrast to the usual breaking processes (3).

Thermal breaking does need a lot of energy and cannot be applied to all emulsions, because of thermal instability of some components. Centrifugation technique has the disadvantage of high mechanical work input and high investment costs; systems containing very small droplets and showing little density difference between the dispersed and continuous phases are unable to be separated by this method.

In contrast to the mentioned usual methods, the electrical emulsion breaking works at room temperature, has no moving parts, low energy input (only low condensator current between the electrodes because of low conductivity organic bulk phase), and it is possible to separate small droplets from the continuous phase.

Theory of Stability

Thermodynamic View. Generally, there are three types of stability for emulsions:
1. thermodynamic stability
2. stability by surface active agents (known as emulsifiers or surfactants)
3. stability by adsorption of colloids

To split a liquid droplet one has to spend surface energy dW by increasing the droplet surface O.

$$dW = \sigma \cdot dO = dG$$

dW ... changing of free energy (surface energy)
σ ... surface tension
dO ... changing of droplet surface
dG ... free energy change

This means that the smaller the droplet there is a greater tendency for two droplets to coalesce into one larger droplet. Therefore, a system of two liquids which exhibit a very low interface tension ($\sigma = 0$) is thermodynamically stable.

Pure water/oil-emulsions are unstable. For this reason, surface active agents (surfactants) are added, which adsorb at the interface between the two immiscible liquids and decreasing the interfacial tension. In this way, the stability of a water droplet in oil will be increased. In addition, there is a second stabilizing effect by steric hindrance which will be explained later (4).

There is another way to stabilize colloids by addition of substances which cover the whole particle or droplet thus hindering the mutual approach of two particles.

DLVO-Theory. The mutual approach of two droplets is a requirement for the coalescence process. Larger droplets approach each other on account of gravitational forces. Smaller droplets show interparticular forces with short distance range, which are responsible for their mutual approach.

Derjaguin, Landau, Overbeek and Verwey explained the stability of dispersions by the superposition of interparticular forces (5). The so called "DLVO-Theory" has been proved by experiments elsewhere (6-9). Principally, there are two forces: attractive Van-der Waals forces (figure 2a) and repulsive forces (figure 2b).

By superposition of these forces, a resulting force will be obtained (Figure 2c). If coalescence should occur, the thermal energy of the particle has to be greater than the energy barrier E (Figure 2) (10). When the thermal energy of the particle is lower, the dispersion only will flocculate, which means that the particles show a mutual approach until a certain distance (between point 1 and 2 in Figure 2).

Overcoming of the energy barrier can also occur by bringing additional energy (thermal energy; mechanical energy: centrifuga-tion, ultrasonic waves, etc.) to the emulsion (11-13).

Water droplets in an organic continuous phase show very low electric repulsion potentials. This means that a pure w/o-emulsion cannot be stable, because the total potential energy is only of the attraction mode.

By addition of surface active agents, a so called "steric hindrance" occurs. In this case, a mutual approach of two particles will be prevented by the lipophilic tails of the surfactants, which can be understood as a mechanical barrier with the same function as the adsorbed colloids as described in the previous section (see Figure 3).

Coalescence Process

The coalescence process can be described by two steps. At first, there is a mutual approach of the drops which is controlled by the rheological properties of the continuous (organic) phase (see Figure 4a). Secondly, a flattening of the droplets appears by the forma-tion of a so called "dimple" (see Figure 4b). The decrease in dis-tance d is determined by the rate of flow out of the continuous phase between the droplets (14,15). A thin film is formed which decreases to a certain critical film thickness, d_{crit}, at which point approach stops (16).

Coalescence of the droplets can only happen if it is possible to break up the thin film. This occurs if surface waves are formed or if external forces are applied. At a certain point, the thick-ness will fall below the critical value and coalescence occurs (17, 18). The influence of this step is given by the interfacial and surface rheological properties such as interface elasticity, inter-face viscosity, type of surfactant, etc. (19-25).

Coalescence in Electrical Fields

The theory of breaking emulsions in electrical fields, especially in a.c. fields, has not been investigated much (26-31). But there are some effects which mainly are responsible for coalescence phenomena (Figure 5; lower figures are microscopic photographs). If a water droplet (containing ions) which is surrounded by an organic phase is exposed to an electric d.c. field, a dipole moment will be in-duced (Figure 5a). Two drops, therefore, show at their adjacent ends electrical charges of opposite signs. This leads to an electr-ical attraction force F_D (see Figure 5b) (32). On the other hand, it can be seen from the distribution of the electrical field lines, that the density of the field lines is increased between two drop-lets. For smaller droplets, this gives rise to attractive forces between them.

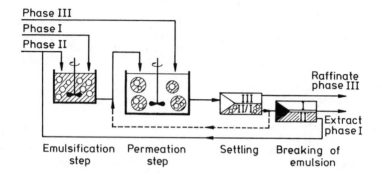

Figure 1.   Flow sheet of a liquid membrane permeation process.

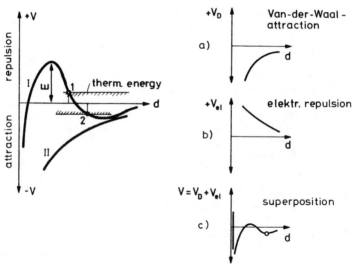

Figure 2.   DLVO-Theory ; attraction and repulsion forces of dispersed
            particles; d: Distance of the centers of two spherical
            particles.

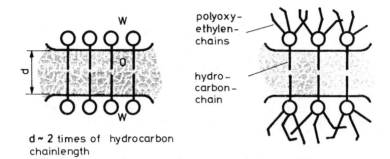

Figure 3.   Steric hindrance; monolayer films of different surfactants between two flattened droplets.

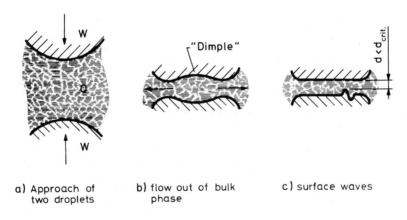

Figure 4.   Film thinning; approach of two droplets and development of a thin film.

As mentioned above, the application of electrical fields deforms
the spherical form of the droplets to an ellipsoid thus decreasing
the distance between two droplets. This deformation can also dis-
turb the thin film between droplet surfaces and lead to coalescence,
especially when using a.c. fields where an oscillating movement of
the droplets also is induced.

Experimental Section

Emulsion preparation: the emulsion was prepared in a stirring vessel
adding the discontinuous water phase dropwise to the continuous phase
under stirring with 5000 rpm. Total stirring time 10 minutes. Volume fraction phase I (discontinuous phase) to phase II (continous
phase) = 1:1.
        Emulsion breaking: an apparatus (see Figure 6) has been deve-
loped to split w/o-emulsion in electrical fields (2). Immediately
after preparation, the emulsion was pumped through the electrical
field in the splitter. After settling down of the greater water
droplets, the unspitted emulsion was recycled through the splitter
again. After a definite circulation time, the emulsion was centri-
fugated at a low centrifugation number (rw^2 = 2000) for two minutes
to get a clear interface between oil-emulsion and emulsion-water
interface.

Materials

Phase I was 5 n sulfuric acid with 15 g CU^{++}/1. Phase II was a
mixture of 60 wt % paraffine thin, 34 wt % Shellsol T, 2 wt % ECA
4360 (EXXON) as surfactant and 4 wt % LIX64N as carrier.
        Table I shows the main parameters which influence emulsion
stability. In addition to the parameters above, there are still
further process parameters.

Effect of Electrical Field Strength. Emulsions with constant phase
ratio were broken in electrical fields of different field strengths.
The applied field was varied from zero up to 1000 V/mm. Figure 7
shows the volume fraction of separated water in relation to the total
emulsion volume. At field strengths lower than 90 V/mm there cannot
be any emulsion breaking. Increasing the field strength leads to a
steep rise in splitting efficiency. From 400 to 1000 V/mm only a
small improvement of separation effect can be observed.

Effect of Splitting Time. The splitting time of an emulsion is shown
in Figure 8. There was an applied electric field strength of 1000
V/mm. Both the volume ratio of separated water and the splitting
time scales are logarithmic. Up to a residence time of 30 seconds,
80% of the emulsion has been broken. With increased breaking time,
there is a smooth change up to a saturation state. After 100 seconds
about 98% of the distributed water is separated from the organic
phase. An extension of splitting time has no more separation effect.

Effect of Residence Time. The influence of varying the residence
time by varying the electrode length on emulsion stability is shown

$$F_D = \frac{K E^2 a^6}{d^4}$$

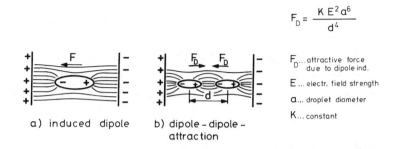

a) induced dipole

b) dipole – dipole –
attraction

F_D...attractive force
       due to dipole ind.

E ... electr. field strength

a... droplet diameter

K... constant

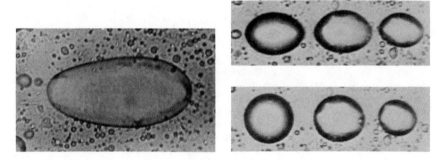

Figure 5.   Interaction forces in electric fields.

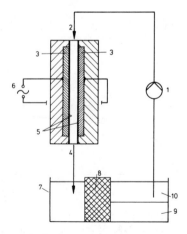

Figure 6.   Principal function of electrical splitting apparatus.
Key: 1, pump; 2, emulsion inlet; 3, electrodes; 4, emulsion
outlet; 5, electrode insulation; 6, high voltage; 7, settler;
8, coalescence aid; 9, water–phase V_I; 10, oil–phase V_{II}.

Table I. Parameters That Influence Emulsion Stability

Phase	Parameter Influencing Stability
All Phases	Applied electrical field strength contact time
Phase II Cont. phase	Viscosity Surfactant concentration Type of surfactant (HLB-value, etc.)
Phase I Discont. phase	Ionic strength
Interface	Surface film properties Interface rheological properties Interface tension
Emulsion	Viscosity Phase ratio $V_I:V_{II}$ Droplet diameter Emulsion preparation Aging

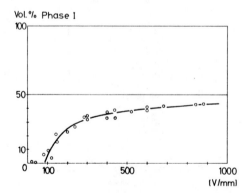

Figure 7. Effect of electrical field strength on emulsion breaking
efficiency; o...vol.% of phase I after splitting for 40
seconds.

in Figure 9. Decreasing the electrode length means decreasing of
the residence time of the emulsion in the electrical field.

When decreasing the electrode length from 160 mm to a few mm -
that means, that the contact time of the emulsion with the electrical
field decreases from about 10 seconds to less than 1 second - the
emulsion breaking coefficient does not decrease very steep.

This experiment shows that there is nearly no influence of contact time on the breaking efficiency. This means, that the coalescence of the droplets occurs very rapidly in the first part of the
electric field up to a certain magnitude of droplet diameter.

In the next part of the electrical field, no further coalescence
will occur.

The effect of electrode length on the electrical current flow
between the electrodes ((3), Figure 6) is also shown in Figure 9.
When pumping air or homogeneous liquids through the splitter, a low
current flow was observed. In the case of pumping emulsion through
the splitter, there was obtained a much higher current flow between
the electrodes. The current flow decreased linearly with decreasing
electrode length.

Effect of Surfactant Type and Concentration. Surfactant concentration and type is of great importance for the stability of thin liquid films and for emulsion stability. Type and concentration of
surfactants are responsible for the degree of lowering the interfacial tension and for the viscoelastic properties of droplet
surface, as well as for the film thickness between two droplets.

Figure 10 shows the volume fraction of the splitted emulsion
after treating in the electrical a.c. field for 30 seconds for
various surfactant concentrations and surfactant types. It is
evident that in the absence of surfactants, there is no stability
and all the water droplets coagulate with coalescence. 50 vol.%
water phase signifies 100% splitting efficiency.

With increasing surfactant concentration, the stability of the
emulsion increases continuously until a certain value. Increasing
surfactant concentration above this point yields no more emulsion
stability. This means that there is a saturation concentration of
adsorbed surfactant molecules at the droplet surfaces corresponding
to the molecular size (12). An increase in bulk phase concentration
has therefore no more effect on emulsion stability.

Span 20 is a SORBITAN-MONOLAURAT with a hydrophobic chain
length of 11 carbon molecules. Span 80 is a SORBITAN-MONOOLEAT with
17 C-molecules at the hydrophobic chain length.

Due to their molecule structures, Span 20 will develop a more
rigid and therefore more stable interface film than Span 80. Because of the higher HLB-value of 816 of Span 20 than 4.3 of Span 80,
the hydrophobic part and therefore the steric hindrance of Span 20
is of higher magnitude than that of Span 80, which fact means higher
stability against coalescence (9,33). With increasing molecular
weight of a homologous series of surfactants, the emulsion stability
decreases (Figure 11).

Viscosity measurements show that with increasing surfactant
concentration even at high concentration the bulk phase viscosity
only increases very smoothly (Figure 12).

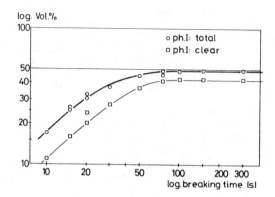

Figure 8.   Effect of breaking time on emulsion breaking efficiency;
             electrical field strength 1000 V/mm.

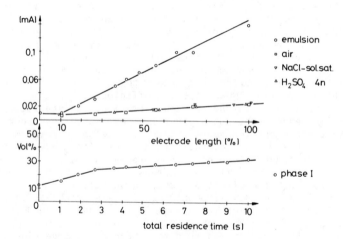

Figure 9.   Effect of electrode length and total residence time on
             splitting efficiency and current flow at an electrical
             field of 1000 V/mm.

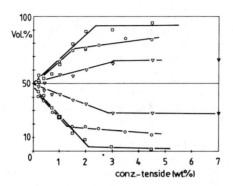

Figure 10. Effect of surfactant type and surfactant concentration on emulsion stability; □ Span 20; o Span 80; ▽ Span 85; breaking time 20's.

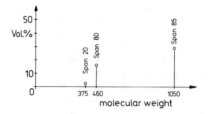

Figure 11. Comparison of surfactants with different molecular weight and molecular structures on their stabilizing effect; o...phase I obtained after 20's splitting by 1000 V/mm electrical field strength; surfactant concentration 3 wt %.

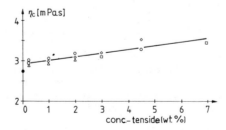

Figure 12. Effect of surfactant concentration on bulk phase viscosity; ◇ ECA 4360, □ Span 20, △ Span 80, o Span 85.

Effect of Continuous Phase Viscosity. Bulk phase viscosity has a
great influence on the approach of two droplets. Figure 13 shows the
influence of continuous phase viscosity as a function of the volume
rate of Shellsol T and paraffine oil as continuous phase. Increasing concentration of Shellsol T decreases the continuous phase
viscosity.
      The emulsions which were obtained with these organic phases
showed an analogous decrease in emulsion viscosity. A mathematical
description of the experimental data even could be obtained from the
organic phase viscosity by an additive factor.

Effect of Continuous Phase Composition. With increasing concentration of Shellsol T in the continuous phase and therefore decreasing
bulk phase viscosity (as shown in Figure 13), the emulsion breaking
efficiency increases (Figure 14). As expected, the mutual approach
of the water droplets will be facilitated in lower viscous continuous
phases.

Effect of Phase Ratio. The effect of phase ratio of discontinuous
phase over continuous phase on the viscosity of the emulsion is
shown in Figure 15. Continous and discontinuous phase composition
and viscosity are constant in each experiment. With increasing
phase ratio, the viscosity of the emulsion increases because of the
increasing of the amount of water droplets.
      The emulsion breaking efficiency was obtained by pumping the
emulsion one time through che splitter and comparing the splitted
water with the overall water content in the unsplitted emulsion.
The increase of emulsion breaking efficiency can be explained by the
shorter distances between the droplets at higher phase ratios and
therefore, the minor mutual approaching time.

Effect of Emulsion Preparation. There are a lot of parameters in
emulsion preparation which have effects on emulsion stability (e.g.
phase ratio, phase I, phase II, stirring time, stirring speed).
Some of them have been explained above, others have already been
investigated (34,35) and others are still under discussion (Kriechbaumer, Thesis; Wacnter, Thesis).
      As an example, I would like to mention the effect of stirring
speed on emulsion stability. In Figure 16, it can be seen that with
increasing stirring speed in a homogenizer, the emulsion can be
broken with less efficiency. This correlates with decreasing
"Sauter diameter" D_p. The droplet diameter distribution of the
emulsion becomes more and more uniform. The smaller the droplets,
the smaller are the mutual attractive forces and the smaller is the
probability of a collision of two particles.

Effect of Aging. With increasing volume fraction of the dispersed
phase, increasing droplet diameter and wider diameter distribution,
the viscosity of a dispersed system increases (36). Unstable emulsions show droplet coalescence by extending the diameter distribution, accompanied by viscosity increasing, an effect, which is
called "aging" (36-38).

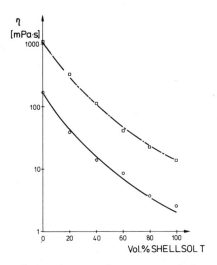

Figure 13. Effect of continuous phase composition on viscosity and emulsion viscosity.   Experimental data: □emulsion, o organic phase; fitted data:— $\ln\eta_0 = \exp[-K_1(\text{vol}\%)+K_2]$ —·— $\ln\eta \ \square = \ln\eta_0 + K_3$.

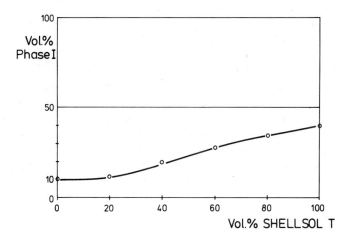

Figure 14. Effect of continuous phase composition on emulsion breaking efficiency.

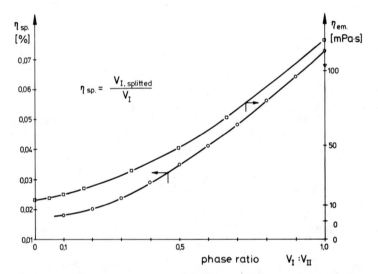

Figure 15. Effect of phase ratio V_I/V_{II} on emulsion viscosity η_{em} and splitting efficiency η_{sp}; V_I...total water content; $V_{I,splitted}$...obtained continuous water phase after splitting.

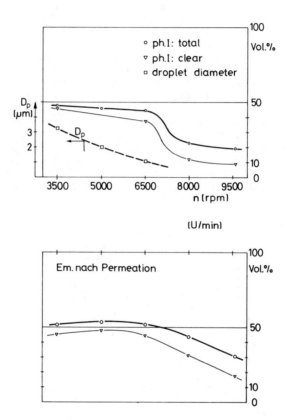

Figure 16. Effect of stirring speed on breaking efficiency and droplet diameter.

Technical Applications

This method can be used to split w/o-emulsions, to separate the
smallest water droplets from motor oil, to clear a turbid organic
phase after solvent extraction (containing water) and crude oil de-
watering, etc.

As an example of an application of the splitter, the continuous
breaking of an emulsion used in the liquid-membrane-permeation
technology is shown in Figure 17.

After the emulsification and permeation step the emulsion enters
the first breaking apparatus. Leaving this step, the emulsion passes
a settler in which the greater water droplets coalesce with the bulk
water phase. The unsplitted emulsion flows in the second splitter
from which it is pumped through the second breaking step, which contains a package of pairs of electrodes. The unsplitted emulsion
will also be recycled through the second breaking step. Clear organic phase can be obtained from the surface, while at the bottom,
the outlet of the water phase is installed.

The breaking efficiency can be seen in the lower part of the
figure. The fresh emulsion contains nearly 14 vol.% of water in the
organic phase. After the first step, it is already reduced to 5.4
vol.% and after the second step there is less than 0.4 vol.% water
in the organic phase.

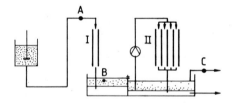

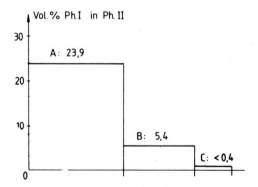

Figure 17. Continuous emulsion breaking; emulsion flow rate: 20 1/h.
     I...first breaking step, II...second breaking step.

Conclusions

It can be seen that there are a lot of parameters influencing emulsion stability. Some of them can be predicted by theory, while others are able to be explained by experimental results. An emulsion breaking device has been developed which enables the investigation of various parameters and different emulsion systems. Thus, on the one hand, emulsion stability can be compared with theory, while on the other hand, conclusions about coalescence processes are obtained.

The scientific investigation will lead to expanding the practical application of electrical emulsion breaking in chemical engineering, chemistry, biology, etc.

Literature Cited

1.   Florence, A.T.; Whitehill, D.   J. Coll. Int. Sci. 1981, 79, 243.
2.   Marr, R.; Bouvier, A.; Draxler, J.; Protsch, M.; Kriechbaumer, A.   Proc. PACHEC 83, 1983, p. 327.
3.   Sharma, M.K. et al.   Ind. J. Techn. 1982, 20, 175.
4.   Tadros, Th. F.   Proc. Int. Symp., 1981.
5.   Verwey, E.J.; Overbeek, J. Th. G.   Ph.D. Thesis, Elsevier, Amsterdam, 1948.
6.   Derjaguin, B.V.   Farad. Disc. Chem. Soc. 1978, 65, 306.
7.   Frens, G.   Farad. Disc. Chem. Soc. 1978, 65, 146.
8.   Goswami, A.K.; Bahadur, P.   Progr. Coll. Polym. Sci., 1978, p. 27.
9.   Srivastava, S.N.; Haydon, D.A.   Proc. Int. Congr. Surf. Act. 4th, 1967, p. 1221.
10. Sonntag, H.; Strenge, K. "Koagulation und Stabilitat disperser Systeme"; VEB Deutscher Verlag der Wissenschaften: Berlin, 1960.
11. Rehfeld, S.J.   J. Coll. Int. Sci. 1974, 46, 448.
12. Rehfeld, S.J.   J. Phys. Chem. 1962, 66, 1966.
13. Vold, R.D.; Groot, R.C.   J. Phys. Chem. 1962, 66, 1969.
14. Schulze, H.J.   Coll. Polym. Sci. 1975, 253, 730.
15. Sherman, P.   "Emulsion Science"; Academic Press: London-New York, 1968.
16. Scheludko, A.; Exerowa, D.   Coll. Polym. Sci. 1974, 252, 586.
17. Ivanov, I.B. et al.   Proc. Sect. 52nd Coll. Surf. Sci. Symp., 1979, p. 817.
18. Gershfeld, N.L.   Molecular Association in Biological and Related Systems, 1965.
19. Prins, A. et al.   J. Coll. Int. Sci., 1967, 24, 84.
20. Mysels, K.J.   J. Phys. Chem. 1964, 68, 3441.
21. Scheludko, A.   Adv. Coll. Interf. Sci. 1967, 1, 392.
22. Vrij, A.   Disc. Farad. Soc. 1966, 42, 23.
23. Lucassen, J.; Luc.-Reynders, E.   J. Coll. Int. Sci. 1967, 25, 496.
24. Matsumoto, S.; Sherman, P.   J. Coll. Int. Sci. 1970, 33, 294.
25. Wasan, D.T. et al.   AIChE Symp. Series No. 192, 1980, p. 93.
26. Stauff, J. "Kolloidchemie"; Springer Verlag: Berlin, 1960.
27. Cho, A.Y.H.   J. Appl. Physics 1964, 35, 2561.
28. Hendricks, Ch. D. et al.   IEEE Transac. Industry Appl., 1977, p. 489.
29. Simonova, T.S.; Dukhin, S.S.   Coll. J. USSR (USA) 1973, 35, 952.
30. Waterman, L.C.   Chem. Eng. Progr. 1965, 61, 51.
31. Waterman, L.C.   Hydrocarbon Proc., 1965. p. 133.

32.   Kriechbaumer, A.; Marr, R.   Chem. Ing. Techn. 1983, 9, 700.
33.   Kriechbaumer, A. Ph.D. Thesis, Technical University, Graz-
      Austria, 1984.
34.   Kopp, G. Ph.D. Thesis, Technical University, Graz-Austria, 1978.
35.   Draxler, J. Ph.D. Thesis, Technical University, Graz-Austria,
      1983.
36.   Gillespie, T.   Proc. Symp. Brit. Soc. of Rheology, 1962, p. 115.
37.   Reddy, S.R.; Folger, H.S. J. Coll. Int. Sci. 1981, 82, 128.
38.   Boyd, J.; Sherman, P.   J. Coll. Int. Sci. 1970, 34, 76.

RECEIVED June 8, 1984

Stabilization of Xylene in Sulfapyridine-Stabilized Water Emulsion
Evaluation and Role of van der Waals and Coulombic Forces

V. K. SHARMA and S. N. SRIVASTAVA

Department of Chemistry, Agra College, Agra 282002, India

Most of the drugs have got the property to promote
relatively stable oil in water emulsions. A study
of a sulphonamide drug sulpha-pyridine reveals that
it yields a stable emulsion. The emulsion behaviours
have been interpreted by studying flocculation, coalescence, electro-kinetic potential, potential
energies and stability factors. These studies were
made as a function of nucleic acids, DNA and RNA
(Deoxy-ribonucleic acid and Ribonucleic acid) which
were chosen in accordance with their biological
significance.

Stability of drug stabilized emulsions plays a role of immense importance in various branches of natural and medical science. However, the effect of nucleic acids on drug stabilized emulsions is
very interesting from the theoretical as well as the practical point
of view.

Some workers attempted physico-chemical interpretations of
emulsion characteristics in a Qualitative manner (1). Previous investigators (2) have studied the aggregation and deaggregation of
globules in hexadecane-water emulsions containing dioctyl sodium
sulpho-succinate as emulsifier and have reported that lower concentrations of surfactant favoured the deaggregation of globules.
Cockbain (3) reported the prevention of coalescence and aggregation
of the dispersed phase of emulsion by the low concentrations of the
cations and anions. Sharma and Srivastava (4,5) studied the effect
of electrolytes and detergents on the emulsions stabilized by the
drugs oxytetracycline hydrochloride and chloramphenicol and observed
the reversible flocculation also at low concentrations of additives.
Upadhyay and Chattoraj (6) studied the adsorption of nucleic acids
at the alumina water interface and noted that the adsorption increases with increasing concentrations of nucleic acids up to some
extent and attains constancy afterwards. Srivastava and Haydon (7)
correlated the emulsion stability with the electrical properties of
the emulsion droplets according to DLVO theory.

0097–6156/85/0272–0399$06.00/0
© 1985 American Chemical Society

From the survey of the above literature, it is concluded that
only a limited work is done on such type of problems. In the present
study, the stability of emulsion has been discussed in the light of
Derjaguin, Landau, Vervey and Overbeek theory (8) using Deoxyribonucleic acid and ribonucleic acid as flocculants for the emulsion
stabilized by the drug sulphapyridine.

Materials and Methods

The emulsifying agent sulphapyridine was obtained from May and Beker,
Bombay. Deoxyribonucleic acid and ribonucleic acid were BDH products. Analar Xylene (BDH) was used as such. Double distilled
water was used in all experiments.
     The emulsion was prepared by suspending 3% by volume of xylene
in 0.02 M solution of sulpha pyridine which was made in N/40 NaOH.
The mixture was then shaken for half an hour and homogenized with
the hand operated stainless steel homogenizer (Fischer Scientific
Co., U.S.A.). For all the estimations, emulsion was prepared under
identical conditions.
     Flocculation was measured by an improved Neubauer model of
haemocytometer. The number of globules with respect to time was
counted with the help of a hand tally counter (Erma, Tokyo) under
the olympus microscope.
     The average size of the emulsion globules as measured by size
frequency analysis of microphotographs with the help of a catheto-
meter was found to be 1.2 μm. (See Figures 1 and 2)
     The electrophoretic measurements were carried out in a rectangu-
lar Northrup-Kunitz type microelectrophoretic cell (9), which was
mounted on the Carl Zeiss Jena microscope, using reversible silver-
silver chloride electrodes. Aqueous solution of KCl having the same
conductance as that of emulsion was filled in electrode chamber to
avoid diffusion. The individual globules were timed for two
divisions of the graticule fitted in the eye-piece of microscope
under constant voltage between Ag-AgCl electrodes. The potential
difference between the two ends of the cell was directly measured by
measuring the voltage across the two electrodes fused at the ends of
cell by vacuum tube voltmeter (10).

Theoretical

The charge on emulsion stabilized by sulpha-pyridine was found to be
negative. The electrokinetic mobility of the emulsion was measured
and the zeta potential was calculated by the Helmholtz equation,

$$\xi = \frac{4\pi\eta}{\varepsilon X} \cdot U \qquad [1]$$

where η and ε are respectively viscosity and dielectric constant of
the solution in the electrical double layer adjacent to the surface,
U is the mobility and X is the applied field strength.
     In the framework of DLVO theory, the forces of attraction and
repulsion were calculated using the equations (11,12):

$$V_A = \frac{Aa}{12 H} \left(\frac{\lambda}{\lambda + 3.45 H} \right) \qquad [2]$$

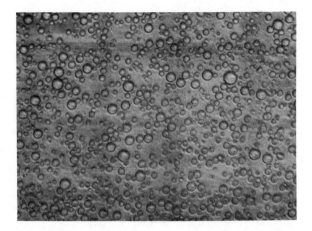

Figure 1.   Microphotograph of sulfapyridine-stabilized emulsion.

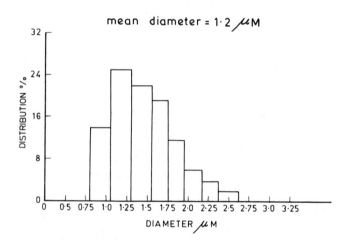

Figure 2.   Particle size distribution curve of sulfapyridine system.

For H < 15 nM

$$V_A = \frac{Aa}{12\ H} \left(\frac{2.45\lambda}{120\ H^2} - \frac{\lambda^2}{1045\ H^3} + \frac{\lambda^3}{5.62 \times 10^4 H^4}\right)$$
$$\dots [3]$$

For H > 15 nM

$$V_R = \frac{\varepsilon a\psi_o^2}{2} \cdot \text{ln.}(1 + e^{-\chi H}) \qquad [4]$$

where, V_A, energy of attraction;
$\quad\quad V_R$, energy of repulsion;
$\quad\quad \varepsilon$, dielectric constant of the medium;
$\quad\quad A$, van der Waals constant;
$\quad\quad a$, particle radius;
$\quad\quad H$, Interparticle distance;
$\quad\quad \chi$, Debye-Huckel parameter;
$\quad\quad \lambda$, Frequency assumed to be 1000 Å;
$\quad\quad \psi_o$, Surface potential, assumed to be zeta potential for the
$\quad\quad$ system having $\chi a \gg 1$.

The total interaction energy (V) may be calculated by summing up these two:

$$V = V_R + V_A$$

In case of reversible flocculation, the temporary existence of monomers and dimers is assumed and the degree of flocculation (D) was calculated from the equation:

$$D = \frac{\text{Number of Doublets}}{\text{No. of singlets + No. of doublets}}$$

Theoretically, D may be calculated by using the equation:

$$D = 4\pi n_o a^3 \int s^2 \exp. (-V/KT) ds \qquad [5]$$

where, n_o is the number of particles;
$\quad\quad a$ is the radius of particles;
and $\quad s = 2 + H/a$.
This value of D differs from the experimentally obtained value of D. Therefore, a graph was plotted between different values of van der Waals constant and theoretical values of D. From this graph, the value of A corresponding to the observed value of D was found.
The rate of flocculation was determined from the equation (13):

$$Vm = \phi / n = K_1 t \qquad [6]$$

Where ϕ is the phase volume ratio of the O/W emulsion and n is the number of drops for mean volume, Vm, at time, t, K_1 is the flocculation rate constant.
Coalescence rate constant (K_2) was estimated with the help of the following equation:

$$K_2 = \frac{2.303}{t} \cdot \log \frac{Nt}{No}$$ [7]

where, Nt and No are the number of drops at time t and t = 0.

Eilers and Korff have shown that the factor which governs the stability of a colloid system is the energy required to bring together two similarly charged particles of the dispersion (14). The following formulae have been derived for the energy (e).

$$E = \xi^2/\chi$$ [8]

where ξ is in millivolts and Debye–Huckel reciprocal distance χ in cm^{-1}. More generally

$$c \; \xi \; (\varepsilon/\chi)^{1/2} = \text{Constant}$$ [9]

or theoretically (15) taking van der Waals constant, A, into account one can determine the conditions for preparing stable systems for which,

$$\frac{2\varepsilon\psi^2}{\chi A} > 1$$ [10]

The stability factor, W, is determined to give clear insight into the emulsion stability with the help of the equation used elsewhere (17):

$$\ln W = \frac{A}{24KT} \left(\frac{Q^2}{1+2\chi a} - 2Q \right) - \frac{3}{2} \ln Q + Z$$ [11]

where $Q = \frac{2}{H/a} + 1$

and $Z = \ln 2\sqrt{\pi} - 1/2 \ln (A/6KT) + (A/24KT)$ [12]

Other notations have their usual significance. Further W may be calculated graphically by plotting Vm against t and the slope of this straight line is = 4π DRϕe $- W_1/KT$ [13]

where, D = $KT/6\pi\eta a$

Here, W_1 should be several KT for stable emulsions. W may also be calculated theoretically from the equation (18):

$$W = e\left(\frac{Vm}{KT}\right)/\chi a$$ [14]

An estimation of van der Waal's constant could be made from surface tension of the material following a simple treatment suggested by Overbeek and give in Vold's paper (19, 20). According to this, the interaction energy per square centimeter of two semi-infinite slabs of material separated by a distance 'a' is given by $V = \frac{-A}{12\pi a^2}$ and since 2 sq. cm. of surface is destroyed when the slabs are brought together. A is given by the equation used by previous workers (21):

$$A = 24 \times \pi \ a^2 \gamma \qquad\qquad\qquad [15]$$

where γ is the surface tension of the material. Assuming that the surface and the bulk material have the same composition, a is the separation of centres of atoms. Considering γ = 70 dynes/cm and a $\simeq$ 20 Å, A for the nucleai acids RNA and DNA is found to be 2.1 x 10^{-14} ergs.

Origin of Charge

The charge on emulsion is found to be negative. The solution of drug was prepared in N/40 NaOH and the pH of the emulsion was 11.32. Due to high alkalinity, the OH^- ions of alkali replace the H^+ of amino group and thus, the emulsion becomes negatively charged. Moreover, when drug solution was prepared in acid, the charge reversed to positive because of the formation of $-NH_3^+$ group from the H^+ ions of acid and amino group of the drug. The structure of sulpha-pyridine drug is given hereunder - (Molecular weight, 249):

Results and Discussion

Graphs were plotted with the number of individual drops (n_1) against time and are shown in Figure 3. These clearly indicate that the number of unassociated drops decreases somewhat faster primarily with an increase in time, but attains almost constancy afterwards, which shows that reversible flocculation is occurring in the system. This is also borne out by the secondary minimum of the curve given in Figure 4. The high values of stability factor W also lead to the conclusion that the flocculation is likely to be slow and of reversible nature. The values of the flocculation rate constant, K_1 (Table I), are of the order of 10^{-15} cm^3 sec^{-1}. This may be assigned to the energy barriers which inhibit the flocculation and change even its nature. Coalescence rate constant (K_2), given in Table I, increases with an increase in the concentrations of RNA and DNA and the values of K_2 are found to be of the order 10^{-5} sec^{-1}, which are more or less equal to that of the macromolecular stabilized emulsion (22). This is presumably due to the considerable strength of the interfacial film of the adsorbed drug, sulpha-pyridine.

To judge whether flocculation or coalescence is the rate determining step ratio of halflives of these two processes was ascertained (Table I), which is greater than unity for the present case, indicating the coalescence is slow and, therefore, is the rate determining step.

The change in zeta potential of the emulsion by addition of different quantities of RNA and DNA was studied. The initial zeta potential of the emulsion was found to be 147 mV. The zeta potential of the emulsion decreases with an increase in concentrations of RNA and DNA (Figure 5). The flocculation sets in when the added nucleic

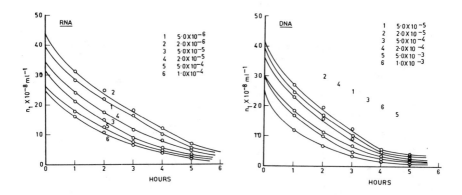

Figure 3.   Plots of individual drops vs. time for sulfapyridine-stabilized emulsion flocculated by RNA and DNA.

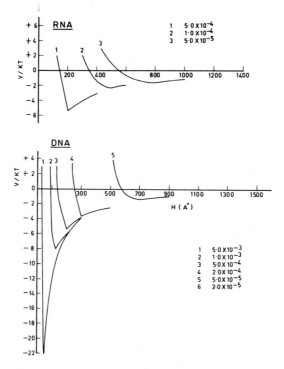

Figure 4.   Elaborated secondary minimum curves of RNA and DNA for sulfapyridine system.

Table I. Kinetic Parameters of the Sulfapyridine-Stabilized
Emulsion in Presence of RNA and DNA

ADDITIVES	CONCENTRATION mol m^{-3}	K_1 x 10^{15} cm^3 sec^{-1}	K_2 x 10^5	$\dfrac{1.47\ K_2}{\eta_o K_1}$
DNA	5×10^{-3}	6.7	9.0	7.5
	1×10^{-3}	4.7	7.0	6.8
	5×10^{-4}	4.6	6.9	7.3
	2×10^{-4}	3.6	6.7	6.6
	5×10^{-5}	3.4	5.2	6.2
	2×10^{-5}	3.0	4.7	3.6
RNA	5×10^{-4}	4.7	5.2	6.2
	1×10^{-4}	4.1	4.8	5.5
	5×10^{-5}	3.8	4.6	5.7
	2×10^{-5}	3.5	4.4	5.6
	5×10^{-6}	2.9	4.3	5.7
	2×10^{-6}	2.6	4.2	5.3

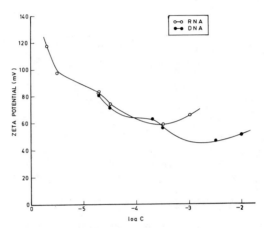

Figure 5.   Plots of zeta potential vs. log concentrations of RNA and DNA for sulfapyridine system.

acids compress the double layer and reduce the zeta potential to a
point, when the repulsive forces decrease considerably and the emulsion droplets can come into close contact. This happens when the
value of the factor ξ^2/χ decreases below 10^{-3}, a condition derived
by Eilers. The relevant data are recorded in Tables II and III.

Nature of flocculation and stability of emulsion is further
studied by calculating the total interaction energy with the help of
equations 3 and 4. The values of V/KT are plotted as a function of
interparticle distance H in Angstrom units, between the two globules
shown in Figures 4 and 6. The data are recorded on Table IV. The
maximum potential energy in case of DNA and RNA was found to be 5412
and 8158, respectively, which shows that the emulsions have high
stability and flocculation cannot occur in the primary minimum. The
height of potential energy barriers between the emulsion droplets
depends upon the zeta potential and the Debye–Huckel parameter, which
varies with the electrolyte concentration. Thus, it was noted that
the height of the energy maxima decreases with an increase in the
concentration of RNA and DNA. This decrease continues till the
repulsion becomes zero and attraction predominates.

Occurrence of flocculation may be explained if secondary minima
aggregation is assumed. This is confirmed by the V/KT vs (H (A))
curves, which indicate that at higher values of H, repulsion becomes
negligible and attraction predominates and emulsion flocculates. It
is also observed that the depth of secondary minimum is more at
higher concentration, 5–20 KT, which is deep enough for reversible
aggregation, while at low concentrations, the depth of the secondary
minima is too shallow to trap the particles.

From the interaction energy curves of Figures 4, 6, and 7 and the
the corresponding data of Table IV, it is evident that the interparticle distance, at which energy is equal to zero, decreases with
increasing concentrations of RNA and DNA, which indicates the
flocculation of the emulsion system.

Van der Waal's constant was calculated with the help of equation
16 and found to be 2.1×10^{-14} ergs. It was also evaluated by
plotting the graph (Figure 8) between theoretical values of degree of
aggregation and van der Waal's constant and by interpolating into the
corresponding curves the values of observed degree of aggregation
calculated haemocytometrically. The values are recorded in Table V.
The mean value of van der Waal's constant obtained is 4.1×10^{-14}
ergs.

The above discussion of emulsion stability is further supported
by the stability factors calculated with the help of equations 11, 12,
and 14. The results are given in Table VI. All values are greater
than 10 KT, the condition derived by Derjaguin for stable dispersions.
These values are higher than the corresponding values of V/KT, because
in this estimation, the effect of van der Waal's interaction is not
taken into account.

Conclusion

Emulsion, stabilized by the chemotherapeutic drug sulpha–pyridine,
is found to be stable and this stability is reasonably sustained even
in the presence of nucleic acids, DNA and RNA. Coalescence is found
to be the rate determining step. Energy barriers are of very high

Table II. Addition of RNA on Sulfapyridine-Stabilized Emulsion

ξ	$C(M)$	χ	ξ^2/χ	$\dfrac{2\varepsilon\xi^2}{A\chi}$	$\xi(\dfrac{\varepsilon}{\chi})^{1/2}$
59	5×10^{-4}	0.43×10^6	8.0×10^{-3}	2.5×10^6	1.96×10^{-6}
66	1×10^{-4}	0.20×10^6	21.0×10^{-3}	7.6×10^6	1.00×10^{-5}
74	5×10^{-5}	0.14×10^6	3.9×10^{-2}	13.6×10^6	2.30×10^{-5}
84	2×10^{-5}	0.08×10^6	8.1×10^{-2}	28.3×10^6	6.80×10^{-5}
98	5×10^{-6}	0.04×10^6	22.3×10^{-2}	77.9×10^6	3.20×10^{-4}
118	2×10^{-6}	0.02×10^6	49.7×10^{-2}	17.3×10^7	9.20×10^{-4}

Table III. Addition of DNA on Sulfapyridine-Stabilized Emulsion

ξ	$C(M)$	χ	ξ^2/χ	$\dfrac{2\varepsilon\xi^2}{A\chi}$	$\xi(\dfrac{\varepsilon}{\chi})^{1/2}$
47	5×10^{-3}	1.4×10^6	1.5×10^{-3}	5.5×10^5	1.4×10^{-7}
51	1×10^{-3}	0.6×10^6	4.1×10^{-3}	1.46×10^6	8.1×10^{-7}
57	5×10^{-4}	0.4×10^6	7.5×10^{-3}	2.6×10^6	1.9×10^{-6}
63	2×10^{-4}	0.2×10^6	1.4×10^{-2}	5.6×10^6	4.9×10^{-6}
71	5×10^{-5}	0.1×10^6	3.6×10^{-2}	12.5×10^6	2.2×10^{-5}
81	2×10^{-5}	0.09×10^6	7.5×10^{-2}	26.3×10^6	7.3×10^{-5}

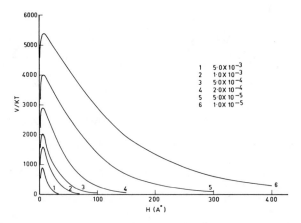

Figure 6. Plots of interaction energy vs. interparticle distance for sulfapyridine-stabilized emulsion with DNA concentrations.

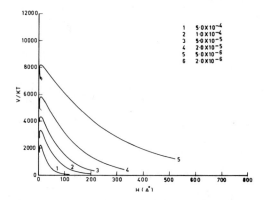

Figure 7. Plots of interaction energy vs. interparticle distance for sulfapyridine-stabilized emulsion with RNA concentrations.

Table IV. Values of Interaction Energies, Interparticles Distance at V = 0, and Depth of Secondary Minimum.

SURFACTANTS	CONCEN-TRATION mol m^{-3}	HEIGHT OF MAXIMUM IN KT	INTER-PARTICLE DISTANCE AT WHICH V = 0	DEPTH OF SECONDARY MINIMUM IN KT
DNA	2×10^{-5}	5412	980–990	0.5
	5×10^{-5}	4010	580–590	1.4
	2×10^{-4}	2942	230–240	3.0
	5×10^{-4}	2048	130–140	5.4
	1×10^{-3}	1591	85–95	8.0
	5×10^{-3}	908	30–40	22.0
RNA	5×10^{-6}	8158	–	–
	2×10^{-5}	5836	–	–
	5×10^{-5}	4384	570–580	1.5
	1×10^{-4}	3357	340–350	2.2
	5×10^{-4}	2267	130–140	5.3

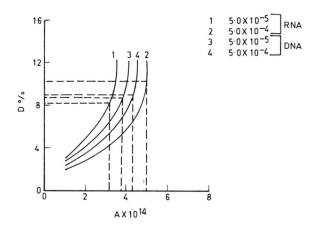

Figure 8. Degree of aggregation (D%) vs. van der Waals' constant for sulfapyridine system.

Table V.   Van der Waal's Constant Values

STABILIZER	SURFAC-TANTS	CONCEN-TRATIONS mol m^{-3}	D% EXPERI-MENTAL	Ax10^{14} ergs	MEAN VALUE OF A in ergs
SULPHA-PYRIDINE	DNA	5x10^{-5}	8.7	3.8	
		5x10^{-4}	9.0	4.3	
	RNA	5x10^{-5}	8.2	3.2	4.1x10^{14}
		5x10^{-4}	10.3	5.0	

Table VI.   Stability Factors of Sulfapyridine-Stabilized Emulsion on Addition of RNA and DNA.

SURFACTANTS	CONCENTRATION mol m^{-3}	Wx10^4	W_1 in KT from plots	Stability Factor in KT (Calc. using Ref. 16)
DNA	5x10^{-3}	3.2	36.7	8.1
	1x10^{-3}	3.9	35.9	9.3
	5x10^{-4}	4.5	36.0	11.8
	2x10^{-4}	5.2	35.6	14.6
	5x10^{-5}	6.0	35.7	18.7
	2x10^{-5}	6.6	35.5	24.0
RNA	5x10^{-4}	5.2	36.4	12.9
	1x10^{-4}	5.7	35.9	15.9
	5x10^{-5}	6.6	35.8	20.4
	2x10^{-5}	7.3	35.7	26.1
	5x10^{-6}	8.1	35.5	35.4
	2x10^{-6}	8.8	35.4	51.8

order. This obviates the possibility of the occurrence of flocculation in primary minima. The possibility of flocculation taking
place in secondary minima is not ruled out. It will, however, be
reversible in nature.

Acknowledgment

The authors are thankful to Professor D. K. Chattoraj for his advice
and help in carrying out the experiments at Jadavpur University,
Calcutta. Thanks are also due to the Council of Scientific and In-
dústrial Research for kindly granting a Junior Research Fellowship
to one of them (VKS).

Literature Cited

1. Clayton, W. "Theories of Emulsions and their Technical Treat-
    ment"; J. Churchill Ltd.: London, 1954; p. 755.
2. Higuchi, W. I.; Okada, R.; Limberger, A. P. Pharm. Sci. 1962,
    51, 683.
3. Cockbain, E. G. Trans. Faraday Soc. 1952, 48, 185.
4. Sharma, M. K.; Srivastava, S. N. Ann. Soc. Chim. Polonium 1975,
    49, 2047.
5. Sharma, M. K.; Srivastava, S. N. Z. Phys. Chemie. Leipizig.
    1977, 2, 258.
6. Upadhyay, S. N.; Chattoraj, D. K. Biochem. Biophys. Acta. 1968,
    561, 161.
7. Srivastava, S. N.; Haydon, D. A. Proc. 4th Intnatl. Cong. Surf.
    Activity., 1967, p. 1221.
8. Vervey, E. W.; Overbeek, T. C. "Theory of Stability of Lyopho-
    bic Colloids"; Amsterdam, 1948.
9. Northrup, J. H.; Kunitz, M. J. Gen. Physiol. 1925, 924.
10. Derjaguin, B; Kussakov, M. Acta Phys. Chim. URSS 1939, 10, 75
    and 153.
11. Schenkel, J. N.; Kitchener, J. A. Trans. Faraday Soc. 1960, 56,
    161.
12. Srivastava, S. N.; Prakash, C. Bull. Chem. Soc. Japan 1967, 40,
    1754.
13. Davies, J. T.; Rideal, E. K. "Interfacial Phenomena"; Academic
    Press: New York and London, 1961.
14. Eilers, H.; Korff, J. Trans. Faraday Soc. 1940, 36, 229.
15. Derjaguin, B. Disc Faraday Soc. 1954, 18, 95.
16. Derjaguin, B. Trans. Faraday Soc. 1940, 36, 203.
17. Daluja, K. L.; Srivastava, S. N. Physic. Chemie. 1970, 173, 244.
18. Daluja, K. L.; Srivastava, S. N. Indian J. of Chem. 1967, 262,
    5.
19. Vold, M. J.; Rathnama, D. V. J. Phys. Chem. 1960, 64, 1619.
20. Vold, M. J. J. Colloid Sci. 1961, 16, 1.
21. Srivastava, S. N. Zeitschr-fur Physikalische Chemie 1966, 237,
    233.
22. Srivastava, S. N. American Chemical Soc. Symp. Series 9-Colloid
    Dispersion and Micellar Behavior, 1975, p. 110.

RECEIVED January 23, 1985

Formation and Stability of Water-in-Oil-in-Water Emulsions

SACHIO MATSUMOTO

Department of Agricultural Chemistry, College of Agriculture, The University of Osaka Prefecture, Sakai, Osaka 591, Japan

This paper reviews a series of the works on the W/O/W emulsions so as to obtain further insights into the factors affecting the formation and stability of W/O/W type dispersion, which can be characterized by the vesicular structure of the dispersed globules consisting of the aqueous compartments separated from the aqueous suspending fluid by the layer of oil phase components. The contents are divided into four sections: methods and conditions for preparing W/O/W emulsions, thinning behavior of oil layers on the surface of the aqueous compartments, water permeability of the thin oil layer, and viscometric method for measuring the stability of W/O/W emulsions.

Emulsions occur so widely in the various fields that a large number of investigations have been made of O/W and W/O systems from the different standpoints of emulsion science and technology.

Recently, reports on the preparations and properties of W/O/W (water-in-oil-in-water) emulsions have come from several laboratories (1-21). Although a variety of phase compositions are employed in different studies, the dispersed globules in the W/O/W emulsions can be characterized by a spherical vesicular structure with single or multiple aqueous compartments (inner aqueous phase), which are separated from the aqueous suspending fluid (outer aqueous phase) by a layer of the oil phase components.

In view of such structural characteristics, some considerable interests will be expected toward the practical use of the W/O/W emulsions, as reviewed by Davis (22). It seems to be necessary, however, to obtain further informations on the formation, dispersion state and stability of W/O/W emulsion systems so as to extend the situation of the emulsion utilization.

It is purpose of this paper to present the experimental results obtained from a series of the works on the W/O/W emulsions in the author's laboratory in order to search further insights into the factors affecting the formation and stability of W/O/W-type dispersion.

0097-6156/85/0272-0415$06.50/0

Formation of W/O/W Emulsions

Two-step Procedure of Emulsification. W/O/W emulsions can be prepared by the method of a two-step procedure. The first step is made
for providing an ordinary W/O emulsion by use of hydrophobic emulsifier. The W/O emulsion is then mixed with an aqueous solution of
hydrophilic emulsifier as the second step procedure for obtaining
W/O/W-type dispersion.

  This technique was originally applied by Herbert (1) in the comparative studies on the role of antigen adjuvants. Engel et al. (2)
also tried to prepare a W/O/W emulsion using the two-step procedure
for examining the intestinal absorption of insulin. Then, the similar works have followed in some fields, especially in the pharmaceutical science (22).

  It was to be made clear that the two-step procedure inevitably
requires to employ the two different types of emulsifiers, i.e.,
hydrophobic- and hydrophilic emulsifiers, as described above. The
author and his colleagues (4) made an attempt to clarify the effect
of the ratio of both emulsifiers on the formation of W/O/W emulsions,
so that they provided a technique for measuring the yield of W/O/W
type dispersion in the freshly prepared sample. That is, the first
step emulsification was made with an aqueous dilute solution of glucose, which distributed uniformly in the aqueous compartments due to
the second procedure. Immediately after the second step procedure,
a certain amount of the freshly prepared sample was dialyzed against
a definite volume of pure water until the system attained dialysis
equilibrium. The quantity of glucose migrated to the pure water was
then determined precisely by means of the Nelson-Somogyi's method
(23). When the quantity of glucose in the pure water was obtained
as a grams/ml, the yield could be given by

$$\text{W/O/W emulsion formation } (\%) = 100 - 100a \, / \, (c/(v_1 + v_2 + v_3)) \qquad (1)$$

where c is the original weight of glucose in the inner aqueous phase
of the dialyzed W/O/W emulsion, and v_1, v_2 and v_3 are the volumes of
the inner aqueous phase, suspending fluid and pure water in the dialyzed system.

  From the data obtained in a series of the experiments, the yield
of the W/O/W-type dispersion is influenced by the concentrations of
the emulsifiers used in both steps of emulsification and by the volume fraction of the W/O emulsion in the final form of the sample, as
shown in Figures 1, 2 and 3. Therefore, the ratio of amount of the
hydrophobic emulsifier to the hydrophilic emulsifier plays an important part in the formation of the W/O/W emulsions. This can be represented by a plot of the yield of W/O/W-type dispersion as a function of the weight ratio of Span 80 (sorbitan monooleate) to Tween 20
(polyoxyethylene sorbitan monolaurate), as shown in Figure 4. The
ratio of both emulsifiers was evaluated from the constitution of each
sample to be tested. Figure 4 suggests that more than 10 times as
much Span 80 as Tween 20 is necessary to obtain 90% or higher yields
of the W/O/W emulsion. The same observation was made with some of
the other samples prepared by use of the various hydrophilic emulsifiers in nonionic type for the second step procedure, as shown in
Figure 5. These results also suggest that the chain length of hydrophobic groups in the emulsifiers relates to the yield of the samples.

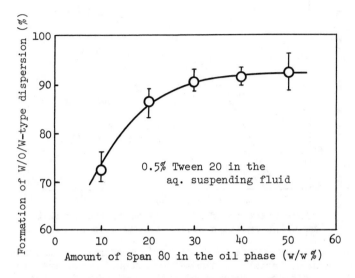

Figure 1. Effect of Span 80 concentration on the formation of the W/O/W-type dispersion. "Reproduced with permission from Ref. 4. Copyright 1976, 'Academic Press'."

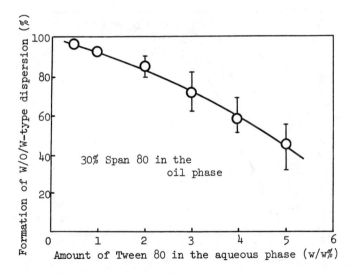

Figure 2. Effect of Tween 20 concentration on the formation of the W/O/W-type dispersion. "Reproduced with permission from Ref. 4. Copyright 1976, 'Academic Press'."

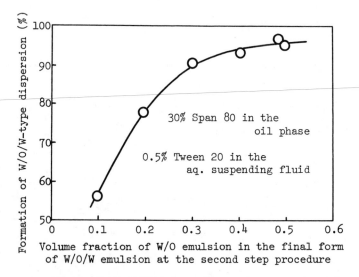

Figure 3. Yield of the W/O/W-type dispersion as a function of
of the volume fraction of the W/O emulsion in the final form
of W/O/W emulsion. "Reproduced with permission from Ref. 4.
Copyright 1976, 'Academic Press'."

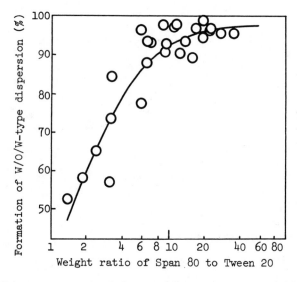

Figure 4. Plot of the W/O/W-type dispersion formation against
the weight ratio of Span 80 to Tween 20 in the final form of
the samples. "Reproduced with permission from Ref. 4. Copyright 1976, 'Academic Press'."

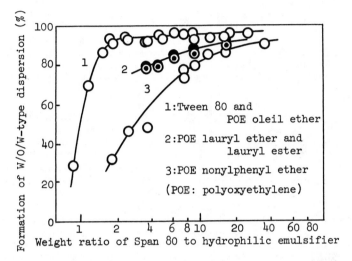

Figure 5. Plot of the W/O/W-type dispersion formation against the weight ratio of Span 80 to the hydrophilic emulsifiers in the final form of the samples. "Reproduced with permission from Ref. 4. Copyright 1976, 'Academic Press'."

One of the effects of hydrophilic emulsifier seems to be attributable to disperse the aqueous compartments into the aqueous suspending fluid from the mass of the W/O emulsion, while hydrophobic emulsifier may contribute to the formation of a rigid adsorbed layer on the surface of the aqueous compartments. On the other hand, some of the molecules of hydrophobic emulsifier are solubilized in the aqueous suspending fluid from the oil layer when the micelle concentration of hydrophilic emulsifier is above a critical level. This process might occur by the interaction of the incorporated micelles with both emulsifiers, increasing pronouncedly with increasing concentrations of hydrophilic emulsifier in the aqueous suspending fluid. Such a phenomenon may play a significant part in the rupture of the oil layer on the surface of the aqueous compartments after the second procedure of emulsification. Thus, the significance of the ratio of both emulsifiers in the yield of the W/O/W-type dispersion might be explained.

Development of W/O/W Emulsion during Phase Inversion. Sherman et al. (24) and Dokic et al.(25) emphasized that the development of a W/O/W type dispersion precedes the thermal induced phase inversion of O/W emulsions. This suggest that the state of multiple structure in emulsions may be generalized as one of the mesophase between O/W and W/O emulsions, and also that there is a possibility of more simplifying the method for preparing W/O/W emulsions.

Recently, it has been found (21) that if the first step procedure described is carried out by introducing an aqueous solution of hydrophilic emulsifier into liquid paraffin containing Span 80, the W/O/W emulsion appears in course of such the first step process of the two step procedure of emulsification due to the phase inversion of W/O emulsion. An assembly for examining the above phenomenon was composed of a pin-mixer and of a peristaltic pump. It was essentially identical to that used for providing W/O emulsions in the first step procedure of the W/O/W emulsion preparation. A well defined volume of a mixture of liquid paraffin and Span 80 was introduced into a vessel of the pin-mixer, and an aqueous mixed solution of glucose and hydrophilic emulsifier was then added successively to the oil phase in the vessel at a rate of 5ml/min by use of a silicone rubber tube connected to the peristaltic pump. The glucose was a necessary component for providing an osmotic pressure gradient between the aqueous compartment and the aqueous suspending fluid in the sample so as to assess the W/O/W emulsion formation in the light of the degree of extent of the oil layer in a unit volume of each sample. The detailes of this method will be described precisely later on in this paper.

The pin-mixer used was a part of a Mixograph, which is an assembly for examining the dough consistency, made by the National Mfg. Ltd., Lincoln, Nebraska. This mixer consisted of two units: an aluminium vessel with 4.2cm depth and 7.5cm inner diameter, and a stirrer unit consisting of a four-pins rotor. The length and diameter of each metal pin were 3.9 and 0.26cm, respectively. The vessel also possessed three metal pins fixed vertically from the bottom of the vessel, while the rotor was driven steadily at 88rpm by means of a speed reducer motor. When the stirrer unit was immersed in the vessel and made to revolve in the oil phase, the latter circulated steadily through the gap between the four-pins rotor and the fixed three pins, thus providing a constant shear force to the system.

The phase inversion of emulsions generally occurs when the dispersed globules are packed very closely in the suspending fluid. It seems that the same thing happened during the mixing procedure in this experiment. The mixing system was initiated by forming an ordinary W/O emulsion for all test systems evaluated. When the volume fraction of an aqueous solution of hydrophilic emulsifier, denoted as ϕ_{aq}, added successively to the mixing system was over 0.7, the continuous phase of the system was substituted by the aqueous phase, as shown in Figure 6. This induced the development of a W/O/W emulsion structure in the system. It then followed that the maximal extent of the oil layer in the newly developed W/O/W emulsion could be observed at around 0.75 for ϕ_{aq}, because an increasing amount of the aqueous suspending fluid containing hydrophilic emulsifier promoted the dispersement of the aggregates of the aqueous compartments.

The appex of the plot in Figure 6C when using Tween 80 (polyoxyethylene sorbitan monooleate) as the hydrophilic emulsifier is much broader than that in Figures 6A or B when using SDS (sodium dodecylsulfate) or CTABr (cetyl trimethyl ammoniumbromide), respectively. This may be brought about by the deficient role of Tween 80 in the dispersement of the aggregates. Additional increments of the aqueous phase, however, caused the rupture of oil layer due to the solubilization of the oil layer components in the micelles of hydrophilic emulsifier in the aqueous phase, as has been described in this paper, so that the extent of the oil layer decreased with increasing amount of the aqueous phase. It was confirmed that the Tween 80 system inverses completely to an ordinary O/W emulsion when the value for ϕ_{aq} is higher than 0.85. Therefore, the development of W/O/W-type dispersion in course of emulsification procedure may be characterized as a mesophase between W/O and O/W-type dispersions during the phase inversion process.

It should be noted that the necessary conditions for obtaining the above phenomenon are not only the closely packed state of the dispersed globules in the W/O emulsion but the presence of a certain amount of hydrophilic emulsifier in the aqueous phase. Figure 7 shows the effect of the molar ratio of Span 80 to the hydrophilic emulsifiers on the formation of W/O/W emulsions, when the volume fraction ϕ_{aq} is fixed at 0.75. The results obtained indicate that the range of the W/O/W emulsion formation as a function of the molar ratio is more or less influenced by the type of hydrophilic emulsifier employed. It is clear that the use of SDS facilitates in a relative manner of the development of a W/O/W emulsion structure, while the formation range of the W/O/W emulsion for CTABr or Tween 80 systems scatters in a comparative narrow region of the molar ratio. In any case, the left hand side of the W/O/W emulsion range in Figure 7 is occupied by the O/W-type dispersion, and the right hand side yields the W/O-type dispersion. Thus, the W/O/W emulsion may be stated as one of the emulsion phases in the phase diagram of the multicomponent systems of oil, water, hydrophobic- and hydrophilic emulsifiers.

Thinning of Oil Layer

It was assumed in an early stage of this study that if the oil phase components distribute uniformly over the surface of each aqueous compartment during the second step procedure in the W/O/W emulsion preparation, the thickness of oil layer on the surface of the

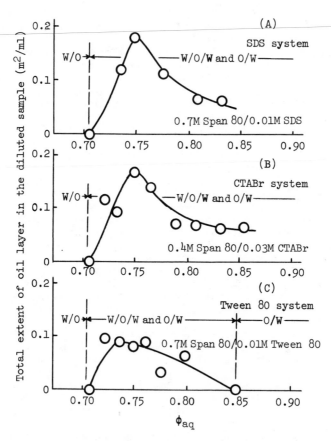

Figure 6. Development of the W/O/W-type dispersion during phase inversion of W/O emulsions. "Reproduced with permission from Ref. 21. Copyright 1983, 'Academic Press'."

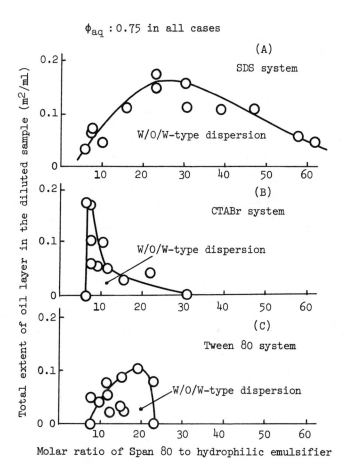

Figure 7. Correlation between the W/O/W-type dispersion formation and the molar ratio of Span 80 to hydrophilic emulsifiers. "Reproduced with permission from Ref. 21. Copyright 1983,'Academic Press'."

compartments can be estimated from the values of diameter and volume
fraction of the dispersed phase in the W/O emulsion prepared by the
first step procedure. An attempt at estimating the thickness of oil
layer according to the above assumption gave 0.13 μm as an example
when the values for the diameter and the volume fraction of the dispersed globules are 2 μm and 0.7, respectively (4).

However, the precise observation of W/O/W emulsions using an
optical microscope has indicated clearly that the thickness of oil
layer is not influenced by the amount of the oil phase components.
That is, the thinning process occured in the oil layer immediately
after the W/O/W emulsion formation, so that the aqueous compartments
were subsequently surrounded by the thin oil layer, as shown in Figure 8. On the other hand, the surplus oil components located hetero-
geneously on the surface of the compartments.

It has been tried to observe the thinning behavior of the oil
layer on the surface of a drop of water using an assembly consisting
of a microsyringe and an optical microscope. The tip of a needle
connected to a microsyringe was immersed in pure water, and a very
small amount of liquid paraffin containing 30% Span 80 was applied to
the tip by use of a small spatula. A drop (about 0.05 cm in diameter,
so that about $6.5 \times 10^{-5} cm^3$ in volume) of pure water was then drawn
from the needle by means of the microsyringe, while the appearance of
the drop surrounded by the oil layer was recorded photomicrographically for a few minutes. Figure 9 shows a typical example of the observations. It is obvious that a large extent of the oil layer on the
surface of the water drop exhibits a thinning process within 1 min
after the preparation of the drop.

The D.L.V.O. theory on the stability of lyophobic colloids (26,
27) can be applied to consider qualitatively the thinning process of
the oil layer in W/O/W emulsions. First, the van der Waals disjoining pressure $\Pi_{VW}(h)$ plays a part in the distance between the both
side surfaces h of the oil layer, as follows:

$$\Pi_{VW}(h) = -H / 6\pi h^3 \tag{2}$$

where H is the Hamaker constant. The pressure appears negatively in
the oil layer, so that the thinning process occurs spontaneously in
the oil layer. The potential energy $V_{VW}(h)$ of the pressure corresponds to the work of the thinning process, expressing as

$$V_{VW}(h) = -\int_{\infty}^{h} \Pi_{VW}(h)\ dh = -H / 12\pi h^2 \tag{3}$$

Second, the electrostatic disjoining pressure $\Pi_{el}(h)$ playing another
part in the thinning process of the oil layer should also be considered, as follows:

$$\Pi_{el}(h) = 64\ nkT\ \psi^2\ e^{-\kappa h} \tag{4}$$

where n is the concentration of ions, kT is the thermal kinetic energy,
ψ is the surface potential, and κ is the constant relating to the value
for n. The potential energy $V_{el}(h)$ of this pressure can be given by

$$V_{el}(h) = 2\int_{\infty}^{h/2} \Pi_{el}(h)\ dh = (64\ nkT / \kappa)\ \psi^2\ e^{-\kappa h} \tag{5}$$

Consequently, the total potential energy V(h) on the disjoining

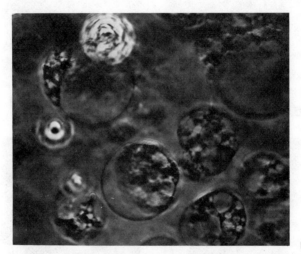

10μm

Figure 8. Photomicrograph of the dispersed globules in W/O/W emulsion prepared by the phase inversion technique.

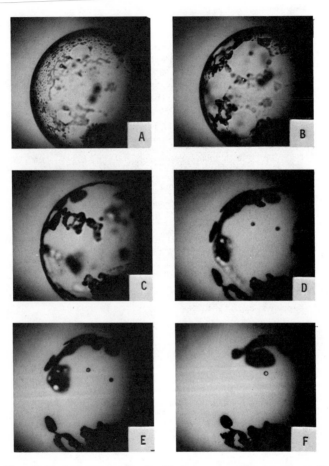

Figure 9. Thinning process of oil layer on the surface of
a water drop taken photomicrographically at 4sec (A), 7sec (B),
16sec(C), 30sec (D), 41sec (E), and 55sec (F) after the
preparation.

pressure of the oil layer is given by the summation of "Equations 3 and 5", as follows:

$$V(h) = V_{el}(h) + V_{vw}(h) = (64 \, nkT / \kappa) \, \psi^2 \, e^{-\kappa h} - H / 12\pi h^2 \tag{6}$$

Therefore, the thinning process of the oil layer observed is explained as a transitional stage in attaining a minimal level of the total potential energy at a given condition.

Water Permeability of Oil Layers

The viscosity of W/O/W emulsions is influenced by the osmotic pressure gradient between the aqueous compartments and the aqueous suspending fluid (12, 15). Although the viscosity change could be observed on either side of increase or decrease according to the drift of the osmotic pressure, the rate of viscosity change in an initial stage of the measurement were proportional to the degree of the osmotic pressure gradients between the two aqueous phases. Therefore, such viscosity behaviors can be analyzed by means of the swelling or the shrinkage of the aqueous compartments due to the osmotic permeation of water through the thin oil layer (15).

It has been tried to measure microscopically the rate of changes in the size of the dispersed globules in a series of the W/O/W emulsions so as to obtain informations on the water permeation coefficient of oil layer (16). The sample emulsions were prepared by the two step procedure using Span 80 in the oil phase and Tween 80 in the aqueous suspending fluid. The two aqueous phases of the compartments and suspending fluid in the original W/O/W emulsions contained an equal concentration of glucose, i.e. there was an isotonic condition between the two aqueous phases in the original emulsions. The original one was then diluted about 200 times with pure water. Since this treatment made dilution of the aqueous suspending fluid of the original sample, the aqueous phase of each compartment had an high osmotic pressure against the suspending fluid across the oil layer depending on the concentration gradient of glucose between the two aqueous phases. Immediately after the dilution, a drop of the diluted emulsion was placed on a glass slide with a small depression in the center (0.05cm in depth and 1.0cm in diameter). The glass slide was then mounted in an optical microscope for recording periodically the changes in the size of the dispersed globules in a fixed place of the microscopic field by means of a photomicrography for about 30min at room temperature. The swelling rate of the compartments under an osmotic pressure gradient was measured with multicompartment globules in many cases, but with single compartment globules if possible. It was confirmed experimentally that the swelling rate is not affected by the number of compartments in a globule, but depends on the magnitude of the osmotic pressure gradients between the two aqueous phases. Figure 10 shows some typical plots of the volume of the globules as a function of aging time after the dilution for the case of a water-in-liquid paraffin-in-water system.

The swelling of the globules is brought by the migration of water to the aqueous compartments in the globules from the aqueous suspending fluid across the oil layer due to the osmotic permeation. The rate of the volume flux of water dv/dt in an initial period of aging can be expressed by the relation used for lipid membrane system (28),

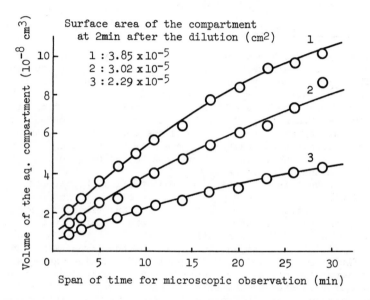

Figure 10.   Changes in volume of the dispersed globules in a
W/O/W emulsion under an osmotic pressure gradient between the
two aqueous phases.   "Reproduced with permission from Ref. 16.
Copyright 1980, 'Academic Press'."

such as

$$dv/dt = L_p \, ART \, (g_2 c_2 - g_1 c_1) \tag{7}$$

where the value for dv/dt can be obtained from the initial slope of
the curve plotted the volume of a globule against time, as shown in
Figure 10, L_p is the hydrodynamic coefficient of the oil layer (29),
A is the surface area of a globule, R is the gas constant, T is the
absolute temperature, g and c are the osmotic pressure coefficient
and concentration of glucose, and subscripts 1 and 2 refer to the suspending fluid side and the compartment side of the oil layer. The
water permeation coefficient of oil layer P_o is defined from
"Equation 7", as follows:

$$(dv/dt) \, / \, \bar{V} = L_p \, ART \, (g_2 c_2 - g_1 c_1) \, / \, \bar{V} = P_o \, A \, (g_2 c_2 - g_1 c_1) \tag{8}$$

where $\bar{V}$ is the partial molar volume of water. Microscopy on the
changes in the size of the dispersed globules in the diluted W/O/W
emulsions gives informations on the volume flux of water dv/dt and on
the surface area of each globule A at any measuring time. Thus, an
evaluation of the water permeation coefficient of the oil layer P_o is
possible for all W/O/W emulsion systems.

Table I. Parameters of the Osmotic Permeability of Water across
the Oil Layer in a Series of Water/Liquid-paraffin/Water
Emulsions prepared by 30% Span 80 in the Oil Phase and
1% Tween 80 in the Suspending Fluid. "Reproduced with
permission from Ref. 16. Copyright 1980, 'Academic Press'."

	A $(10^{-5} cm^2)$	dv/dt $(10^{-10} cm^3 \, sec^{-1})$	P_o $(10^{-4} cm \, sec^{-1})$
n-heptane	5.54	1.29	7.73
	4.78	1.21	8.40
	2.38	0.59	8.27
			(av. 8.13)
n-decane	2.83	0.54	6.35
	1.81	0.33	6.01
	1.08	0.20	6.18
			(av. 6.18)
kerosene	3.96	0.61	5.10
	2.46	0.39	5.23
	1.96	0.33	5.55
			(av. 5.29)
liquid paraffin	3.85	0.71	6.13
	3.02	0.43	4.78
	2.29	0.28	4.11
			(av. 5.01)

Table I summarizes some examples of the values for A, dv/dt and
P_o of the W/O/W emulsions prepared from various hydrocarbons. The
water permeation coefficient of the oil layers obtained under the
various conditions consequently scatters within a range of values
from 2×10^{-4} to 8×10^{-4} cm sec^{-1}, although it has a tendency to

decrease slightly as the carbon number of the hydrocarbons used in the oil phase increases. It seems that the mechanism of water permeation is not affected by the kind of hydrocarbon and by the degree of osmotic pressure gradient, but that the rate of extent of the oil layer due to the swelling of the globule may be restricted by the consistency of the oil phase.

The values for P_o of the oil layers in the W/O/W emulsions appear a little smaller than that of lipid membrane systems, for which values ranging from 10^{-3} to 10^{-4} cm sec^{-1} have been reported ([30-37]). When these permeation coefficients are compared with each other, the mechanism of water permeation across the oil layers in W/O/W emulsions may be considered as similar to that of the lipid membrane systems. That is, the structure of oil layers in the W/O/W emulsions will be in the main bimolecular layers of hydrophobic emulsifier. During the first step emulsification in the two-step procedure or the early stage of the one-step procedure, the molecules of hydrophobic emulsifier, e.g. Span 80, form an adsorbed film on the surface of the dispersed aqueous globules, and then another adsorption of the emulsifier molecules occurs on the surface of the aqueous compartments at the second step procedure or the phase inversion process. Immediately after that, a large part of the oil phase components existing between the two adsorbed films start to move to a definite place on the surface of the compartments or to disperse into the aqueous suspending fluid due to the total disjoining pressure of the oil layer as described. It is well known that many hydrophobic amphiphiles form a lamellar liquid crystal phase with water in a region of the smectic mesophase ([38], [39]). Such phase will make possible the formation of a stable bimolecular layer of the amphiphilic molecules on the surface of the aqueous compartments in the W/O/W emulsions.

Stability of W/O/W Emulsions

The W/O/W emulsion stability may be phenomenologically understood as being brought about by the durability of the oil layer as each dispersed globule has a vesicular structure, which consists of single or multiple aqueous compartments separated from the aqueous suspending fluid by a thin oil layer. If a rupture occurs in an oil layer, the compartment disappears instantaneously, and then the aqueous phase of the compartment is mixed with the suspending fluid. Therefore, the volume fraction of the aqueous compartments in the W/O/W emulsion gradually decrease with increasing number of the rupture of oil layer, and this leads to the decrease of viscosity of the system, because the viscosity of disperse systems depends strongly upon the volume fraction of the dispersed phase. In view of the above, it will be possible to estimate the stability of the W/O/W emulsions by following the changes in the bulk viscosity of the emulsions with time ([6]).

Figure 11 shows how the apparent viscosity at 9.31 sec^{-1} of shear rates changed with time for a series of water-in-olive oil-in-water emulsions containing different additives in the aqueous compartments ([19]). The results can be divided into two groups according to whether they produced a continuous increase in viscosity with aging time until a steady value is reached (saccharides as additive), or an increase followed by a decrease (salts and organic acids as additive). When saccharides were added, the viscosity increased for 2 hr after the preparation of the samples as the volume fraction of the dispersed

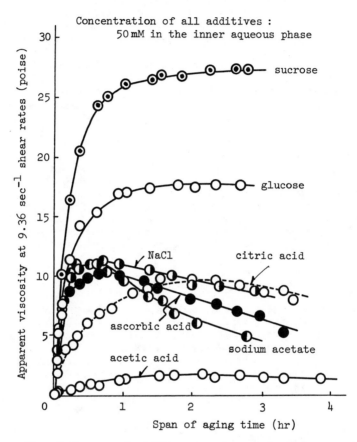

Figure 11.   Influence of additives on changes in the apparent
viscosity of water/olive-oil/water emulsions within a short
span of aging.   "Reproduced with permission from Ref. 19.
Copyright 1981, 'Food and Nutrition Press, Inc.'."

aqueous compartment was increased. This was due to the migration of
water under the osmotic pressure gradient between the two aqueous
phases. The viscosity was then subsequently maintained for at least
a month. Its value, and the rate at which it was achieved, depended
on the concentration of saccharide used. Using salts or organic acids
as additives the rate of viscosity increase and subsequent decrease
depended on the kind of the electrolytes. In the presence of organic
acids they had broken down the W/O/W-type dispersion after four days,
but when salts were added the process continued for at least a month.

On the other hand, it has been tried to assess the stability of
W/O/W emulsions from the view point of the total extent of the oil
layers separating the two aqueous phases in a unit volume of the
sample (17). The viscosity of the W/O/W emulsions can be expressed by
the Mooney's relation (40) in a Newtonian flow region (6, 15), as
follows:

$$\ln \eta_{rel} = \alpha \, (\, \phi_o + \phi_w) \, / \, (1 - \lambda \, (\, \phi_o + \phi_w)) \tag{9}$$

where η_{rel} is the relative viscosity of the sample, ϕ_o and ϕ_w are the
volume fractions of the oil phase and the aqueous compartments in the
sample, and α and λ are the constants. The values for these constants
were examined experimentally with the various W/O/W emulsions as 2.5
for α and 1.5 for λ (15), so that the volume fraction of the aqueous
compartments ϕ_w can be calculated from the viscosity using "Equation
9", as follows:

$$\phi_w = ((1 - \lambda \phi_o) \ln \eta_{rel} - \alpha \phi_o) \, / \, (\alpha + \lambda \ln \eta_{rel}) \tag{10}$$

When the concentration of glucose in the compartments is higher than
that in the suspending fluid, an increase of the emulsion viscosity is
observed. It is due to the increase of the volume fraction ϕ_w caused
by the osmotic permeation of water across the oil layer. Using
"Equation 10" one can obtain information on the water flux $d\phi_w/dt$ by
measuring the changes in the viscosity of the W/O/W emulsion under an
osmotic pressure gradient, so that the total extent of the oil layer
in a unit volume of the sample, denoted as $\bar{A}$, can be evaluated (see
"Equation 8") by

$$\bar{A} = (d\phi_w/dt) \, / \, P_o (g_2 c_2 - g_1 c_1) \, \bar{V} \tag{11}$$

This is the principle for evaluating the durability of the oil layer
in W/O/W emulsions. The phase inversion technique in preparing W/O/W
type dispersion described was also assessed by the extent of oil layer
using this method.

Measurement was made with a small part of the original W/O/W emulsion over one month period, while the two aqueous phases in each original emulsion contained an equal concentration of glucose so as to
provide an isotonic condition between the two aqueous phases. During
aging period a small part of the original sample was diluted periodically with pure water for each time measurement in order to produce a
concentration gradient of glucose between the two aqueous phases.
This dilution was also efficient to provide the Newtonian flow fluid
to the diluted sample. Immediately after the dilution, measurement
was commenced to follow the changes in the viscosity of the sample at
a fixed shear rate for about 20 min.

Table II summarizes the values of $\bar{A}$ for a series of the W/O/W emulsions against each aging period of time (17). Figure 12 also shows the plot of the values for $\bar{A}$ vs. aging time for the W/O/W emulsions provided by the phase inversion technique (21). Within the experimental error, the extent of the oil layers in the W/O/W emulsions evaluated is not influenced by aging time, so that the dispersion state of these samples can be assessed as stable. In the Tween 80 system in Figure 12, however, the extent of the oil layer decreased about 50% of that in the freshly prepared sample in an initial stage of aging, and then reached a stedy state. Optical microscopic observation suggests that the decrease of the oil layer extent in the Tween 80 system was brought by the formation of aggregates among the dispersed globules during the initial stage of aging. This was clarified by since the aggregation formation caused a decrease in the extent of the oil layer separating the two aqueous phases.

Table II Changes in Total Extent of Oil Layer ($\bar{A}$) in a Unit Volume of W/O/W Emulsions during Aging at 25°C. "Reproduced with permission from Ref. 17. Copyright 1980, 'Academic Press'."

aging period (days)	Hydrocarbon used for the oil phase			
	n-heptane $\bar{A}$ (cm^2)	n-decane $\bar{A}$ (cm^2)	kerosene $\bar{A}$ (cm^2)	liquid paraffin $\bar{A}$ (cm^2)
0	95.1	129.4	155.2	101.5
1	93.3	129.4	135.8	104.5
3	72.8	130.6	135.8	116.4
5	85.8	126.9	—	101.5
7	93.3	133.1	140.3	—
9	—	—	—	101.5
10	85.8	124.4	137.3	106.0
30	69.1	110.6	143.7	107.8

On the other hand, Figure 11 indicates that the dispersion state of W/O/W emulsions becomes to be unstable in the presence of electrolytes in the aqueous phase. External appearance of the W/O/W emulsions also indicates that the various organic acids such as acetic acid, citric acid, ascorbic acid, etc. are the efficient ingradients to rupture the oil layer, but that the addition of salts promotes the aggregation formation among the dispersed globules more than the rupture of the oil layer.

The details about any mechanism of these phenomena are not fully investigated yet.

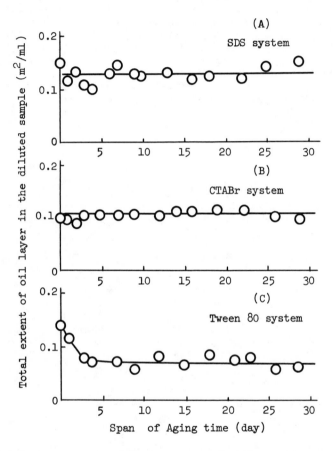

Figure 12. Stability of the oil layer in W/O/W emulsions prepared
by the phase inversion technique. "Reproduced with permission
from Ref. 21. Copyright 1983, 'Academic Press'."

Literature Cited

1. Herbert, W. J. Lancet 1965, 2, 771.
2. Engel, R. H.; Fahrenbach, M. J. Nature (London) 1968, 219, 856.
3. Davis, S. S.; Purewal, T. S.; Burbage, A. S. J. Pharm. Pharmacol. 1976, 27(Supplement), 60.
4. Matsumoto, S.; Kita, Y.; Yonezawa, D. J. Colloid Interface Sci. 1976, 57, 353.
5. Davis, S. S.; Burbage, A. S. J. Colloid Interface Sci. 1977, 62, 361.
6. Kita, Y.; Matsumoto, S.; Yonezawa, D. J. Colloid Interface Sci. 1977, 62, 87.
7. Matsumoto, S.; Kohda, M.; Murata, S. J. Colloid Interface Sci. 1977, 62, 149.
8. Sheppard, E.; Tcheurekdjian, N. J. Colloid Interface Sci. 1977, 62, 564.
9. Kita, Y.; Matsumoto, S.; Yonezawa, D. J. Chem. Soc. Japan 1977, 748.
10. Kita, Y.; Matsumoto, S.; yonezawa, D. J. Chem. Soc. Japan 1978, 11.
11. Matsumoto, S.; Ueda, Y.; Kita, Y.; Yonezawa, D. Agri. Biol. Chem.(Tokyo) 1978, 42, 739.
12. Matsumoto, S.; Kohda, M. In "Food Texture and Rheology"; Sherman, P., Ed.; Academic: London, 1979; p.437.
13. Panchal, C. J.; Zajic, J. E.; Gerson, D. F. J. Colloid Interface Sci. 1979, 68, 295.
14. Pilman, E.; Larsson, K.; Tornberg, E. J. Disp. Sci. Technol. 1980, 1, 267.
15. Matsumoto, S.; Kohda, M. J. Colloid Interface Sci 1980, 73, 13.
16. Matsumoto, S.; Inoue, T.; Kohda, M.; Ikura, K. J. Colloid Interface Sci. 1980, 77, 555.
17. Matsumoto, S.; Inoue, T.; Kohda, M.; Ohta, T. J. Colloid Interface Sci. 1980, 77, 564.
18. Florence, A. T.; Whitehill, D. J. Colloid Interface Sci. 1981, 79, 243.
19. Matsumoto, S.; Sherman, P. J. Texture Stud. 1981, 12, 243.
20. Tomita, M.; Abe, Y.; Kondo, T. J. Pharmaceutical Sci. 1982, 71, 332.
21. Matsumoto, S. J. Colloid Interface Sci. 1983, 94, 362.
22. Davis, S. S. Chemistry and Industry 1981, 683.
23. Somogyi, M. J. Biol. Chem. 1952, 195, 19.
24. Sherman, P.; Parkinson, C. Prog. Colloid Polym. Sci. 1978, 63, 10.
25. Dokic, P.; Sherman, P. Colloid Polym. Sci. 1980, 258, 1159.
26. Derjaguin, B. V.; Landau, L. Acta Physicochim. URSS 1941, 16, 633.
27. Verwey, E. J.; Overbeek, J. Th. G. "Theory of Stability of Lyophobic Colloids"; Elsevier, Amsterdam, 1948.
28. Cass, A.; Finkelstein, A. J. Gen. Physiol. 1967, 50, 1765.
29. Kedam, O.; Katchalsky, A. Biochim. Biophys. Acta 1958, 27, 229.
30. Mueller, P.; Rudin, D. O.; Tien, H. T.; Westcott, W. C. In "Recent Progress in Surface Science"; Danielli, J. F.; Pankhurst, K. G. A.; Riddiford, A.C., Ed., Academic: New York, 1964, Vol. I, p.379.

31. Hanai, T.; Haydon, D. A. J. Theoret. Biol. 1966, 11, 370.
32. Huang, C.; Thompson, T. E. J. Mol. Biol. 1966, 15, 539.
33. Bangham, A. D.; De Gier, J.; Greiville, G. D. Chem. Phys. Lipids 1967, 1.225.
34. Tien, H. T.; Ting, H. P. J Colloid Interface Sci. 1968, 27, 702.
35. Reeves, J. P.; Dowben, R. M. J. Membrane Biol. 1970, 3, 123.
36. Fettiplace, R. Biochim. Biophys. Acta 1978, 513, 1.
37. Nagle, J. F.; Scott, Jr. H. L. Biochim. Biophys. Acta 1978, 513, 236.
38. Larsson, K. Soc. Chem. Ind. London Monogr. 1968, No.32, p.8.
39. Friberg, S.; Larsson, K. In "Advances in Liquid Crystals"; Brown, G. H., Ed.; Academic: New York, 1976; Vol. II, P. 173.
40. Mooney, M. J. Colloid Sci. 1951, 6, 162.

RECEIVED December 26, 1984

Use of Macroemulsions in Mineral Beneficiation

BRIJ M. MOUDGIL

Center for Research in Mining and Mineral Resources, Department of Materials Science and
Engineering, University of Florida, Gainesville, FL 32611

Macroemulsions of water insoluble reagents can
enhance the coating efficiency because of large
surface area of the droplets coming in contact
with solid particles. However, stability of oil
droplets by fine particles is encountered in these
operations, and causes problems in recycling of
the water immiscible reagents. Applications of
emulsifying the reagents in achieving desired
solid-solid separation by electronic ore sorting,
emulsion flotation, liquid-liquid extraction, and
froth flotation are discussed. Role of surface
charge of the particles and interfacial tension in
spreading of fluids on solids is presented.

Macroemulsions are extensively used in coating of different
substrates. In such cases, generally the coating material is
insoluble or sparingly soluble in the bulk solvent, thereby,
limiting the area of contact between the coating material and
the substrate. Emulsification of the insoluble fluid using a
surface active agent helps in overcoming this limitation.
Pesticides, paints, and road surfacing by bitumen or tar are
some of the examples in which emulsification of the coating
medium is necessary to obtain an effective coating of the
substrate. It should be noted that in all these cases, a
nonselective coating is obtained. However, selectivity of the
coating is essential in applications of oil in water
macroemulsions in solid/solid separation operations encountered
in mineral processing. A discussion of this aspect of
macroemulsions in coating of mineral particles is presented in
this paper.

0097-6156/85/0272-0437$06.00/0
© 1985 American Chemical Society

Background

Reagent coating process involves collision between oil droplets
and the particle, and spreading of the fluid over the solid
substrate. Collision between droplets and substrates can be
enhanced by agitation or imparting other mechanical forces.
Spreading of the fluid during the time it is in contact with the
substrate is influenced also by surface chemical forces such as
electrostatic charge and steric interactions. In both these
cases adsorption of surfactant molecules at various interfaces
can have a major influence on the spreading of droplets on solid
substrates. Adsorption of surfactant molecules will modify the
electrokinetic properties of the oil droplets and solid surface,
thereby resulting in more fruitful collisions. On the other
hand, surfactant adsorption at the liquid-liquid interface can
reduce interfacial tension which favors spreading of the fluid
over the solid surface.
     In correlating electrophoretic mobility of the bitumen
droplets with the rate of spreading of the droplets on quartz
particles, Lane and Ottewill (1) determined that spreading
occured only below the HTAB (hexatrimethylammonium bromide)
concentration of 0.5 mM at which the amount of surfactant
adsorbed on quartz was equivalent to a vertically oriented
monolayer. They suggested that at the lower concentrations, the
surfactant molecules pre- ferentially adsorb on the quartz
surface, thus, depleting the bitumen-water interface and
destabilizing the emulsion. Effect of surfactant concentration on emulsion droplet coating is shown schematically in
Figure 1.
     In achieving solid-solid separation in mineral
beneficiation, selective coatings of water insoluble or
sparingly soluble reagents are required in electronic ore
sorting, froth flotation, and emulsion flotation operations.
Use of macroemulsion in achieving the selective coating for
these applications is discussed below.

Electronic Ore Sorting

Electronic sorting of minerals based on the surface chemical
differences involves achieving a selective coating of a coloring
or fluorescent dye on the desired mineral component. The fluorescent dye may or may not be water soluble. In the case of
water soluble dyes, it is essential to make only the desired
mineral component water wetted leaving the other particles water
unwetted or vice versa. In such cases applications of macroemulsion technology in achieving the desired coating are
limited. On the other hand, when the dyes are water insoluble,
an effective coating of the dye may only be achieved by macro-
mulsion technology which uses an intermediatory reagent compatible with the dye and the solid substrate. For example, in
separation of limestone from quartz to obtain a coating of fluoranthene on limestone particles, the intermediatory chemical is
fatty acid (oleic acid) which adsorbs on limestone pieces but

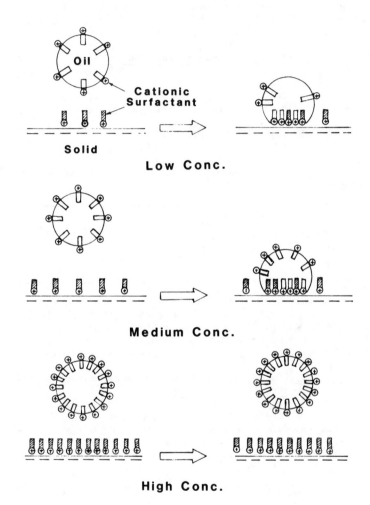

Figure 1. Effect of surfactant concentration on spreading of an oil droplet on a solid surface.

not on quartz particles. It should be noted that oleic acid is
sparingly soluble in water and, therefore to achieve an
efficient coating, the dye is first dissolved in an oil and/or
oleic acid, which is agitated with water to form a oil in water
macroemulsion. After a suitable coating of the fluoranthene on
limestone is achieved, the coated and uncoated particles are
separated using the Oxylore sorting machine.(2) Results of such
a separation are presented in Table 1.

Table I. Separation of Limestone from Quartz by Fatty Acid
          Coating

```
           Feed:    Limestone and Quartz Mixture
Coating Emulsion:   Fluoranthene Dye, Oil and Fatty Acid
                    Mixture
   Particle Size:   1.3-10.0 cm
```

Sample	Limestone Content (%)	Limestone Recovery (%)
Feed	47.5	100
Concentrate	94.5	93.7

however, if only the quartz particles are to be made fluorescent, the dye is dissolved in a beta amine which has limited
solubility in water.(2) The amine is then added to water, and
an emulsion is made by agitating the mixture. Results of
limestone separation from quartz are presented in Table II.

Table II.   Silica Content of Limestone after Separation by
            Electronic Ore Sorting

```
           Feed:    Limestone and Quartz Mixture
Coating Emulsion:   Fluoranthene Dye, and Armeen L-9
                    (Beta Amine - Armak Chemicals)
   Particle Size:   1.3 - 10.0 cm
```

Sample	Silica Content (%)
Feed	25.1
Concentrate	1.1

In another example of separating coal from slate, fluoranthene is dissolved in decylalcohol which is then dispersed into water. Results of separation of fluorescent dye coated coal particles from uncoated slate pieces using the Oxylore sorting machine are presented in Table III. (3)

Table III.   Separation of Coal From Slate by Decyl Alcohol
             Coating

                    Feed: Coal (87%) and Slate (13%) Mixture
         Coating Reagent: Fluoranthene in Decyl Alcohol
           Particle Size: 10.0 cm

Sample	Ash (%)	Pyritic Sulfur (%)	Thermal Valve (BTU/lb)	Coal Recovery (%)	BTU Recovery (%)
Feed	17.3	0.2	11339	100	100
Concentrate	5.0	0.1	13384	96	98

Emulsion Flotation

In emulsion flotation technique, a neutral hydrocarbon oil, together with a surface active agent function as collectors for specific minerals. In this process as shown in Figure 2, hydrophobic mineral particles and oil droplets form an aggregate which is then floated out of the pulp by air bubbles. The mixture of oil and surfactant is either emulsified (O/W) prior to addition to the pulp, or in the pulp itself. The effect of neutral oil in emulsion flotation is not yet completely known. According to Fahrenwald (4) and Karjalahti (5) the neutral oil coats the hydrophobic particles which then form an agglomerate. Livshits and Kuzkin (6) on the other hand have reported that hydrophobicity of the particles is unaltered in the presence of neutral oil, and that the added oil reduces the induction time of mineral particles in contact with the air bubbles. Lapidot and Mellgren (7) concluded that addition of the neutral oil improved the dispersion of the collector and also increased the hydrophobic nature of the particles. It has been reported that the flotation activity of nonpolar oils depends on their viscosity and dispersity. (8) It is to be expected that there is an optimum droplet size of emulsions which is a function of particle size and mineral chemistry. This aspect of emulsion flotation, however, has not been systematically studied. Emulsion flotation has been employed in the treatment of ores containing, iron, manganese, molybdenum and titanium oxide. (9-11) A major role of emulsion of fatty acids in the flotation of Florida phosphate rock also cannot be

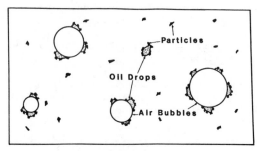

Emulsion Flotation

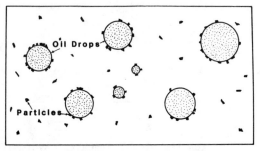

Liquid-Liquid Extraction and Flotation

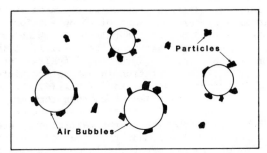

Froth Flotation

Figure 2.   Schematic of emulsion flotation, liquid–liquid extraction and flotation, and froth flotation.

ruled out. In this case, probably ionization of the fatty acid provides the required surfactant molecule for emulsification of the fuel oil which is used as a co-collector.

McCarrol (9) reported that recovery of manganese ore increased from 70 to 84% when an emulsion of oil, soap and a surfactant (Ornite-S) was used as the collector as compared to a mixture of only soap and fuel oil.

Burkin and Bramley (10) observed that less than 1% coal floated in 3 minutes when fuel oil was used as the collector. On the other hand, 100% of the coal floated when fuel oil was emulsified using a nonionic surfactant, Lissapol NDB. It should be noted that in both cases the droplet size of the fuel oil was maintained at 7 microns. The increased efficiency of coal flotation in the presence of oppositely charged surfactant was attributed to lowering of the zeta potential, which indicated adsorption of surfactant on coal particles, and thereby increased tendency of the fuel oil to spread on surfactant coated particles. These investigators suggested that if the kinetic energy of approach between particles and droplets is less than the energy of repulsion, collision between them does not lead to the spreading of oil on coal. If the energy barrier is smaller than the kinetic energy to overcome it, collision between coal particle and droplet occurs but does not yield a hydrophobic coal particle. Flotation of coal occurs only when the energy barrier is much smaller than the kinetic energy available, and a surface active agent is present in the system.

Liquid-Liquid Extraction

This process involves extraction of fine particles from an aqueous phase into an oil phase. The effectiveness of this technique, as shown in Figure 2, is based on the stability of emulsion droplets with solid particles. If a particle is partially wetted by two immiscible liquids the particle will concentrate at the liquid-liquid interface. The thermodynamic criteria for distribution of solids at the interface of two immiscible liquids is the lowering in the interfacial free energy of the system when particles come in contact with two immiscible liquids. (12) If γ_{sw}, γ_{wo} and γ_{so} are the interfacial tensions of solid-water, water-oil and solid-oil interfaces respectively, and if $\gamma_{so} > \gamma_{wo} + \gamma_{sw}$ then the solid particles are preferentially dispersed within the water phase. However, if $\gamma_{sw} > \gamma_{wo} + \gamma_{so}$, the solid is dispersed within the oil phase. On the other hand, if $\gamma_{wo} > \gamma_{so} + \gamma_{sw}$, or if none of the three interfacial tensions is greater than the sum of the other two, the solids in such case will be distributed at the oil-water interface.

The surface charge of the solid particles has been reported to play an important role in the recovery process. Maximum recovery is achieved when the particles exhibit net zero surface charge.(13-15) Lai and Fuerstenau (16) have reported that ultrafine (0.1 µm) alumina particles can be extracted from aqueous suspension into iso-octane (oil phase) through the adsorption of sulfonates. These investigators reported that sulfonate molecules control the hydrophobicity of the alumina particles and also act as emulsifying agent, thus, forming a large oil-water interfacial area. Maximum amount of alumina was extracted into the oil phase when the contact angle exceeded 90°, and the electrokinetic potential was zero. Under these conditions, the particles have a net zero charge and, therefore, can transfer into the oil phase without experiencing repulsion due to electrostatic forces.

One of the major limitations of the liquid-liquid extraction and flotation process is the breaking of the stable emulsions to recover the separated solids, and to recycle the oil phase. Some of the methods examined to break the emulsions include filtration, modifying the phase volume ratio, centrifugation and sedimentation. An efficient and economic solution to this problem is yet to be developed.

Froth Flotation

In this process, as shown in Figure 2, hydrophobic particles attach to the air bubbles and rise to the top of the cell where they are removed by skimming. Separation by froth flotation is based on selective hydrophobicity of the particles. The surfactant molecules which selectively adsorb on the particles are mostly water soluble. In cases where the reagents are water insoluble, an efficient coating is achieved by emulsifying the reagent. Brown and co-workers (17) have reported a marked increase in the flotation of complex low-grade Michigan phosphate ore when the collector (fatty acid-fuel oil mixture) was emulsified using oil soluble petroleum sulfonate. The potential of applying the emulsification technology in froth flotation has not been investigated to any extent.

Summary

It is evident from the above discussion that an efficient coating of a water insoluble or sparingly soluble reagent can be achieved through emulsification. Adsorption of the emulsifying agent (surface active agent) on the solid substrate reduces the surface charge and oil-water interfacial tension, which leads to the spreading of reagent on the solid substrate. Reagent coating on the desired substrate can be achieved by selecting an emulsifying agent which will adsorb also on the specific solid-water interface, e.g. by charge attraction or by chemical bonding. The enhanced effectiveness of emulsion coatings in case of water insoluble reagents can be attributed to increased area of contact between the substrate and the reagent droplets.

Acknowledgments

The author wishes to acknowledge the College of Engineering-EIES (COE Funds) for partial financial support of this work.

Literature Cited

1. Lane, A.R.; Ottewill, R.H. in "Theory and Practice of Emulsion Technology," Smith, A.L. Ed.; Academic: New York, 1976; p. 157.
2. Moudgil, B.M. US Patent 4 208 272, 1980.
3. Moudgil, B.M.; Messenger, D.F.; US Patent 4 208 273, 1980.
4. Fahrenwald, A.W. Mining Congress J. 1957 43, 72-74.
5. Karjalahti, K. Trans. Inst Mining & Met (London), 1972, 81, C219-C226.
6. Livshits, A.K.; Kuzkin, A.S. Tsvetn. Metal, 1964, 37, 76-77.
7. Lapidot M.; Mellgren, O. Trans. Inst Mining & Met. (London), 1968, 77, C149-C165.
8. Glembotskii, V.A.; Dmitrieva, G.M.; Sorokin, M.M. "Non-polar Flotation Agents,"; Israel Program for Scientific Translations, Jerusalem, 1970, p. 45.
9. McCarroll, S.J. Mining Engineering, 1954 (March), 289-293.
10. Burkin, A.R.; Bramley, J.V. J. Appl. Chem. 1961, 11, 300-309.
11. Hoover, R.M.; Molhotra, D. In "Flotation-Gaudin Memorial Volume," Fuerstenau, M.C., Ed.; AIME: New York, 1976, Vol. I, p. 485.
12. Green, E.W.; Duke, J.B. Trans. SME-AIME 1962, 223, 389-393.
13. Shergold, H.L.; Mellgren, O. Trans. SME-AIME, 1970, 247, 149-159.
14. McKenzie, J.M.W. Trans SME-AIME, 1969, 244, 393-400.
15. McKenzie, J.M.W. Trans SME-AIME, 1970, 247, 202-247.
16. Lai, R.W.M.; Fuerstenau, D.W. Trans. SME-AIME, 1968, 241, 549-556.
17. "Characterization and Beneficiation of Phosphate-Bearing Rocks from Northern Michigan," U.S. Bureau of Mines, RI 8562, 1981.

RECEIVED December 18, 1984

Kinetics of Xylene Coagulation in Water Emulsion Stabilized by Phthalylsulfathiazole

S. RAGHAV and S. N. SRIVASTAVA

Department of Chemistry, Agra College, Agra 282002, India

Kinetics of coagulation of Phthallylsulfathiazole
stabilized xylene in water emulsion in the presense
of some cationic detergents is examined. Kinetic
parameters such as rate constant of flocculation,
coalescence and creaming have been evaluated. Studies
were also made to determine zeta potential values and
stability factor W_1 in the presence of same additives
with the help of some equations. Temperature effect
was also calculated by determining coalescence rate
constant at different temperatures. With the help
of these kinetic parameters particle loss mechanism
or coagulation has been delineated. This approach
explained the kinetics of coagulation i.e., which
process is the rate determining one and also expose
the comparative account of effect of the detergents
used in the systems under investigation.

Alkylpyridinium or ammonium halides constitute a group of cationic
detergents with well known surface properties. The kinetics of
emulsion breaking or in general coagulation, comprises flocculation and coalscence. It has been reported (1) that cations and
anions at low concentrations retard the coalescence and association of dispersed phase of emulsions.

Sharma and Srivastava (2,3) have determined emulsion stability from the rate of coagulation as well as by measuring the zeta
potential with the variation of concentration of cationic detergents
and electrolytes. Recently some expressions have been proposed by
Reddy et al (4-6) for the calculation of Brownian and sedimentation
collision frequencies and creaming. However, their application to
our experimental data is not possible because it involves some
adjustable parameters i.e., a_i and a_j, particle radii. The
turbidity technique (5,7) has been used to determine the stability
of suspensions by assuming that turbidity is directly proportional
to particle concentrations. Recently coalescence rate constants of
detergents stabilized polar oil/water emulsion have also been
evaluated (8).

0097-6156/85/0272-0447$06.00/0
© 1985 American Chemical Society

The present paper deals with kinetics of coagulation of
Phthallylsulfathiazole stabilized xylene in water emulsion in the
presence of some cationic detergents. Rate of flocculation, rate of
coalescence and rate of creaming have been determined. To estimate
the stability of the present systems their zeta potentials have been
measured and stability factors calculated. Temperature effect on
the system was also studied.

Materials and Methods

The emulgent Phthallylsulfathiazole, an important drug, was a gift
from May and Backer, Bombay (India). The cationic detergents
Cetyl Pyridinium Bromide (CPB), Cetyl Pyridinium Chloride (CPC),
Cetyl Trimethyl Ammonium Bromide (CTAB), Lauryl Pyridinium Chloride
(LPC) used in the system were obtained from British Drug House Lts.,
Poole, England. The oil phase consisting of xylene was also of
B.D.H. (A.R.) quality. Corning glass apparatus was used throughout
the experimental work.

Emulsion

The emulsions were prepared by suspending 4.0% by volume of xylene
in 0.01 M aqueous solution of the emulgent phthallylsulfathiazole
(alkaline, pH 10.48) and 0.01 M KCl. The mixture, after being
hand shaken for 15 minutes, was homogenized three times in a stainless steel homogenizer (Scheer Sci.Co., Chicago). The method is
explained in the theoretical part. All chemicals were used without
any further purification.

Theoretical

Rate of flocculation was determined by counting the number of
particles haemocytometrically using an improved Neubauer model.
Hand tally counter (Erma, Tokyo) was used to count the numbers of
associated and unassociated globules under Olympus microscope. The
average size of emulsion droplet was found to be 1.15 μm from the
size frequency analysis of microphotograph (9). Particle size and
its distribution have been calculated and presented in Figure 1 and
Figure 2.

From Smoluchowski's (10) theory we have

$$\frac{1}{n} - \frac{1}{n_o} = 4\pi \, DRt \qquad\qquad [1]$$

Where n_o and n are number of singlets present initially and at time
t, D is the diffusion coefficient and R = 2a, a being particle
radius. $\emptyset$ is the phase volume ratio ($\emptyset$ = 0.04 in our system). If
Vm is mean drop volume of n drops after time t, we have −

$$Vm = \emptyset/n \qquad\qquad [2]$$

combining [1] and [2]

$$Vm = \emptyset/n_o + 4 \, DR\emptyset t \qquad\qquad [3]$$

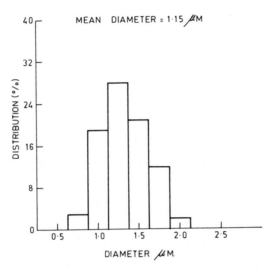

Figure 1.   Plot for particle size distribution of phthalyl-sulfathiazole-stabilized system.

Figure 2.   Microphotograph of phthalylsulfathiazole-stabilized emulsion.

As the phase volume ratio remains the same during coalescence, we may have

$$\emptyset = nVm = n_o Vm_o$$

Where Vm_o is mean drop volume of initial drops, hence,

$$Vm = Vm_o + 4\pi DR\emptyset t = K_f t \qquad\qquad [4]$$

Where K_f is Smoluchowski's flocculation rate constant.

For determination of the stability factor W_1 eq. [4] may be modified as

$$Vm = Vm_o + 4\pi DR\emptyset t \ e^{-W_1/kT}$$

Plot of Vm against time will be a straight line. The slope of this curve = $4\pi DR\emptyset t \ e^{-W_1/kT}$. Using Einstein equation $D = kT/6\pi\eta a$, W_1 values were calculated (11).

Coalescence occurring in the system was measured tubiditimetrically. Turbidity is found to decrease with time for dilute emulsions. As droplets coalesce, there is a net decrease of oil surface area and an increase of average globule diameter. This has been shown (12) that the turbidity for a dilute emulsion as measured in Klett-Summerson photoelectric colorimeter is directly proportional to the surface area of the globules. Thus for Phthallylsulfathiazole stabilized emulsion it is possible to relate the ratio of number of droplets at any time t and that at time = 0 (N_t/N_0) to Klett readings, provided that the size distribution of emulsion droplet does not vary much. If R_t and R_0 are the Klett readings at any time t and t = 0 and r_t and r_0 are average volume surface radiiiat respective times, then

$$R_t \propto 4\pi r_t^2 \ N_t \qquad\qquad [5]$$

$$R_o \propto 4\pi r_o^2 \ N_o \qquad\qquad [6]$$

if the total volume of oil is constant, then

$$4/3 \quad \pi r_t^3 \ N_t = 4/3 \ \pi r_o^3 \ N_o \qquad\qquad [7]$$

From equations [5] to [7]

$$\log \ N_t/N_o = 3 \log R_t/R_o \qquad\qquad [8]$$

For turbidity measurements, diluted emulsions were used to avoid multiple scattering effects (13) which may produce non-linearity between the photomicrographically determined surface area and turbiditimetrically determined one.

The electrophoretic mobilities were determined in a rectangular closed cell of Northrup Kunitze type (14) in an air thermostat. The usual precautions were taken.

As the deserved droplets had large radii compared with the Debye Huckel parameter $(1/\chi)$, the zeta potentials were calculated from the following equation:

$$\xi = 4\pi\eta U/\varepsilon \; X \tag{9}$$

Where η and ε are respectively the viscosity and the dielectric constant of the solution in the electrical double layer adjacent to the surface, U is the mobility and X is the applied field strength. In the present cases η and ε were taken as equal to their bulk values. The temperature for mobility measurements was at $25° \pm 0.1°C$.

RESULT AND DISCUSSION

Charge on Emulsion

Phthallylsulfathiazole stabilized xylene in water emulsion was found to be negatively charged. Since this compound is acidic in nature, the -COOH group ionises to furnish negatively charged carboxylate anions which impart the same charge on emulsion droplets at alkaline pH 10.48 by virtue of the adsorption of the emulgent.

R-remaining part of compound

Effect of Cationic detergents on flocculation

Different concentrations of all the four detergents were used to study the flocculation occurring in the system. Unassociated and associated (monomers and dimers) globules were counted at various intervals of time. Individual oil globules (n_1 or monomers) against time in minutes were plotted. Here only one typical graph (Fig. 3) obtained from $CTAB$ has been presented. Initially particle concentration was 2.39×10^9/ml. It is clear from Fig. 3 that the number of unassociated drops first decreases fastly and then it is slow. Later on it attains almost constancy which indicates reversibility of flocculation occurring in the system. Fig. 4 shows effect of different concentrations of different detergents. It is quite evident from this figure that the flocculation decreases as concentration increases. In the case of LPC the rate constants increase with increase in concentration. Rate constant K_f was of the order of 10^{-14}, 10^{-14}, 10^{-13}, and 10^{-13} for CPB, CPC, CTAB, and LPC, respectively, These low values of the rate constants are indicative of reversible nature of flocculation which in fact occurs in the secondary minimum.

Effect of Cationic detergents on Coalescence

There are three processes by which the number of oil drops in an emulsion is decreased. These are Brownian flocculation, sedimentation flocculation and creaming. But it should be noted that if the absorbed film strength is quite high, flocculation may not necessarily result in coalescence. It is also important to note that flocculation which may be due to any of above three reasons is reversible, but coalescence which follows flocculation is irreversible.

All the emulsions follow a first order kinetic equation of the form:

$$N_t = N_o \exp(-K_c t) \tag{10}$$

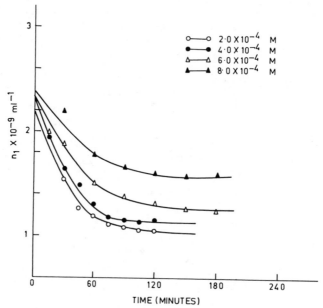

Figure 3. Plots of monomer against time for the system flocculated by CTAB.

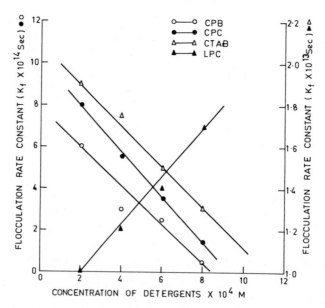

Figure 4. Plots of flocculation rate constant (K) against concentration of detergents.

Where K_c is the first order rate constant. On addition of increasing amounts of the cationic detergents the corresponding doalescence rate constants of the emulsion decrease rapidly except in the presence of LPC which increases the rate of coalescence. This is depicted in Fig. 5, which is a plot of rate constants vs. concentration of different detergents. This figure presents a clear picture of comparative study of effects of all detergents taken in our system. It is also interesting to note that the rate constants in the presence of surfactants are of the order of 10^{-5} s-1, 10^{-5}s-1, 10^{-4}s-1 and 10^{-4}s-1 for CPB, CPC, CTAB and LPC respectively. This means the stability of the emulsion increases with the chain length. From Figs. 4 and 5, it is obvious that rate constants of coalescence and rate constants of flocculation follow the same trend, so it can be assumed that flocculation is resulting in coalescence.

It is evident from the date recorded in Table I that at higher temperature coalescence rate constant is higher, that is in order of $K_{c30°} < K_{c40°} < K_{c50°}$. The temperature effect for all detergents was found to be in the order LPC < CTAB < CPC < CPB. This implies that in the presence of CPB the coalescence rate constant of the emulsions increases enormously (e.g. four to six times for every 10°C rise of temperature) with the rise of temperature whereas in the presence of LPC the increase in the rate is not that pronounced.

From the coalescence rates of different temperatures it has also been possible to calculated some kinetic and thermodynamic parameters such as energy of activation (E_a), enthalpy change ($H°$), entrophy change ($S°$) and free energy change ($F°$) from the equation:

$$RT \ln K_c = -\Delta F° = -\Delta H° + T\Delta S° \qquad [11]$$

within the formalism of flocculation/coalescence rate theory assuming that the intermediate floc of the following overall process of coalescence may be taken to be identical with the transition state or activated complex of the Transition State Theory of Chemical Kinetics.

$$\begin{array}{ccc} \text{O} & & \\ & \xrightarrow{\text{Flocculation}} & \text{OO} & \xrightarrow{\text{Coalescence}} & \text{O} \\ \text{O} & & \text{Transition} \\ & & \text{State floc} \end{array}$$

Different parameters reflected in Table II are within the expected range of values. The positive activation entropies denote that the entropy of the transition state floc is greater than the entropy of the separate globules. The stability preventing coalescence from the flocculated intermediate state may be due to the metastability of thin films against rupture which in their turn offer steric hindrance owing to the close packed ordered layers of the emulgent at oil/water interface. The intervening water films between the two droplets of the transition state floc may play the same sort of the role. In the wake of coalescence, rupture of the said films occur and the orderliness disappears and so the entropy increases.

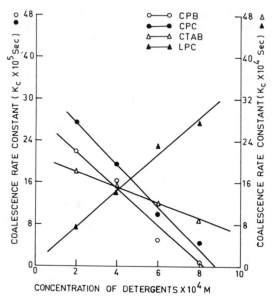

Figure 5. Plot of coalescence rate constant (K_c) against detergent concentration.

Table I.   Effect of Temperature on Coalescence Rate Constant for the System

Detergent (2.0 x 10^{-4}M)	$K_{c30°C}$ Sec.	$K_{c40°C}$ Sec.	$K_{c50°C}$ Sec.
CPB	2.2×10^{-4}	9.4×10^{-4}	1.3×10^{-3}
CPC	2.7×10^{-4}	4.7×10^{-4}	5.2×10^{-4}
CTAB	1.8×10^{-3}	2.4×10^{-3}	2.9×10^{-3}
LPC	7.5×10^{-4}	9.8×10^{-4}	1.1×10^{-3}

Effect on zeta potential

Figure 6 is a graphic representation of the zeta potential data as a function of log molar concentration of different cationic surfactants. At lower concentration, zeta potential is high while on addition of additive it decreases slowly. This decrease is sharp at higher concentrations. On continuous addition of surfactant, the values first become zero and then reversal of charge takes place.
It is very important to note the lower is the zeta potential, the greater the flocculation. As it increases, charge also increases. As charge increases, electrostatic repulsion (Coulombic force) increases and hence flocculation decreases in the system under investigation. The amount of detergent required to reverse the charge is in the order LPC > CTAB > CPC > CPB (Table III).

Effect of Creaming

Brownian flocculation is dominating in the system as the particle diameter is approximately equal to $1 \mu m$ (i.e. $1.15 \mu m$). Particles under Brownian motion collide first and then coalesce to form larger size particles. Simultaneously these coalesced particles are creaming out due to the differences in the densities of the particles and the continuous phase. In Figures 7 and 8 per cent creaming is plotted against time in hours. In this experiment phase volume ratio was changed from 0.04 to 0.4 to measure creaming data accurately. It is obvious from these figures that the per cent creaming is faster initially but later on it is very slow as the coalescence becomes slow. It was noticed that per cent creaming rate follows the same trend and order as in the case of both flocculation and coalescence.

Kinetics of Coagulation of the system under investigation

One may conclude from the figures and the data that the emulsions are most stable in the presence of higher concentrations of CPB while with LPC reverse is the case. So emulsion stability was found to be in the order CPB > CPC > CTAB > LPC. Zeta potential values also support this type of behavior as its values are lower for LPC and higher for CPB. This comparative stability is also reflected from the stability factors W_1 of the system recorded in Table IV. Stability factor is greater for CPB and lower for LPC, e.g. for 8.0×10^{-4} M concentration of CPB and LPC, its values were found to be 11.96 and 8.44 (in kT) respectively. This order clearly indicates that the adsorption of these detergents on emulsion globules is in accordance with their chain length. The curves in Figs. 4 and 5 also shed light on the behavior of CPB and CPC which differ only in the head group, the former being more efficient.
To find whether flocculation or coalescence is the rate determining step in the system under investigation, the ratio of the half lives of these two processes was determined. Half life of flocculation is given by equation $1/K_f n_0$ whereas that of coalescence is $1/1.47 K_c$ (assuming it is kinetically of first order). The relative ratio is given by the ratio of the two half lives (i.e. $1.47 K_c/n_0 K_f$) (11). Flocculation or coalescence is more rapid according as this ratio is greater or less than unity. Since in the present case it

Table II. Kinetic Thermodynamic Parameters of the
Coalescence Process

Detergent $(2.0 \times 10^{-4}M)$	E_a (1)	$- F°$ (1)	$H°$ (1)	$S°$ (2)
CPB	4.6	−5.0	−27.3	73.9
CPC	6.2	−4.9	−10.4	18.1
CTAB	4.2	−3.8	−4.0	4.6
LPC	3.7	−4.3	−4.9	2.4

(1) $Kcal\ deg^{-1}\ mol^{-1}$ (2) $Cal\ deg^{-1}\ mol^{-1}$

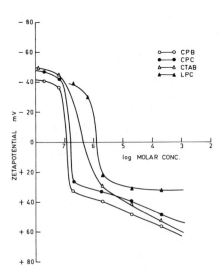

Figure 6. Plots of zeta potential as a function of log molar
concentration with different detergents.

Table III. Values of Zeta Potential for the System Flocculated
By Different Cationic Detergents

Concentration of Detergents		Zeta Potential (in mV)
CPB	2.0×10^{-8} M	40
	6.0×10^{-8} M	37
Charge	2.0×10^{-7} M	+35
Reversal	2.0×10^{-6} M	+39
	2.0×10^{-5} M	+48
	2.0×10^{-4} M	+56
CPC	2.0×10^{-8} M	48
	6.0×10^{-8} M	42
Charge	2.0×10^{-7} M	+31
Reversal	2.0×10^{-6} M	+33
	2.0×10^{-5} M	+39
	2.0×10^{-4} M	+48
CTAB	2.0×10^{-8} M	49
	6.0×10^{-8} M	36
Charge	2.0×10^{-7} M	+32
Reversal	2.0×10^{-6} M	+40
	2.0×10^{-5} M	+47
	2.0×10^{-4} M	+51
LPC	2.0×10^{-7} M	40
	2.0×10^{-6} M	30
Charge	2.0×10^{-5} M	+22
Reversal	6.0×10^{-5} M	+24
	2.0×10^{-4} M	+32
	6.0×10^{-4} M	+32

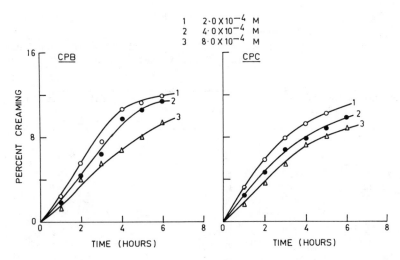

Figure 7.   Creaming % vs. time plots for phthalylsulfathiazole system.

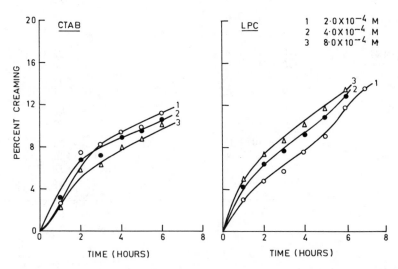

Figure 8.   Creaming % vs. time plots for phthalylsulfathiazole system.

Table IV.   Kinetic Parameters for the System Stabilized
By Phthallylsulfathiazole

Conc. of Detergent $\times 10^{-4}$ M	Coalescence Rate Constant (K_c)	Flocculation Rate Constant (K_f)	Creaming Rate Constant (Per Cent/ Hour)	Stability Factor (kT)
CPB	$K_c \times 10^5$ Sec^{-1}	$K_f \times 10^{14}$ Cm$^3 \times$ Sec^{-1}		
2.0	22.1	6.0	2.2	9.5
4.0	16.5	3.1	1.7	10.2
6.0	5.0	2.5	–	10.3
8.0	1.2	0.5	1.4	12.0
CPC				
2.0	27.5	8.0	2.4	9.2
4.0	19.5	5.5	1.3	9.6
6.0	9.0	3.5	–	10.0
8.0	4.5	1.5	1.6	10.9
CTAB	$K_c \times 10^4$ Sec^{-1}	$K_f \times 10^{13}$ CM$^3 \times$ Sec^{-1}		
2.0	18.2	1.9	3.7	8.3
4.0	15.5	1.7	3.4	8.4
6.0	12.0	1.5	–	8.6
8.0	8.5	1.3	2.9	8.7
LPC				
2.0	7.5	1.0	0.9	9.0
4.0	14.0	1.2	3.1	8.8
6.0	23.0	1.4	–	8.6
8.0	27.5	1.7	3.8	8.4

is greater than unity so coalescence is slower and hence it determines the rate of coagulation of these systems.

Conclusion

The emulgent, phthallylsulfathiazole provides highly stable type of emulsions. Even in the presence of cationic surfactants, these emulsions were found to be reasonable stable. Effect of detergents on the stability was in the order CPB < CPC < CTAB < LPC, which is compatible with their chain length. The rate determining step of coagulation kinetics of the system under investigation is the coalescence as this is slower than flocculation.

Acknowledgments

Authors are grateful to Prof. D. K. Chattoraj, Head, Department of Food Technology and Biochemical Engineering, Jadavpur University, Calcutta, for providing us laboratory facilities and for his valuable advice. We are also thankful to C.S.I.R., India for financial support as Junior Research Fellowship to one of us (S.R.).

Literature Cited

1. Cockbain, E. G. Trans Faraday Soc. 1952, 48, 185n.
2. Sharma, M. K.; Srivastave, S. N. Ann. Soc. Chim. Plonorum. 1975, 49, 2047.
3. Sharma, M. K.; Srivastava, S. N. Z. Phys. Chemie, Leipsig. 1977, 258, 2s, 235.
4. Reddy, S. R. Ph.D. Thesis, The University of Michigan, Michigan, 1980.
5. Reddy, S. R.; Fogler, H. S. J. Phys. Chem. 1980, 84, 1570.
6. Reddy, S. R.; Melik, D. H.; Fogler, H. S. J. Collid Interface Sci. 1981, 82, 116.
7. Friedlander, S. K. "Smoke, Dust and Haze, Fundamentals of Aerosol Behavior"; Wiley: New York, 1977.
8. Dass, K. P.; Chattoraj, D. K. Colloids and Surfaces. 1982, 5, 75.
9. Sharma, M. K.; Srivastava, S. N. Agra Univ. J. of Res (Sci.). 1974, 10, 153.
10. Smoluchowski, M. V. Zeit. Physik Chem. 1917, 92, 129.
11. Srivastava, S. N. Progr. Colloid and Polymer Sci. 1978, 63, 41.
12. Ghosh, S.; Bull, H. B. Arch. Biochem. Biophys. 1962, 99, 121.
13. Sherman, P. "Emulsion Science"; Sherman, P., Ed.; Academic Press: London and New York, 1968; 159.
14. Northrup, J. H.; Kunitz, M. J. Gen. Physiol. 1925, 925.

RECEIVED January 23, 1985

Colloidal Stability: Comparison of Sedimentation with Sedimentation Flocculation

D. H. MELIK[1] and H. S. FOGLER

Department of Chemical Engineering, The University of Michigan, Ann Arbor, MI 48109

We have developed simple analytical equations one can
use to estimate the overall stability of a
polydisperse colloidal system undergoing simultaneous
creaming and gravity-induced flocculation (note: our
approach can be easily extended to include other
particle loss mechanisms, such as shear-induced
flocculation and Brownian flocculation). As an
example of this analysis, we have studied the relative
particle loss rates of creaming and gravity-induced
flocculation under conditions of negligible
electrostatic repulsion. From this study we have
determined that the total particle concentration, the
particle size, and the particle size ratio are the
most sensitive operating parameters in controlling the
stability of a colloidal system undergoing
simultaneous gravity-induced flocculation and
creaming.

In quiescent media, polydisperse colloidal dispersions are broken by
the coupled mechanisms of creaming, Brownian flocculation, and
sedimentation (gravity-induced) flocculation (see Figure 1). In
creaming, particles either rise or sediment out of the system as a
result of the density difference between the particles and the
suspending medium. Encounters between particles occur because of
random Brownian motion, frequently resulting in particle aggregation
if the interparticle interactions are favorable. Gravity-induced
flocculation arises from the differential creaming rates of large
and small particles. The large particles sweep out the slower
moving small particles in their path, often resulting in particle
aggregation if the interparticle interactions are favorable. In the
case of emulsions, coalescence of the particles can occur after
flocculation.

In order to rigorously analyze the behavior of these particle
loss mechanisms on colloidal stability, one must solve the

[1]Current address: The Procter & Gamble Company, Miami Valley Laboratories, Cincinnati, OH 45247

1. Creaming:

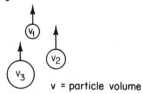

v = particle volume

2. Brownian Flocculation:

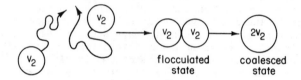

3. Sedimentation Flocculation:

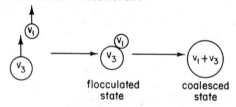

Figure 1. Colloidal breakage mechanisms in quiescent media

convective-diffusion equation which describes the distribution of particles around a central or reference sphere (1). Depending on the particle sizes and the net gravitational force, five important regimes can be identified:

(i) Both Brownian motion and interparticle forces (in particular, London van der Waals attractive and electric double layer repulsive) are negligible, but the effects of differential creaming are considerable. Under these conditions flocculation never occurs due to the hydrodynamic resistance of the fluid as the particles approach each other. However, orbital pairs do exist (2,3).

(ii) Brownian motion is negligible, but the effects of interparticle forces and differential creaming are significant (4).

(iii) Brownian motion is appreciable. Brownian motion, interparticle forces, and differential creaming must be taken into account. This case requires the complete solution of the convective-diffusion equation; a task which has yet to be accomplished.

(iv) Brownian motion and interparticle forces are significant, whereas the effects of differential creaming are moderate (2).

(v) The effects of creaming are negligible, but Brownian motion and interparticle forces are important (5-9).

The various flocculation models which are valid in the different regimes described above allow one to compute the particle/particle collision rate for any given particle sizes, chemical and physical condition. From the magnitude of this collision rate, one can estimate a colloidal system's stability in cases (iv) and (v). However, in cases (ii) and (iii), both flocculation and creaming will be important in the colloidal breaking process. Consequently, in order to determine whether a colloidal system will be stable in these two cases, we have to determine the net rate of particle loss due to both creaming and flocculation.

In this paper we propose a simple procedure whereby one can estimate the overall stability of a given colloidal system undergoing simultaneous creaming and flocculation. Since case (iii) has not been solved yet, we will focus our attention only on case (ii). As an example, we use this procedure to compare the net rates of particle loss due to creaming and gravity-induced flocculation when electrostatic repulsion is negligible.

Theory

For a polydisperse system of particles undergoing both creaming and gravity-induced flocculation, the time evolution of the particle size distribution, and hence the time rate of change of the total particle concentration, can only be determined from the solution of the governing population balance equations (10,11). For the special case of two different sized particles, this system of balance equations reduces to:

$$\frac{\partial N_1}{\partial t} = -\frac{\partial}{\partial x_1}(u_1 N_1) - G_{12}N_1 N_2 \qquad (1a)$$

$$\frac{\partial N_2}{\partial t} = -\frac{\partial}{\partial x_1}(u_2 N_2) - G_{12} N_1 N_2 \tag{1b}$$

$$\frac{\partial N_{12}}{\partial t} = -\frac{\partial}{\partial x_1}(u_{12} N_{12}) + \frac{1}{2} G_{12} N_1 N_2 \tag{1c}$$

where u_1 and u_2 are instantaneous creaming velocities for particles
of radius a_1 and a_2, respectively, u_{12} the instantaneous creaming
velocity for a doublet comprised of one particle of radius a_1 and
one of radius a_2, N_i and N_{ij} the respective particle concentrations,
G_{12} the gravity-induced collision kernel for particles of size a_1
and a_2, t the time, and x_1 the cartesian coordinate axis parallel to
the direction of gravity. The creaming terms can be approximated by
the following relationship (12,13):

$$\frac{\partial}{\partial x_1}(u_1 N_1) \simeq \frac{u_1 N_1}{h} \tag{2}$$

(with analogous expressions for the other creaming terms), where h
is the height of the suspending medium.

Since we are most interested in determining the stability of a
colloidal system, an analysis of Equation (1) is only required for
initial times. This is because if the system is unstable at t=0,
then it will be unstable for t > 0. Noting that as t → 0, $N_1 \to N_{01}$,
$N_2 \to N_{02}$, and $N_{12} \to 0$, and making use of Equation (2), for short
times Equation (1) reduces to:

$$\frac{dN_1}{dt} = -\frac{u_1 N_{01}}{h} - G_{12} N_{01} N_{02} \tag{3a}$$

$$\frac{dN_2}{dt} = -\frac{u_2 N_{02}}{h} - G_{12} N_{01} N_{02} \tag{3b}$$

$$\frac{dN_{12}}{dt} = \frac{1}{2} G_{12} N_{01} N_{02} \tag{3c}$$

Therefore, the total initial rate of particle loss due to creaming
is given by

$$R_{Cr} = \frac{1}{h}\{u_1 N_{01} + u_2 N_{02}\} \tag{4}$$

and the initial net rate of particle loss due to gravity-induced
flocculation is given by

$$R_{Gr} = \frac{3}{2} G_{12} N_{01} N_{02} \tag{5}$$

Equations (4) and (5) provide a basis for not only comparing the net rate of particle loss due to both creaming and flocculation, but the sum $(R_{Cr} + R_{Gr})$ allows one to determine the initial stability of a given colloidal system. It should be noted that the analysis presented above could easily be extended to completely polydisperse systems by summing Equation (3) over all pairwise combinations of particle sizes $a_1, a_2, \ldots, a_\infty$.

The rest of this paper will be devoted to studying the relative rates of creaming and gravity-induced flocculation, that is

$$\mathcal{R} = \frac{R_{Cr}}{R_{Gr}} = \frac{2(u_1 N_{01} + u_2 N_{02})}{3hG_{12} N_{01} N_{02}} \tag{6}$$

for the case of negligible electrostatic repulsion.

The gravity-induced collision kernel for spheres of radius a_1 and a_2 is given by (4):

$$G_{12} = \pi \alpha_{Gr} (u_{02} - u_{01})(a_1 + a_2)^2 \tag{7}$$

where u_{oi} is Stokes creaming rate for a particle of radius a_i, and α_{Gr} the gravity-induced flocculation capture efficiency which, for the case of negligible repulsion, can be satisfactorily computed from (4):

$$\alpha_{Gr} = f(\lambda) N_G^{-0.20} \tag{8}$$

with λ being the particle size ratio ($\lambda = a_1/a_2$), and $f(\lambda)$ is equal to 0.11 for $\lambda = 0.2$ and 0.17 for $\lambda = 0.5$. The dimensionless gravity parameter N_G describes the relative importance of gravitational to interparticle attractive forces and is given by:

$$N_G = \frac{\pi g \Delta \rho a_2^4}{3A} (1+\lambda)^2 (1-\lambda^2) \tag{9}$$

where g is the local acceleration of gravity, $\Delta\rho$ the density difference between the particles and the suspending medium, and A the system Hamaker constant. Stokes creaming rate is given by:

$$u_{oi} = \frac{2g \Delta \rho a_i^2}{9 \mu_f} \tag{10}$$

where μ_f is the viscosity of the suspending fluid.

The instantaneous creaming velocities for moderately concentrated systems can be expressed in the following form (3):

$$u_1 = u_{01}(1 - S_{11}\phi_1 - S_{21}\phi_2) \tag{11a}$$

$$u_2 = u_{02}(1 - S_{12}\phi_1 - S_{22}\phi_2) \tag{11b}$$

where ϕ_i is the particle volume fraction for a particle of radius a_i (note: $\phi_1 + \phi_2 < \simeq 0.1$), and S_{ij} the first order corrections due to particle/particle interactions (surface and hydrodynamic). The particle volume fractions are related to the particle concentrations through:

$$\phi_i = \frac{4}{3}\pi a_i^3 N_{oi} \tag{12}$$

If we assume that the creaming rates are only significantly affected by hydrodynamic interactions, then the coefficients S_{ij} can be computed from Batchelor and Wen's ($\underline{3}$) results. Using a simple linear regression, the following empirical expressions are obtained:

$$S_{11} = S_{22} = 6.55$$

$$S_{12} = 3.42\exp(0.497\lambda) \quad (0.25 \leq \lambda \leq 4.0, \; r^2 = 0.998)$$

$$S_{21}(\lambda) = S_{12}(\lambda^{-1}) \tag{13}$$

where r is the computed correlation coefficient.

The assumption that hydrodynamic forces dominate the effect of interparticle forces on the instantaneous creaming velocities is reasonable when one considers the following rationale. If the particles are close enough for the interparticle forces to be significant, the particles will undergo what can only be described as a sedimentation flocculation encounter and, therefore, would be accounted for in R_{Gr}.

By combining Equations (4)-(7) with Equations (10)-(13), one finds that the ratio of creaming to sedimentation flocculation rates $\mathcal{R}$ can be computed from:

$$\mathcal{R} = \frac{2f(a_2, N_{02}, \lambda, R_N)}{3\pi H \alpha_{Gr}(1-\lambda^2)(1+\lambda)^2 R_N} \tag{14}$$

where

$$f(a_2, N_{02}, \lambda, R_N) = \lambda^2 R_N \left\{ \frac{1}{a_2^3 N_{02}} - 14.3\exp(.497/\lambda) - 27.4\lambda^3 R_N \right\}$$

$$+ \frac{1}{a_2^3 N_{02}} - 14.3\lambda^3 R_N \exp(.497\lambda) - 27.4 \tag{15}$$

with $H = h/a_2$ and $R_N = N_{01}/N_{02}$

Results and Discussion

Of the various parameters which affect the particle loss ratio $\mathcal{R}$ defined in Equation (6), under conditions of negligible electrostatic repulsion, we have found that the total initial particle concentration N_{TOT} (where $N_{TOT} = N_{01}+N_{02}$), the size of the larger particle a_2, and the particle size ratio λ give the most interesting results. The effect of each of these parameters is discussed next.

Figure 2 shows the effect of the total particle concentration N_{TOT} on the relative rates of creaming to gravity-induced flocculation. (Note: $\nu=2\lambda_L/(a_1+a_2)$ is the dimensionless electromagnetic retardation parameter in which the characteristic London wavelength of the atoms, λ_L, is typically 10^{-5} cm. The parameter N_R is the dimensionless electrostatic number with $N_R=0$ indicating the absence of electrostatic forces. See reference 4 for complete details.) As expected, the higher the particle concentration the more significant the process of sedimentation flocculation becomes. What is interesting to note is the sensitivity of $\mathcal{R}$ to changes in N_{TOT}. For low particle concentrations, a 10-fold increase in N_{TOT} results in approximately a 10-fold decrease in $\mathcal{R}$. However, for higher particle concentrations this is not the case. Increasing N_{TOT} from 10^9 to 4×10^9 particles/cm^3 results in a 6.7-fold decrease in $\mathcal{R}$, indicating that the higher particle concentrations are more sensitive to changes in N_{TOT} than the lower particle concentrations. The presence of a minimum in Figure 2 is also to be expected. For small values of R_N, $N_{02} \gg N_{01}$ and for large values of R_N, $N_{02} \ll N_{01}$. In each of these extremes the effect of gravity-induced flocculation is reduced since there is either not enough collectors (a_1) for the given number of particles (a_2), or there are too many collectors. In either case, creaming begins to dominate the colloidal breaking process. When $N_{02} \simeq N_{01}$, the effect of gravity-induced flocculation attains its maximum contribution to the breaking process.

Figure 3 shows the effect of increasing particle size on the relative rates of creaming and gravity-induced flocculation. As the particle size increases, the rate of flocculation increases faster than the creaming rate of the particles. This is a result which, at first, one wouldn't expect. According to Equations (4), (10), and (11), the total rate of particle loss due to creaming R_{Cr} is approximately proportional to a_2^2 for a constant particle size ratio λ. On the other hand, combining Equations (5) and (7)-(10), one

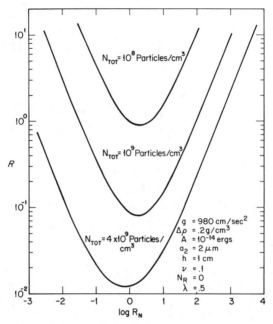

Figure 2. Effect of increasing total particle concentration on the
relative rates of creaming and gravity-induced
flocculation

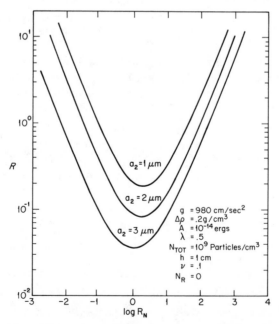

Figure 3. Effect of increasing particle size on the relative rates
of creaming and gravity-induced flocculation

finds that $R_{Gr} \propto a_2^{3.2}$ for constant λ.   Therefore, the ratio R is
approximately proportional to $a_2^{-1.2}$, indicating that as the
particle size increases, the rate of gravity-induced flocculation
increases more rapidly than the rate of creaming.

     Figure 4 shows how changes in the particle size ratio affect
the relative rates of creaming and gravity-induced flocculation.
When the concentration of smaller particles is less than the
concentration of larger particles ($R_N < 1$), the larger of the two
particle size ratios reduces the rate of gravity-induced
flocculation as compared to the creaming rate of the particles.   On
the other hand, when the concentration of the smaller particles is a
bit greater than that of the larger particles ($R_N > \approx 3.2$), the
situation is reversed.   Smaller particle size ratios favor gravity-
induced flocculation over creaming more than larger particle size
ratios.

<u>Summary</u>

We have presented a simple procedure whereby one can estimate the
stability of a colloidal system undergoing simultaneous creaming and
gravity-induced flocculation.   This procedure is by no means
restricted to only this case.   One can easily take into account
other particle loss mechanisms, such as shear-induced flocculation
or Brownian flocculation.   What is required in these cases are the
appropriate particle/particle collision kernels, which can be
computed by solving the governing convective-diffusion equation.

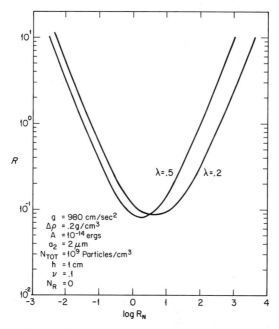

Figure 4.   Effect of the particle size ratio on the relative rates
            of creaming and gravity-induced flocculation

Literature Cited

1. Melik, D.H.; Fogler, H.S. J. Colloid Interface Sci. 1984, (In press).
2. Wacholder, E.; Sather, N.F. J. Fluid Mech. 1974, 65, 417.
3. Batchelor, G.K.; Wen, C.-S. J. Fluid Mech. 1982, 124, 495.
4. Melik, D.H.; Fogler, H.S. J. Colloid Interface Sci. 1984, (In press).
5. Derjaguin, B.V.; Muller, V.M. Dokl. Akad. Nauk. S.S.S.R. (English Translation) 1967, 176, 738.
6. Spielman, L.A. J. Colloid Interface Sci. 1970, 33, 562.
7. Roebersen, G.J.; Wiersema, P.H. J. Colloid Interface Sci., 1974, 49, 98.
8. Feke, D.L.; Prabhu, N. 1982 AIChE Annual Meeting, Los Angeles, Ca. (Nov. 14-19).
9. Valioulis, I.A.; List, J. Adv. Colloid Interface Sci. 1984, 20, 1.
10. Reddy, S.R.; Melik, D.H.; Fogler, H.S. J. Colloid Interface Sci. 1981, 82, 116.
11. Reddy, S.R.; Fogler, H.S. J. Colloid Interface Sci. 1981, 82, 128.
12. Friedlander, S.K. "Smoke, Dust, and Haze: Fundamentals of Aerosol Behavior"; Wiley: New York, 1977; pgs. 282-286.
13. Reddy, S.R.; Fogler, H.S. J. Colloid Interface Sci. 1981, 79, 105.

RECEIVED December 26, 1984

30

Rheology of Concentrated Viscous Crude Oil-in-Water Emulsions

YEIN MING LEE, SYLVAN G. FRANK, and JACQUES L. ZAKIN

Department of Chemical Engineering, Ohio State University, Columbus, OH 43210

Economic pipeline transport of viscous crudes as concentrated oil-in-water emulsions has been demonstrated in at least two commercial pipelines. The present study was undertaken to learn more about the rheological characteristics of concentrated emulsions and the effect of such variables as emulsion formulation and preparation techniques, aging, and crude oil viscosity on emulsion properties.

Two crude oils were used, a California crude with a viscosity of 24 poise at 25°C and a Canadian crude of 164 poise. Both could be emulsified by the addition of NaOH which reacted with the acids present in the crude. A series of oil-in-water emulsions containing 60% (by volume) of oil were prepared. Concentration of NaOH and NaCl and mixer speed were varied. Emulsion stability was measured as was particle size distribution and viscosity and the effect of aging on the latter two. Emulsions of the heavier crude had viscosities about 600 times smaller than the crude viscosity.

With no salt present, moderately stable emulsions of the lower viscosity oil could be prepared at NaOH concentrations as low as 3.0 x 10^{-5} moles NaOH/gram oil. In the presence of 1.0% NaCl in the water, emulsions with NaOH contents above 5.0 x 10^{-5} were less stable than those with lower NaOH contents. In general average particle diameters decreased as NaOH concentration increased, with slightly lower particle sizes when <1.0% NaCl was present. Viscosities of the fresh emulsions were lowest at the lowest NaOH contents with or without salt.

Aging generally caused increase in the average particle diameters and reduction in the viscosities of salt-free emulsions, and increase in the average particle diameters and increase in the viscosities of those containing 1.0% NaCl.

Equations proposed by Sherman for predicting viscosities from apparent volume fractions and particle diameters were useful in analyzing the effects of formulation and preparation variables and aging on emulsion viscosities.

The immobilization of water by strong charges on the dispersed oil droplets or by entrapment by flocculation or by the trapping of

0097-6156/85/0272-0471$06.00/0

small water droplets inside the oil droplets all may contribute to
increase in the value of the apparent volume fraction of oil.
Emulsion viscosity is very sensitive to this quantity.

The presence of small amounts of NaCl lowered the viscosity of
two emulsions of the lower viscosity crude (3.0×10^{-5} and 4.0×10^{-5}
moles NaOH/gram oil) by 13% presumably by reducing the charge on the
oil droplets and the immobilized water. Both minima occurred at about
the same Na^+ concentration (NaCl concentrations of 0.40% and 0.25%,
respectively).

Significant reserves of heavy viscous crude oil exist in the United
States, in Venezuela, and in Canada. These crudes are becoming of
increasing importance as lower viscosity crudes are depleted, and the
production of heavy crudes is expected to increase significantly in
the next ten to twenty years. Because of their high viscosities (up
to 10^5 centipoise at $25°C$), transport of these crudes cannot be
accomplished by the usual pipeline methods. A number of techniques
for their transport in pipelines have been proposed including:
dilution with lighter oils, preheating with subsequent heating of the
pipeline, preheating and insulating the pipeline, injecting a water
sheath around the viscous crude, and encapsulation (3, 24, 29). Each
of these has serious practical or economic drawbacks as described, for
example, in References (3) and (16).

Transport as concentrated oil-in-water emulsions is another
promising technique first described for an Indonesian pipeline in 1963
(6). It has also been used in a 13-mile long, 8-inch diameter
pipeline by Getty Oil in California. Marsden (7-11) Sifferman (25)
and others (18, 26) have described specific techniques for oil-in-
water transport for particular applications. Little research has been
done, however, on concentrated high viscosity crude oil-in-water
emulsions to determine their rheological characteristics, the
variables controlling their rheology, and their potential for use for
crude transport.

In addition to the application to pipeline transport of heavy
crudes, there is also considerable interest in downhole emulsification
for heavy crude oil production (2, 13, 27, 28). Here aqueous solutions
of surfactants are added to the tubing-casing annulus of wells
producing heavy oils and water. Oil-in-water emulsions of relatively
low viscosity are formed resulting in increased production rates.

This paper reports results on one phase of a program of study of
concentrated oil-in-water emulsions, namely the effect of several
composition variables on the apparent viscosity of concentrated (60
volume percent) crude oil-in-water emulsions. In particular, the
effects of NaOH content, NaCl content, emulsion preparation technique,
crude oil viscosity and aging of emulsion were studied.

Experimental

Two crude oils were used; one was a California crude donated by Shell
with density of 0.953 gm/cm^3 and viscosity of 24 poise at $25°C$, the
other was a St. Lina (Canada) crude donated by Amoco Production
Company with a density of 0.987 gm/cm^3 and viscosity of 164 poise at
$25°C$. All of the experiments described here were run at 60% volume
fraction of oil.

It was observed by S. Westfall (30) that stable oil-in-water emulsions of the Shell crude could be prepared by addition of NaOH which reacted with the acids present in the crude. This was also found to be true for the St. Lina crude. Thus, surfactant emulsifiers were not needed and the emulsions studied here were prepared using 1.0 to 8.0 x 10^{-5} moles NaOH per gram of crude oil. The procedures used were:

(1) Weigh the oil to $\pm$ .05 gram and add sufficient volume of the aqueous phase to provide a 60% by volume 45.0 ml mixture. Preheat the crude oil to 65°C to reduce the viscosity and facilitate the mixing of oil and water phases.

(2) Using an Omni Mixer containing the proper amount of heated oil, add the water and NaOH solution and stir at 9,000 rpm for 180 seconds. The mixer speed was varied between 7,000 and 11,000 rpm in one set of experiments.

(3) After homogenization, the emulsion was cooled to room temperature.

(4) In a number of experiments, the NaCl content of the water phase was varied to determine its effect on the apparent viscosity of the emulsion. NaCl contents ranged from 0.1% to 2.0%.

(5) For pH measurements, the pH meter was calibrated beforehand and the value for the emulsion was measured.

(6) To determine the emulsion stability, 60 ml test tubes were filled with 45 ml of emulsion and separation of a lower layer was monitored up to 10 days. Stable emulsions showed only one or two milliliters of separation in 10 days; unstable emulsions showed 15 or more milliliters of separation after 10 days.

(7) For particle size analysis, one drop of concentrated emulsion was added to 25 ml of double-distilled water. Particle size analyses were made using a "Micro-Computerized Elzone/ADC-80 XY", which is operated by passing the particles through an orifice and measuring conductivity changes. A detector can count the number of particles of each size range. The number of particles in each of 128 divisions of the size range is read and a histogram and various averages can be obtained. Approximately 8000 particles were measured for each sample.

Contamination will result in significant error, so the final dilution water (50:1) and the 0.9% salt water used for increasing the conductivity for this analysis must be triple filtered through 0.22 micron filter paper to obtain particle-free water. This water was checked by passing it through the orifice before using it and observing any particle size reading. Nitric acid, detergent, toluene and acetone were used to clean the orifice. The emulsion sample for the test was so dilute that it was almost as clear as water ensuring that no two drops would pass through the orifice simultaneously as this would result in an over-estimate of the particle size and in possible blockage of the orifice. Each sample was analyzed at least three times.

(8) Viscometric measurements were made with a Haake Rotovisco MV 1 system with a rotor of diameter 40.08 mm, and length of 60 mm and a cup of 42.00 mm inside diameter. The shear stress was detected using a Head 50. Before measurements, calibrations were made with standard oils to obtain the instrument constant to convert stress readings to shear stress. Shear rate could be varied over the range of 50.74 to

685 reciprocal seconds. The temperature of the test parts of the
Haake was controlled at $25^{\circ}C$. Low shear (50 seconds^{-1}) measurements
were made first and shear rate was increased in steps to 685
seconds^{-1}. Readings were taken at steady state.

Theoretical

In describing the viscosity of an emulsion, the volume fraction of
the dispersed phase is the most important parameter. A model
suggested by M. Mooney (15) in 1951 for solid suspensions and
emulsions with highly viscous dispersed phase described the relative
viscosity as a function of volume fraction and a coefficient, k,
called the self-crowding factor.
    S. Matsumoto and P. Sherman (12) extended the Mooney model to
microemulsion systems. They tried to relate the coefficient, k, to
average particle size and developed an empirical correlation. The
numerator of 2.5, in the original equation is replaced by the
variable, a, which is a function of the ratio of the viscosity of the
dispersed phase to the continuous phase, and is predicted by Taylor's
hydrodynamic theory. The value of a can vary from 1.0 to 2.5
depending on the circulation of fluid within the particles. The
value of 2.5 holds for solid particles and very viscous dispersed
liquid phases such as those used in these experiments. The relative
viscosity, η_R is expressed as:

$$\eta_R = \exp\left[a\phi/(1-k\phi)\right], \qquad \phi = \phi_o \quad \phi_s \qquad\qquad [1]$$
$$\text{where } a = 2.5\ (P+0.4)/P+1.0);\ P = \eta_i/\eta_o$$
$$k = 1.079 + \exp\ (0.01008/Dm) + \exp\ (0.0029/Dm^2)$$

ϕ_o is the nominal volume fraction of dispersed phase and ϕ_s is volume
fraction of the dispersed phase solubilized in the micelles of excess
emulsifier, η_i is viscosity of the dispersed phase, η_o is viscosity
of the continuous phase, and Dm is the mean particle size. Since k
values predicted are always greater than one, this relationship is
not suitable for macroemulsions where k values can be less than one (5).
    C. Parkinson et al. (17) considered the effect of particle size
distribution on viscosity. They studied suspensions of poly(methyl-
methacrylate) spheres in Nujol with diameters of 0.1, 0.6, 1.0 and
4.0 microns with different volume fractions and with different
particle size combinations to determine the influence of size
distribution on the viscosity. Each particle size gave a certain
contribution to the final viscosity based on the volume fraction and
the hydrodynamic coefficient obtained from the empirical equation for
that particle size. The contributions were expressed in the same
form as in Mooney's model, and the viscosity was calculated from the
product of each term.

$$\eta_R = \prod_{i=1}^{\eta} \exp\ (2.5\phi_i/(1.0-k_i\phi_i)) \qquad\qquad [2]$$

    Thus, both volume fraction and particle size affect viscosity.
For a given volume fraction, the smaller particle dispersions gave
higher viscosity. This may be due to more particle-particle
interactions because of the larger interfacial area. Parkinson et
al., varied the fraction of small particles (0.1 micron) in 11% by
volume suspensions. Increasing the concentration of the 0.1 micron

particles, above 25% volume fraction, increased the viscosity. However, when the volume fraction of 0.1 micron particles was less than
25% of the total in the 11% suspension, increase in the 0.1 micron
volume fraction resulted in a decrease in viscosity. They concluded
that at low fraction of small particles, they behaved like ball
bearings between the larger particles. This may explain why emulsions
with broad particle size distributions have lower viscosities than
those with narrow particle size distributions and the same average
particle size (4).

Sherman (19-22) did a series of studies of the rheological
properties of emulsions, which include oil-in-water (O/W) and water-
in-oil (W/O) at both high and low shear rates. The W/O system was a
water in Nujol emulsion stabilized by 1.5% (by weight) sorbitan
monooleate and the O/W system was Nujol in water stabilized by
Aerosol OT or sodium oleate. Both systems showed higher viscosity
at low shear rates than at high shear rates probably due to
flocculation effects. The W/O system showed a decrease in viscosity
with aging presumably due to coalescence. For O/W systems viscosity
increased to a maximum after three or six days (even though the mean
particle size increased continuously), then fell sharply to a value
which changed only slowly with time.

Sherman concluded that in the O/W systems a small fraction of the
continuous phase was immobilized by the dispersed phase either by
attractive forces or by flocculation. Therefore, the apparent volume
fraction was greater than the actual volume fraction of that component.
The equations he used to describe the viscosities of these emulsions
further extended the approach of Mooney and took account of the
particle diameter.

$$\eta_R = \exp{(C/Am - 0.15)}$$
$$Am = Da \left((\phi_{max}/\phi_a)^{1/3} - 1.0\right) \qquad [3]$$

where Am is a mean distance of separation such that further increase
in the shear rate causes no change in viscosity and C is a function of
particle diameter, Da.

$\log C = 0.153 \, Da - 1.01$ for Da greater than 2 microns,
ϕ_{max} is 0.7403, the maximum volume fraction for uniform spheres, and
ϕ_a is the apparent volume fraction of the dispersed phase.

These equations relate the relative viscosity to particle size
and apparent volume fraction and are good for solid spheres in fluid
media as well as for emulsions. It should be noted that these
relations were derived from data for monodispersed systems with
diameters greater than one micron. While polydisperse systems which
have the same average diameter as monodisperse systems show lower C
values and hence lower viscosities (23), it is difficult to establish
an equation to describe the relation between Da and C for non-uniform
spheres. Therefore, the relation reported by Sherman for uniform
spheres has been assumed in the present work permitting values of ϕ_a
to be estimated from experimental values of η_R and D_m. Relative
viscosity as a function of particle diameter with volume fraction as a
parameter computed from Equation 4 is shown in Figure 11. Note that
viscosity is very sensitive to small changes in ϕ_a and that this model
does not predict a monotonic decrease in emulsion viscosity as particle
size increases but rather a minimum value followed by a gradual
increase. At higher values of ϕ_a, i.e., 0.62, the rise is quite steep.

Results
A. The Effect of NaOH Concentration on Shell Crude Emulsion Properties
    (1) Stability
Emulsions were prepared with the Shell crude and distilled water with
sodium hydroxide concentrations varying from 1.0×10^{-5} to 8.0×10^{-5}
moles NaOH/gram oil. The emulsions were very stable and no
separation into two phases occurred when NaOH concentration was above
4.0×10^{-5}. At 4.0×10^{-5} and lower NaOH concentrations, some
separation occurred after one or two days but even at 3.0×10^{-5}
moderate stability was observed. Microscopic examination showed that
the top layer contained a high fraction of oil with large particle
sizes predominating and the bottom layer contained a high fraction
of water with the small particle size oil droplets predominating.
The oil-rich layer eventually coagulated to form an oil layer; this
type of separation was described as mixed-particle-size emulsion by
Tadros and Vincent (1). The addition of sodium chloride to the
water used in making the emulsions gave faster emulsion separation
especially for high NaOH (5×10^{-5} or greater) concentration
emulsions. When the NaCl content exceeded 2.0%, no homogeneous
emulsion could be obtained as separation occurred immediately.
    (2) pH value
pH values of fresh emulsions increased as NaOH concentration
increased as shown on Fig. 1. The presence of NaCl caused a drop in
the pH value of fresh emulsions of the same NaOH concentration at all
NaCl concentrations (Figure 2).
    (3) Particle Size
Average particle diameter of fresh emulsion increased as NaOH concentration decreased as shown on Fig. 3 for NaCl-free emulsions.
Average particle diameters ranged from 2.4 to 3.7 microns. Addition
of NaCl resulted in a decrease in average particle size of fresh
emulsion at low NaOH concentration as shown on Fig. 4.
    (4) Viscosity
All of the emulsions appear to be Newtonian in the range of shear
rates studied (Fig. 5). No significant change in the viscosity of
the fresh emulsions was observed at NaOH concentrations above $6.0 \times$
10^{-5} moles NaOH/gram oil, but the viscosity did decrease as NaOH
concentration decreased, particularly below 4×10^{-5}, as shown in
Table I and Fig. 5.
    The addition of increasing amounts of NaCl gave a minimum in
apparent viscosity (685 seconds^{-1}) which occurred at concentrations
of 0.25% and 0.4% for 4.0×10^{-5} and 3.0×10^{-5} moles NaOH/gram oil
(fresh) emulsions (see Table II and Fig. 6.). Of particular interest
is the fact that the NaCl concentrations giving the lowest apparent
viscosities for the 3.0 and 4.0×10^{-5} NaOH emulsions were close to the
lowest pH values (see Figure 6). At higher NaCl concentrations
(about 0.75% or higher) apparent viscosities of the emulsions were
greater than the salt-free emulsion (see Table I and Figure 7).

B. The Effect of Aging on Shell Crude Emulsion Properties
    (1) Stability
Some separation into two layers was observed for all samples after
ten days aging, but the rate of separation and the amount of
separation were quite different for different formulations. Addition
of NaCl made the emulsions more unstable and in many cases separation

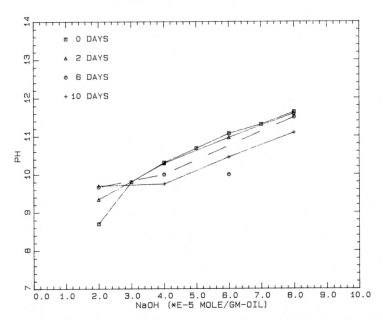

Figure 1. Effect of NaOH Content and Aging on pH Values for 60% Shell Crude Emulsions.

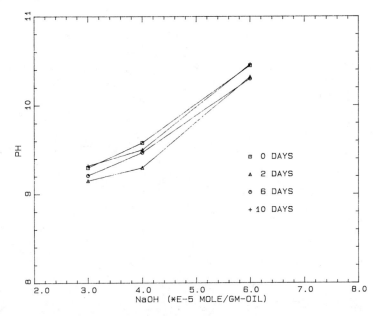

Figure 2. Effect of NaOH Content and Aging on pH Values for 60% Shell Crude Emulsions (NaCl=1.0%).

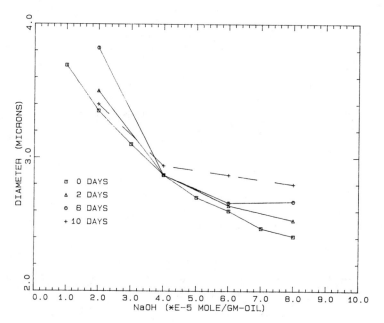

Figure 3. Effect of NaOH Content and Aging on the Average
Diameters of 60% Shell Crude Emulsions.

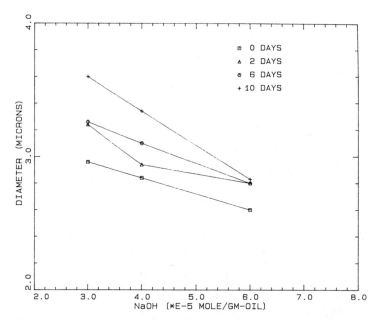

Figure 4. Effect of NaOH Content and Aging on the Average
Diameters of 60% Shell Crude Emulsions (NaCl = 1.0%).

Table I.   Effect of NaOH, NaCl and Preparation Technique
on Emulsion Properties

moles NaOH/ gm-oil x 10^5	Stirring Speed rpm	NaCl %	Apparent viscosity(cp) 685 seconds^{-1}	Dm microns	Apparent Volume ϕ_a
8.0	9000	0	20.7	2.41	0.6135
7.0	9000	0	21.3	2.47	0.6148
6.0	9000	0	20.9	2.60	0.6148
5.0	9000	0	20.0	2.70	0.6136
4.0	9000	0	19.6	2.87	0.6130
3.0	9000	0	19.0	3.10	0.6115
2.0	9000	0	16.5	3.35	0.6053
1.0	9000	0	15.1	3.69	0.6000
6.0	9000	1.00	25.6	2.60	0.6211
4.0	9000	1.00	23.0	2.84	0.6182
3.0	9000	1.00	20.2	2.96	0.6140
4.0	7000	0	19.8	3.10	0.6128
4.0	9000	0	19.6	2.87	0.6130
4.0	11000	0	22.3	2.80	0.6173

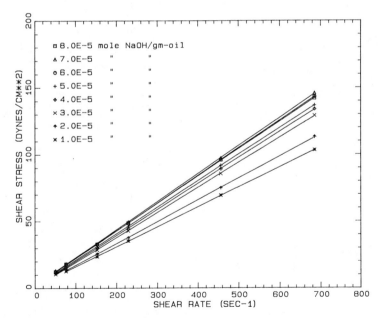

Figure 5.   Effect of NaOH Content on Viscosity of 60% Shell
Crude Emulsions.

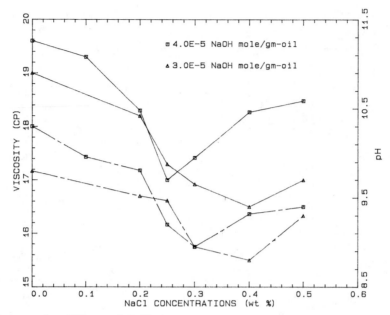

Figure 6. Effect of NaCl Content on Viscosity and pH for 60% Shell Crude Emulsions.

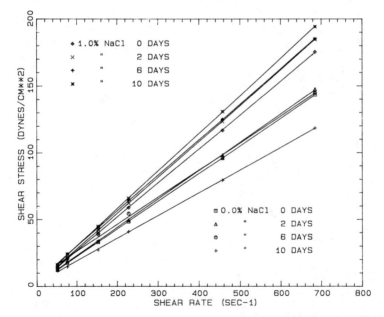

Figure 7. Effect of NaCl Content and Aging on Viscosity of 60% Shell Crude Emulsions (6.0 x 10^{-5} moles NaOH/gram oil).

occurred within a few hours and in some cases immediately. The
influence of NaCl on stability is actually more significant for high
NaOH systems than for low NaOH systems. For example, emulsions containing 6 x 10^{-5} moles NaOH/gram oil with 1.0% NaCl are less stable
than emulsions made with 3.0 or 4.0 x 10^{-5} moles NaOH/gram oil and
1.0% NaCl.

(2) pH Value

As shown on Fig. 1, pH values decreased during aging for the emulsions
made with NaOH concentrations above 3.0 x 10^{-5} moles NaOH/gram oil
presumably due to further completion of the reaction of the NaOH. 2.0
x 10^{-5} emulsion shows the opposite effect but this emulsion is quite
unstable and the results are suspect. For emulsions with NaCl, pH
values decreased during the first two or three days and then
increased as shown in Fig. 2.

(3) Particle Size

For emulsions made with very high NaOH concentration, the average
particle size increased more rapidly with time than for the emulsions
with NaOH concentrations of 4.0 x 10^{-5} moles NaOH/gram oil as shown
in Fig. 3. The unstable 2 x 10^{-5} emulsion also showed larger
increases in diameter with time. The average particle sizes of
emulsions with 1.0% NaCl increased most rapidly with time at low NaOH
contents (Fig. 4). Since in these cases larger particles migrated to
the upper separated layer within one or two days, the average particle
sizes for the lower layers reported here are lower than the overall
averages.

(4) Viscosity

Generally speaking, the viscosity of the moderate to high NaOH content
emulsions without salt increased slightly the first two to six days
and then decreased (see Fig. 7). For the emulsions made with low
NaOH content, the viscosity of the lower layer increased but
separation was so great that the aging results are not reliable. For
the emulsions with salt, the increase of viscosity was continuous for
the ten days as shown in Fig. 7.

C. Effect of Mixing Speed on Emulsion

Different mixing speeds were used to see the effect of mechanical
forces on the emulsion properties. As expected, at high speed
(11,000 rpm), average particle size obtained was smaller than at
low speed (7,000 rpm) and the viscosity was higher, as expected (see
Table I). The lowest viscosity was not observed at the lowest mixing
speed, however, but at 9,000 rpm. This surprising result is
discussed later.

D. Effect of the Nature of Crude Oil

Two crude oils were used in these experiments: Shell California oil
(2400 cp at 25°C) and Amoco St. Lina (Canadian) oil (16,400 cp at
25°C). For typical emulsions, the Shell oil (3.0 x 10^{-5} moles NaOH/
gram oil) gave small average particle size (3.1 microns) and narrow
size distribution while the Amoco oil (NaOH = 3.0 x 10^{-5} moles NaOH/
gram oil) gave large average particle size (7.3 microns) and broad
size distribution using the same preparation techniques (see Fig. 8).
Furthermore, the viscosities of the California crude emulsions were

around 21 cp and those of the St. Lina emulsions were around 30 cp as
shown in Fig. 9 for two formulations of each. The difference may be
due to the difference in viscosity of the two crude oils, to differences in the average particle size, or to different apparent oil
volumes as discussed below.

Discussion

For pipeline transport, highly stable emulsions are not necessary as
some redispersion will occur due to turbulence in the pipe and to
passage through booster pumps. Three days down time is the normal
design criterion (25)(29). Thus, for example, the stability of Shell
crude emulsions without salt prepared by a technique equivalent to
that used here with about 3.0 to 4.0 x 10^{-5} moles NaOH/gram oil would
be more than adequate and higher NaOH concentrations are not needed.
While the addition of NaCl had a greater adverse effect on stability
at higher NaOH concentrations, even at 3.0 or 4.0 x 10^{-5} concentrations the presence of 1.0% NaCl would make the emulsions only
marginally suitable. For high salt concentrations in the water,
improved formulations would be necessary. For the St. Lina crude 2.0
to 3.0 x 10^{-5} moles NaOH/gram oil would be adequate in salt-free
emulsions. (The St. Lina crude emulsions without NaCl were unstable
when NaOH content reached 4.0 x 10^{-5} moles NaOH/gram oil. Stable
emulsions of this crude could not be prepared with 1% NaCl.)
    For emulsions containing no salt the relative viscosities
increased and the average particle sizes decreased with increasing
NaOH content. However, once a sufficient quantity of NaOH was
presented to form stable emulsions (4.0 x 10^{-5} moles NaOH/gram of oil
for the Shell crude) further increases in NaOH content gave only
small increases in apparent viscosity and moderate decreases in
particle size.
    The effect of oil viscosity on initial emulsion viscosity is not
clear from these experiments. The St. Lina crude is about six times
as viscous as the California crude. The apparent viscosity of the
lower viscosity St. Lina Crude emulsion (2 x 10^{-5} moles NaOH/gram oil)
is less than 50% greater than the lowest viscosity moderately stable
California crude emulsion (4.0 x 10^{-5} NaOH). The average particle
size of the St. Lina emulsion is 7 microns while that of the Shell
crude emulsion is about 3 microns (see Figure 8). Since particle
sizes, particle size distributions and types of oil are different, no
conclusions can be drawn about the influence of oil viscosity. There
is, however one fact which should be emphasized, namely that
viscosities 600 times lower than that of the crude were observed for
60% St. Lina crude emulsions.
    The presence of NaCl has an important effect on initial
viscosities. As seen in Figure 6 for Shell crude emulsions containing 3.0 and 4.0 x 10^{-5} moles NaOH/gram oil, the addition of NaCl
first lowers the viscosities which go through minima and then rise
with further addition of NaCl. At the optimum NaCl contents,
viscosities were about 13% lower than the salt-free emulsions.
    At lower Na+ concentration, the high negative charge (zeta
potentials of salt-free emulsion droplets are 115 mv and above (5))
immobilizes a water layer around the droplets, thus effectively
increasing the apparent volume fraction. Addition of Na+ reduces the
net charge and some of the immobilized water is released giving a
lower apparent volume fraction and lower apparent viscosity.

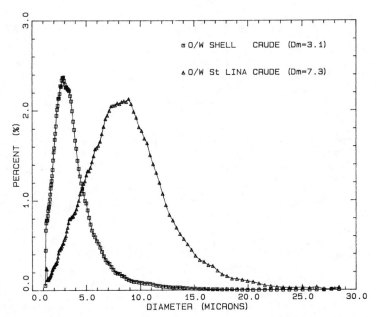

Figure 8. Particle Size Distributions of Two Crude Oil Emulsions (Shell Crude 3.0 x 10⁻⁵ moles NaOH/gram oil, St. Lina Crude 3.0 x 10⁻⁵ moles NaOH/gram oil).

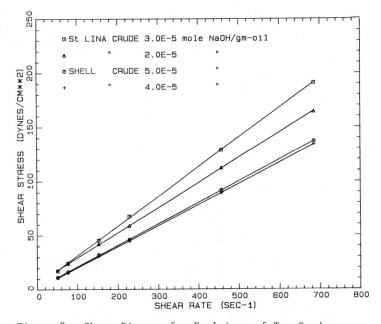

Figure 9. Shear Diagram for Emulsions of Two Crudes.

Beyond the minimum viscosity, the increase in viscosity with addition of NaCl is believed to be related to further reduction in the net (negative) charges on the dispersed droplets. Reduction of the net charge on the oil particles promotes particle-particle interaction leading to flocculation which (as discussed below) will promote viscosity increase. Hence the total Na+ content from NaOH and from NaCl acts to cause an increase in viscosity above the minimum. The location of the minima varied with the initial NaOH content of the emulsion but occurred at almost the same total Na+ concentrations (Fig. 10).

At the same total Na+ concentration, the viscosities of emulsions made from 3.0×10^{-5} mole NaOH/gram oil are always less than those from 4.0×10^{-5}. This may be due to the fact that 4.0×10^{-5} emulsions have a higher concentration of surfactant and smaller particle sizes than 3.0×10^{-5} emulsions. These two factors induce higher apparent volume fractions as shown on Table II.

These effects can be illustrated by estimating values of ϕ_a from the Sherman model [3] from values of relative viscosity and average particle diameter. In Table II we can see the influence of NaCl concentration on ϕ_a for the two emulsions shown in Figure 6. At the NaCl concentrations of minimum viscosity, values of ϕ_a also show minima.

Figure 11 shows all of the fresh emulsion data obtained in this study compared with Sherman's model. Values of ϕ_a for these emulsions ranged from 0.60 to 0.62. The 1.0% NaCl emulsions generally had the highest relative viscosities and the highest ϕ_a values; the salt-free emulsions were intermediate; and the emulsions with NaCl contents below 0.5% had the lowest values. The latter were for NaOH levels of 3.0 and 4.0×10^{-5} moles NaOH/gram oil for which viscosity minima were observed.

The factors affecting ϕ_a are complex. Under different conditions water immobilization by attractive forces, trapping of water particles within the dispersed oil droplets (photomicrographs indicated that in some cases water particles were dispersed in oil droplets), trapping of water by flocculation, and release of water when oil particles coalesce all come into play. Since relative viscosity is so sensitive to ϕ_a, small changes in any of these effects with composition or with time can cause significant viscosity changes.

In the experiment in which mixing speeds were varied, the average particle diameter decreased and apparent volume fraction increased with increasing rpm. The apparent viscosity for the 9000 rpm emulsion, however, was slightly lower than that for the 7000 rpm. So, as shown by the Sherman model, the viscosity depends on both particle size and apparent volume fraction. Information on both are required to predict the viscosity of a system.

Conclusions

1. Stable concentrated oil-in-water emulsions of two viscous crudes were made by adding NaOH which reacts with acids present in the oils.
2. NaOH requirements for stable emulsions and the emulsion properties are specific to the nature of crude. High NaOH concentrations do not give the most desirable emulsion characteristics.
3. Emulsions at least 600 times lower in viscosity than viscous crudes at 60% oil concentration can be made. It is not clear how oil viscosity affects emulsion properties.

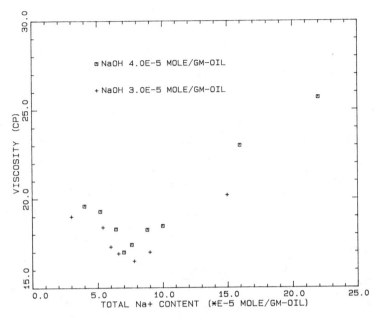

Figure 10. Effect of Total Sodium Ion Content on Viscosity for Two 60% Shell Crude Emulsions with NaCl.

Table II. Effect of NaCl Content on Emulsion Properties

molesNaOH/ gm-oil x 10^5	NaCl %	Visco. (cp)	Dm microns	ϕ_a
4.0	1.50	25.7	3.10	0.6212
4.0	1.00	23.0	2.84	0.6182
4.0	0.50	18.5	2.64	0.6109
4.0	0.40	18.3	2.70	0.6105
4.0	0.30	17.4	2.70	0.6088
4.0	0.25	17.0	2.73	0.6080
4.0	0.20	18.3	2.78	0.6107
4.0	0.10	19.3	2.80	0.6125
4.0	0.00	19.6	2.87	0.6130
3.0	1.00	20.2	2.96	0.6140
3.0	0.50	17.0	2.73	0.6080
3.0	0.40	16.5	2.78	0.6069
3.0	0.30	16.9	2.82	0.6079
3.0	0.25	17.3	2.94	0.6086
3.0	0.20	18.4	2.99	0.6107
3.0	0.00	19.0	3.10	0.6115

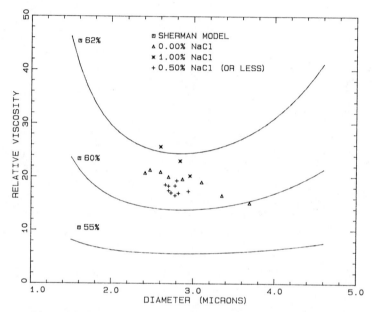

Figure 11.   Relative Viscosity vs. Particle Diameter for 60% Fresh
Shell Crude Emulsions.

4. Average particle size and particle size distribution and salt concentration are other variables which affect emulsion viscosity. Viscosities can be lowered by proper control of these variables.

5. The use of equations proposed by Sherman for predicting emulsion viscosities from apparent volume fractions and particle diameter of the dispersed phase is useful in analyzing the effects of formulation and preparation variables and aging on emulsion viscosities.

Literature Cited

1. Encyclopedia of Emulsion Technology, P. Becker, ed., Chapter by R.F. Tadros and B. Vincent, Marcel Dekker, New York (1983).

2. Boyer, A.H., SPE paper 2676, presented to Society of Petroleum Engineers, Denver meeting (1969).

3. Ells, J.W., and V.R.R. Brown, J. Inst. Pet., 57, 175-183 (1971).

4. Fruman, D.H. and J. Braint, "Investigation of Rheological Characteristics of Heavy Crude Oil in Water Emulsions," presented to International Conference on the Physical Modelling of Multi-Phase Flow, Coventry, England, April (1983).

5. Lee, Y.M., "Rheological Studies of Concentrated Oil-in-Water Emulsions," M.S. Thesis, The Ohio State University, Columbus, 1984

6. Lamb, M.S. and W.C. Simpson, Proc. Sixth World Petroleum Congress, Section VII, 23-33 (1963).

7. Marsden, S.S., and R. Raghaven, Paper 72-Pet-42 presented to Petroleum Div., ASME, Sept. (1972).

8. Marsden, S.S., and S.C. Rose, U.S. Patent 3,670,752 (1972).

9. Marsden, S.S., Paper No. SPE 4359 presented to Society of Petroleum Engineers, Denver Meeting (1973).

10. Marsden, S.S., and P. R. Hooker, U.S. Patent 3,926,203 (1975).

11. Marsden, S.S., Paper No. SPE 8296 (1979).

12. Matsumoto, S. and P. Sherman, J. Colloid Interface Sci., 30, 525 (1969).

13. McClaflin, G., C. Clark, and T.R. Sifferman, SPE Paper 10094, presented to Society of Petroleum Engineers, San Antonio Meeting (1981).

14. Mewis, J. and A.J.B. Spaull, Adv. Colloid Interface Sci, 6, 197 (1976).

15. Mooney, M., J. Colloid Sci., 6, 162 (1951).

16. Myers, R.W., J. Pet. Tech., 890-894, June (1978).

17. Parkinson, C., S. Matsumoto, and P. Sherman, J. Colloid Interface Sci, 33 (1), 150 (1970).

18. Seymour, E.V., U.S. Patent 3,530,310 (1970).

19. Sherman, P., J. Colloid Interface Sci., 24, 97 (1967).

20. Sherman, P., J. Colloid Interface Sci., 24, 107 (1967).

21. Sherman, P., J. Colloid Interface Sci., 27, 282 (1968).

22. Sherman, P., J. Physical Chem., 67, 2531 (1963).

23. Sherman, P., Proc. 4th Intern. Congr. Rheol., 3, 605 (1965).

24. Sifferman, T.R., J. Pet. Tech., 1042-1050, August (1979).

25. Sifferman, T.R., U.S. Patent 4,265,264 (1981).

26. Simon, R., and W.G. Poynter, U.S. Patent 3,519,006 (1970).

27. Simon, R., and W.G. Poynter, J. Petrol. Technol., 20, 1349 (1968).

28. Steinborn, R., and D.L. Flock, presented at 33rd Annual Meeting of the Petroleum Society of CIM, Calgary, Alberta, Canada, June (1982).

29. Uhde, A., and G. Kopp, J. Inst. Pet., 57, 63-73 (1971).

30. Westfall, S.A., personal communication, 1980.

RECEIVED June 8, 1984

INDEXES

Author Index

Subject Index

Production by Anne Riesberg
Indexing by Karen McCeney
Jacket design by Pamela Lewis

Elements typeset by Hot Type Ltd., Washington, D.C.
Printed and bound by Maple Press Co., York, Pa.